Fundamentals and practices in colouration of textiles

Fundamentals and practices in colouration of textiles

J N Chakraborty

WOODHEAD PUBLISHING INDIA PVT LTD
New Delhi ● Cambridge ● Oxford

Published by Woodhead Publishing India Pvt. Ltd.
Woodhead Publishing India Pvt. Ltd., G-2, Vardaan House, 7/28, Ansari Road
Daryaganj, New Delhi – 110002, India
www.woodheadpublishingindia.com

Woodhead Publishing Limited, Abington Hall, Granta Park, Great Abington
Cambridge CB21 6AH, UK
www.woodheadpublishing.com

First published 2010, Woodhead Publishing India Pvt. Ltd.
© Woodhead Publishing India Pvt. Ltd., 2010

Woodhead Publishing India Pvt. Ltd. ISBN 13: 978-81-908001-4-3
Woodhead Publishing India Pvt. Ltd. EAN: 9788190800143

Woodhead Publishing Ltd. ISBN 10: 1-84569-788-4
Woodhead Publishing Ltd. ISBN 13: 978-1-84569-788-4

Typeset by Sunshine Graphics, New Delhi
Printed and bound by Replika Press Pvt. Ltd.

Contents

Preface

Colouration of textiles is an art and less complicated than actually thought; just needs due attention to ascertain look, quality and consistency in product. Each textile is coloured with specific colourants through its own technology. Thorough realization of basics of colouration is the basic need to achieve that.

This book is an attempt to explain basics of colouration with a match in practical application precisely but in detail. Effective technical excellence can be achieved in dyehouses only if the dyer has adequate skill to run the shop-floor with a grip over it. It is common in all dyehouses to face typical new problems on every new day, which adds on to the daily experience and certain part of dyed textile comes out as faulty dyeing. Fundamental knowledge is a tool to reduce these faults, save money and makes the dyer capable of handling critical situations in dyeing. With implementation of ISO, tremendous pressure has been felt to produce quality products and to remain in rhythm with highly competitive global market.

It was realized since long to have a text book in chemical processing of textiles which can cater needs of industries, academics and to shape potential technicians. To materialize it, dye and chemical manufacturer's recommendations have been introduced to make the book practically healthy. However to limit its volume, textile fibres of prime importance only have been focused in detail. This is purely a text book; references supported at places are to help readers to know more if they desire so. Chapters have been made precise and compact; even a small chapter covers related area with full depth.

Any suggestion towards upgrading this book will be duly acknowledged with thanks from its readers to make it more meaningful and reader friendly.

June, 2009   J N Chakraborty

Acknowledgements

This book 'Fundamentals and Practices in Colouration of Textiles' spreads over a large spectrum related to colouration, basically nurtured and shaped by researchers throughout the world. Author is thankful to them for their valuable contribution to script this book.

Author duly acknowledges with thanks the permission extended by Society of Dyers and Colourists (Bradford), American Association of Textile Chemist and Colorist (USA), Melliand Textilberichte GmbH, The Textile Institute (Manchester), IIT (Delhi), ATIRA and NITRA (India), to reproduce required texts, figures and tables from their published journals and books.

Author also expresses thanks to NISCAIR (Delhi), Colour Publications (Mumbai), The ITJ (Mumbai), PHI Ltd (Delhi), Sevak Publications (Mumbai), The TRJ (USA) for their kind permission to reproduce some material from their respective publications.

Author duly acknowledges permission extended by globally reputed manufacturers of dyes, chemicals and apparatus, viz. Atlas (USA), Atul Ltd (India), Dystar (Germany), ICI (UK), Jaysynth Dyechem Ltd (Mumbai) to incorporate their technical recommendations in this book to make it practically so meaningful. Figure 21.1 was originally reproduced from ICI manual but the organisation could not confirm the same. The dyeing technology with Indigo developed by BASF was later transferred to DyStar.

Some reproductions have been made from 'Dyeing & chemical technology of textile fibres' by E R Trotman, 6th Edition published in India by B I Publications, Delhi. When communicated, the publisher refused to grant permission but asked to approach Charles Griffin Co. Ltd, UK. Trying best, author could not find contact details of the latter. Author had sent letters and e-mails to ITS Publishing (International Textile bulletin, Switzerland), Verlag Konradin Verlag Robert Kohlhammer GmbH (Textile Praxis International, Stuttgart), Oliver & Boyd (UK), Thomas Nelson & Sons Ltd (UK) but in vain. Author draws attention of these publishers to communicate to include their name in its forthcoming edition, if they desire so.

This book was composed throughout a long period, during which author had misplaced a very few research papers after inclusion of specific table or figure and could not recollect the source. Author requests various publishers, researchers and manufacturers to please communicate him with proof to make required correction in the next edition of this book.

Author extends thanks to Woodhead Publishing India for their interest to publish this book.

June, 2009   J N Chakraborty

Introduction to dyeing of textiles

Abstract: Dyeing is a process of thorough colouration of textiles, and its success lies on type and extent of pre-treatment imparted to develop good absorbency and whiteness. Other factors, viz. pH, form of textile, type of fibre, formulation of dyeing recipe, preparation of dye solution, liquor ratio, selection of machinery, etc. too play a crucial role in the process to develop levelled shades with least effort and cost. This chapter highlights these basic concepts to be kept in mind for effective implementation of technology of dyeing.

Keywords: Pre-treatment, absorbency, dyeing, dye yield

1.1   Importance of pre-treatment

Textiles are dyed in three different forms: fibre, yarn and fabric (knitted/woven); and the methods include either of exhaust/batch or padding. Thorough colouration in exhaust method involves a sequential mechanism, e.g. migration of dye from bath to fibre surface, adsorption of dye on the surface of fibre, diffusion of dye inside followed by fixation. It is essential to confirm complete absence of impurity in fibre to ensure formation of levelled shades. If impurities are hydrophilic in nature, an efficient wash prior to dyeing removes these and facilitates effective deposition of dye on surface of fibre (the first step). Absence of impurity inside fibre keeps fibre pores clean to give a passage to dye to migrate inside. In fact, the impurities are mostly hydrophobic, hinders surface deposition, adsorption and diffusion. The final result is an uneven dull light shade in spite of higher dose of dye with very poor wash and rubbing fastness. If dyeing technique is changed to padding, affinity of dye does not play crucial role as dye is forcibly placed on textile avoiding the problem associated with the first two steps, i.e. migration of dye from bath to fibre and surface deposition, but presence of impurities at the interior will suppress adsorption as well as diffusion both, i.e. irrespective of dyeing method, presence of impurities is bound to resist formation of levelled and fast shades. Further, a pale shade can only be formed on a good white ground but different textiles possess own inherent colours requiring bleaching. This has necessitated pre-treatment of all textiles to acquire required absorbency and whiteness prior to dyeing.

1.2   Type of impurities

Impurities present in various textiles are of two types: natural and added. Natural fibres like cotton, wool and silk, etc. are grown under natural conditions, during which these acquire natural impurities like dirt, dust, minerals, fatty matters, etc. while man-made fibres like polyester, nylon and acrylic, etc. do not possess natural impurities but pick up some impurities like machine oil, grease, etc. during various processes or externally added-like spin finish and are called added impurity. All natural fibres also collect little added impurity during mechanical processing imparted, and obviously these retain collectively more impurity requiring a drastic pre-treatment for effective removal to develop a wetting time around 5 s when a drop of water is allowed to fall on scoured cotton to disappear inside within this time. Man-made fibres retain little added impurity, and a mild pre-treatment is adequate.

1.3   Type of pre-treatment

The pre-treatment process is not same for all fibres due to difference in their properties, variation in type and percent of impurity content. Cotton is singed first to remove protruding fibres, which otherwise hinder formation of brilliant shade due to enhanced scattering of light. Cotton warp is sized to protect it from damage during weaving but requires its subsequent removal to allow diffusion of chemicals and dyes. Presence of natural as well as added impurities, which are hydrophobic in nature, impair absorbency; and the wettability remains too poor to exercise any successful dyeing. This necessitates scouring of cotton textiles with an aim to improve absorbency with least fibre damage – a purely purification process. Scoured cotton is bleached to make it whiter – essential to build up brilliant shades, and finally bleached cotton is mercerized to improve its lustre and chemical reactivity or dyeability. If cotton is to be marketed as full bleach where absorbency is of least importance, thorough scouring may be avoided but double bleach is to be introduced instead.

Wool retains grease, stain, suint, etc., which get hardened during singeing and pose problem in removal during scouring. It is carbonized first with acid to eliminate cellulosic impurities followed by removal of stain with the help of a solvent; either singeing or cropping is imparted to burn or cut protruding fibres respectively (Peters, 1967; Karmakar, 1999). If cropping is to be imparted and not singeing, then it may be done at any stage before dyeing. However, fabric may get distorted during cropping, and hence singeing is preferred (Sule, 1981). Woollen yarn is preferred in knitting and is not sized. Fabric structure is stabilized through crabbing or blowing to avoid distortion followed by mild scouring to remove impurities at 55–60°C; H_2O_2 bleaching is imparted for required whiteness.

Silk consists of sericin and fibroin; sericin forms the outer layer as protective coating while fibroin remains at the interior and is the fibrous part. Silk yarn is made of filaments and do not require sizing, obviously singeing and desizing both are omitted. A well-controlled degumming with very mild alkaline soap at boil is imparted to remove sericin effectively and to make the fibre highly absorbent. Degumming of silk with boiled-off liquor which comes from the spent bath from previous degumming process, is the most efficient method of degumming. Bleaching is done with H_2O_2 at 70°C.

Man-made fibres like polyester, nylon and acrylic retain mainly added impurities. Thermoplastics like polyester and nylon are not singed at the initial stages as protruding fibres may melt to form molten beads trapping impurities as well as sizing materials inside and hindering its removal during subsequent treatments. Beads formed accept more dye causing specky dyeing. It is better to start pre-treatment with desizing; singeing may be imparted in post-colouration stages. Both of the fibres are scoured in mild alkali followed by chlorite bleaching ($NaClO_2$); presence of chlorine in sodium chlorite reduces elastomeric nature of nylon, and peracetic acid bleach is the most preferred option. Polyester is preferably dry heat set for structural stabilization while nylon is steam set. Acrylic, on the other hand, is not sized as the main use is knitting; mild scouring, chlorite bleaching and stabilization by autoclaving at 105°C with steam to develop bulkiness completes pre-treatment. However, a modified sequence comes into effect when different textiles are blended; the sequence of pre-treatment depends on share of component fibres in that blend, impurity content of component fibres, whether any of the components were previously passed through any pre-treatment, and treatment of one component should not have negative effect on another component.

1.4   Structure of fibre

In dyeing, dyebath conditions are adjusted as per the technology for application of a specific dye. In some cases, a fibre may be dyed with a specific dye but in exchange for some damage in it, e.g. dyeing of wool and silk with sulphur, reactive hot brand and vat dyes. The highly alkaline pH maintained during dyeing at high temperature partially disintegrates these fibres causing fall in tensile strength. That is, in spite of having the required technology, a fibre can not be dyed with few methods on technical grounds. Wool, silk and nylon having acid groups in chain are not dyed with basic dyes due to poor light fastness, but acrylic is dyed with basic dye to develop excellent light fastness. Nylon and acrylic, being man-made fibres are not dyed with disperse dye as the dye produces poor light fast shades on nylon and only light shades on acrylic in spite of adding more and more dye. Wool, silk, nylon and acrylic are dyed with acid/metal complex and basic dyes respectively. Another aspect in dyeing is availability of free volume in fibres as it determines how much dye the fibre can accommodate *in situ*. This, in turn, ascertains openness of fibre as well as rate of dyeing, permissible size of dye molecule that can be used for a specific dye, polarity in fibre, etc. This is important because two fibres possessing affinity for a single dye when dyed as blend, the dye uptake will be in proportion to their openness and affinity for dye. Another factor is why any dye can not be applied on any fibre, viz. why disperse dye is not applied on cotton and direct dye on polyester, the hindrance must be known to suggest the most effective method of dyeing for a given fibre.

1.5   Absorbency, whiteness and pH of textiles

Natural textile for dyeing must have required absorbency, whiteness (reflectance factor) and pH. Various chemical pre-treatment processes as well as dyeing are carried out at specific pH; if there is match in working pH of both, it does not become essential to go for neutral pH concept of textile to start dyeing, e.g. dyeing of mercerized cotton with vat dyes does not require necessarily a fabric of neutral pH; in such a case, neutralization of mercerized cotton can be omitted and may directly be sent for dyeing, thus reducing time and cost both. The pH of man-made fibres must be neutral as these are invariably dyed in acidic pH.

Well-pre-treated natural textiles possess good absorbency which in turn enhances rate of dyeing through better surface deposition of dye followed by its diffusion. The latter occurs at a faster rate if fibre canals or pores are cleaned up effectively during pre-treatment; inadequate or uneven absorbency throughout textile causes patchy dyeing. Under industrial conditions, huge amount of textile is pre-treated in a single operation with probable uniform packing and right control over parameters to ensure desired absorbency and whiteness. Processing of such a huge assignment in a single operation poses problem to get uniform results. Poor and varying absorbency disturbs wetting and dye uptake. In contrast, dyeability of man-made textiles depends on its rate of opening with temperature for a given dye–fibre system. Bleaching is a must to produce light shades but may be omitted if deep shades are to be produced.

1.6   Form of textile in dyeing

An important factor in dyeing to produce levelled shades is that the textile in any form should have homogeneity in itself. If the textile is natural fibre, whole of the lot must have the same fineness and maturity to form thorough shades, while man-made fibres should have uniformity in fineness, orientation as well as monomer or reactant composition. If the dyed fibre is to be transformed to yarn and/or fabric, little variation in dyeing do not pose any such problem as blending of fibre in spinning ensures good mixing. Once grey fibre is converted to yarn, migration of fibre is arrested and uneven shade can not be rectified. Conversion of such dyed yarn to fabric may improve the situation. Dyeing of textiles in fabric form is most typical and any unlevelled shade formed can not be so easily rectified; the situation becomes so complicated in some cases that the dyed fabric is ultimately converted to black shades. Width of fabric pieces to be joined must have identical width and stitching should be free from faults to obtain levelled shades (Gokhale and Modi, 1992).

1.7   Formulation of dyeing recipe

Formulation of instrumental dyeing recipe and maintenance of optimal dyeing conditions are also the key to produce levelled shades. The recipe for a dye–fibre system should remain unchanged to reproduce a specific hue and shade. Dyeing recipe is mostly formulated with computer colour-matching system which works on the basis of stored data of dyed textiles at different concentrations for a specific dye. The stored data also helps the dyer to know cost, fastness, etc. for each recipe. To maintain quality, stock of dyes used in the database as well as required chemicals must be adequate and dyer must remain adhered to the established recipe and conditions; adding any foreign colour may spoil the shade. Even a dye can not be substituted with a second one having the same colour index specification as the latter does not take care of the extent of dilution of various dyes after their synthesis causing variation in share of component colour in recipe. An established recipe with dyeing conditions and the dyed sample must be well-documented for reproduction of that very shade, the so-called concept of ISO-9002.

1.8   Liquor ratio in dyeing

Liquor ratio may be defined as volume of liquor to be taken in dyebath in proportion to weight of textile. If a piece of textile weighing 'y' g is to be dyed at a liquor ratio of 1:5, volume of water would be '5y' ml. Aqueous dyeing processes are highly influenced by volume of water in dyebath.

The higher the affinity or strike rate of dye for fibre, a higher liquor ratio would ensure formation of levelled shades. Dyes with higher affinity if applied from a lower liquor ratio bath are bound to form unlevelled shades. This is because the effective concentration of dye molecules in bath increases on lowering liquor ratio – obviously strike rate of dye increases with ultimate formation of faulty shade as a few dye molecules try to occupy the same point on fibre simultaneously. Increase in liquor ratio dilutes the bath, and spacing between dyes get increased; any two dye molecules can not reach same point on fibre surface at the same time thus producing levelled shades, e.g. dyeing of cotton with vat dyes. However, each drop of water in industries costs a lot and machines have specific capacities. A bath of higher liquor ratio also increases waste-water load, more use of chemicals and heating of bath causes wastage of energy. Thus, a compromise is to be made between possible lower liquor ratio and highest possible dye uptake. If the dye has intermediate affinity, liquor ratio may be moderate, e.g. dyeing of cotton with sulphur and direct dyes, while reactive dyes possesses negligible affinity with chances of hydrolysis and are preferred to apply from a low liquor bath (Chakraborty, 1991).

Effect of liquor ratio is not as important in dyeing acrylic with basic dyes as dyeing is controlled through either control over temperature or adding retarding agents in bath. Nylon, wool and silk are dyed with acid or metal complex dyes, the higher the liquor ratio, the higher the formation of zwitter ions and better the dye uptake. Polyester is dyed with disperse dye, and the liquor ratio does not play crucial role as such because dye goes to solution very slowly beyond 90°C and the same enters *in situ* fibre at higher temperature.

1.9   Selection of machinery

Use of a specific dyeing machine also influences quality of dyeing. A woven cotton fabric can be dyed in a winch or a jigger. If the fabric is finer and chances of damage exist, use of jigger is to be avoided as the fabric is dyed in jigger in open width and under tension. On the other hand, open width processing in jigger causes effective dye–fibre contact with a levelled shade; the appearance of dyeing remains free from crease marks. Dyeing in a winch reduces risk of damage but does not form the bright shade that is produced in jigger. Polyester fabrics, specifically suiting and shirting, can be dyed in jet dyeing as well as HTHP beam dyeing machines. Dyeing in jet dyeing machines is a quicker and efficient process; a well-levelled shade is produced with a serious problem of crease marks on fabric, which even poses trouble in removal during finishing. Dyeing in beam dyeing machine is associated with wrapping of fabric on beams with optimal and consistent tension, other wise moiré effect (faulty dyeing) is produced which can only be corrected by redyeing the fabric in a jet dyeing machine. Fabrics dyed in beam dyeing machine do not form crease marks rather develops a fuller handle.

1.10   Dye uptake and dye yield

Efficiency of a dyeing process is assessed by the extent of colouration of fibre using two different terms: dye uptake and dye yield. The first one is a quantitative approach and the second term is qualitative. Dye uptake can be defined as 'grams of dye per 100 g of dyed textile' while dye yield is the ratio of co-efficient of absorption and co-efficient of scattering termed as K/S, known as Kubelka–Munk ratio. It is purely a number with no units and is measured by colour matching system. Dye yield is also known as colour yield, dye strength or colour strength. Dye yield is the coloured property of the surface of a textile and does not include the dyeing status at the interior. To get higher K/S using lesser dye, the dyeing process is to be controlled to achieve only a surface dyeing, where dyes mostly remain on the surface of textile. For example, if two identical samples are dyed with same amount of same dye, dyeing process can be varied for these two to get two dyed samples with two different K/S values; the sample where diffusion is more will show lesser K/S and where little or no diffusion is there shows a higher K/S. In both the cases, dye uptake might be the same. As different colour matching systems use different setting for colour measurement, K/S does not have a universal value and rather varies from one instrument to other, and K/S obtained from one system can not be directly compared with that obtained from another system of some other make. Assessment of dye uptake is made with a UV-Visible spectrophotometer by dissolving the dyed sample in a suitable solvent and to get the extinction or absorption value. The method is laborious and time-consuming. Suitable solvent is required which can either solubilise the textile or the dye; during solubilisation, dye structure must remain intact and λ_{max} unchanged. In practice, reflectance value (R) of the dyed sample is measured by the colour matching system which is converted to K/S using the relationship:

$$K/S = (1 - R)^2/2R;$$

the higher the reflectance value, the lower the K/S.

1.11   Sequencing dyeing programme

Smooth running of a dye house is influenced by uninterrupted back feed of pre-treated material. Various types of textiles in various forms are dyed in machines; the flow of back feed must be smoothened to ensure running of machines at maximum possible capacity. Sequencing of goods for dyeing is to be programmed based on sales demand, timing of back feed of specific pre-treated textile for dyeing, availability of dye/chemicals at ready stock, and minimum washing of machines after each cycle. The latter plays a crucial role to reduce time gap between dyeing of two successive batches. Dyeing in a machine may be started with lightest shade and successive batches if possible, may be with same dye or dye combination to omit washing. Programming of dyeing schedule for an entire day is really a tough job for the management to get maximum output.

1.12   Selection of dyeing method

Dyeing can be performed in three ways: exhaust or batch, semi-continuous and continuous method. For successful implementation of exhaust method, an optimum batch size is preferred as dyeing of several batches, may be in same machine, results variation in shade and hue for a given consignment. Exhaust method is applicable for any form of textile: fibre, yarn or fabric. In contrast, dyeing of longer length or relatively larger amount of textile for a given hue and shade is dyed better in pad-roll-steam, pad-exhaust, etc., which are semi-continuous methods to have same hue and shade throughout. Coarse fabrics, e.g. suiting, canvasses, etc. are better dyed in pad-exhaust technique. The fabric is padded with pad liquor followed by dyeing in jigger with the rest pad liquor. If the fabric would have been dyed in exhaust method, it would take long time to dye thoroughly requiring more time for diffusion and distribution of dye. Thorough dyeing becomes impossible if the fabric becomes coarser and compact, the latter inhibits diffusion of dye at the interior during exhaust dyeing; padding followed by exhaust may be a solution for such fabrics. Continuous dyeing methods are to dye huge length of fabric in a uniform shade without any variation. This is done by padding followed by development or padding–drying–curing/steaming sequence. Dyes with good affinity are selected for exhaust dyeing whereas dyes with poor affinity are preferred for padding.

1.13   Dye saturation value of fibres

Every fibre has its own dye saturation value up to which shades can be produced, and beyond that it can not accommodate more dye. The value depends on whether the dye is retained by the fibre physically or through chemical bonding. In the first case, free accessible volume inside fibre plays the role and accommodates definite amount of dye in it. In case of chemical attachment, dyes may occupy free volume inside fibre and reacts with that, at the same time a little more dye reacts with the reactive group of fibre projected on its surface. Knowledge of saturation value of a fibre is important keeping in view the maximum shade depth that can be produced on it. Redyeing of deep shades, even to a black fails sometimes due to non-availability of accessible volume inside; the latter is inversely proportional to draw ratio of the man-made fibre during spinning (Vickerstaff, 1950).

1.14   Chemical processing technology

Wide spectrum of chemical processing of textiles may be divided into a few areas, viz. dye chemistry, colouration technology, colour technology, chemical technology, etc. Dye chemistry ascertains synthesis, chemical structure, solubility, stability, fastness and other properties of dyes; colouration technology predicts the technique of its efficient application on a specific fibre and how to produce the shade having better fastness through after-treatments if there is deficiency the way to rectify it etc. Colour technology deals with colour aspects, viz. primary, secondary and tertiary colours, colour mixing, recipe prediction, etc., also known as colour management. In spite of having stored data of various dyes, computer colour-matching system asks dyer to enter probable colour combination to form a composite colour, and it can not formulate these by itself. So the dyer must have knowledge of mixing of colours. In light theory, lights of all three primary colours (red, blue and green) are mixed to get white, while black means absence of any colour; in textile colouration, a white textile having no colour on it looks white and dyeing of it with three primary colours (red, yellow and blue) at a specific ratio develops a black with no specific λ_{max}.

Chemical technology deals with chemistry and suitability of a chemical for a specific dyeing process, its working mechanism, etc. If the required chemical is short in stock, then it is either to be replaced by a second one thus keeping production schedule undisturbed, or it is to be prepared immediately. For example, acetic acid is very important in most of dyeings; absence of it may be managed by citric or formic acid. Sodium acetate is another chemical, which may be prepared in shop floor by reacting NaOH with CH_3COOH in cold water. The percent strength of the solution is to be calculated based on total volume, and recipe is to be formulated accordingly.

1.15   Prediction of recipe for mass dyeing

To develop a specific shade, trials are run on smaller samples in the chemical laboratory. On confirmation of the desired shade produced, the recipe and conditions with a piece of dyed sample is passed on to the shop-floor level for mass dyeing. However, required reduction in concentration of dye and chemicals is to be made in the laboratory recipe to make it effective for mass dyeing. This is because in laboratory practice, a small sample is dyed in higher liquor ratio to keep it immersed in bath which is not so in mass scale dyeing implying a low liquor dyeing process. In mass scale dyeing, a higher dye uptake is obtained with even lesser dye. An overall reduction up to even 40% may be exercised to convert a laboratory recipe to a shop-floor level one, though the accuracy in it varies from dye to dye.

1.16   Preparation of dye solution

During preparation of a dyebath, solid dye particles must be pasted with little Turkey red oil (T R oil) to prepare dye slurry followed by addition of water to prevent floating of dye particles on the upper layer of solution and will not be available for dyeing. This is a must if the dye is water soluble; addition of T R oil may be omitted in other cases as the dye being insoluble in water gets precipitated. Addition of more T R oil may increase solubility of dye and reduce affinity as well as substantively necessitating least possible application for better dye uptake.

1.17   Documentation of process details

To reproduce a given hue and shade for a specific dye–fibre system, once carried out dyeing process requires being well-documented with details of dyeing parameters and any other comments. This eliminates hit and trial in subsequent dyeing processes and reduces extent of redyeing, addition, etc.; also laboratory trial is by-passed. Detailed documentation and strict adhesion to the norms introduces consistent quality of processed textiles, the so-called ISO-9002.

1.18   References

CHAKRABORTY J N (1991) 'Exact Volume of water to be used in Jigger dyeing', *Textile Dyer Print*, **24**, 23, 31–33.

GOKHALE S V AND MODI J R (1992) *Process and Quality Control in Chemical Processing of Textiles*, India, Ahmedabad Textile Industries Research Association.

KARMAKAR S R (1999) *Chemical Technology in the Pretreatment Processes of Textiles*, Amsterdam, Elsevier.

PETERS R H (1967) *Textile Chemistry*, Vol-II, London, Elsevier.

SULE A D (1981) 'Scientific and practical aspects of finishing polyester-wool blended fabrics', Blended textiles, Paper presented at 38th All India Textile Conference, Mumbai, 18–20th Nov, The Textile Association (India), 321–353.

VICKERSTAFF T (1950) *The Physical Chemistry of Dyeing*, London, Oliver and Boyd.

Abstract: Colouring materials used in dyeing of textiles possess several features; the dyer must have basic knowledge of those for effective handling to produce hues and shades, their mode of attachment with fibres and related fastness properties. All colouring materials are not suitable for all fibres rather depend on chemistry of the latter. This chapter deals with definition of various colourants, their classification, general chemical structure, suitability for different textiles, ionic nature, solubility in water, nomenclature, colour index specification, interpretation of commercial name, etc.

Keywords: Chromophore, dye, pigment, lake, colour index

2.1   Types and definition

Colouring materials are mainly of three types, viz. dyes, pigments and lakes (ingrain dyes). A dye has three parts in its structure – chromophore, chromogen and auxochrome – and is soluble in a specific medium under certain conditions. Chromophore is an unsaturated group that absorbs light and reflects it at specific angle to give the hue, e.g. azo, keto, nitro, nitroso, thio, ethylene, etc.; chromogen retains chromophore and plays a crucial role to determine the final hue and its affinity for fibre, fastness, stability, etc. while auxochrome is a substituted acidic or basic group in dye structure to intensify depth of shade, e.g. –OH, –COOH, SO_3H, $–NH_2$, $–NH(CH_3)$, etc. (Shenai, 1987; Finar, 1975). Further addition of substituents to dye structure deepens the shade and extent of deepening varies with increase in molecular weight of dye. Following are examples of various parts of a dye.

- Chrysophenine G, C I Direct Yellow 12, C I 24895, is a direct dye in which chromophores are –N=N– and –CH=CH–, auxochrome – OC_2H_5 on both sides and chromogen is the part containing three chromophores but excluding the auxochromes as depicted in Fig. 2.1 (Colour Index, 1987).

11

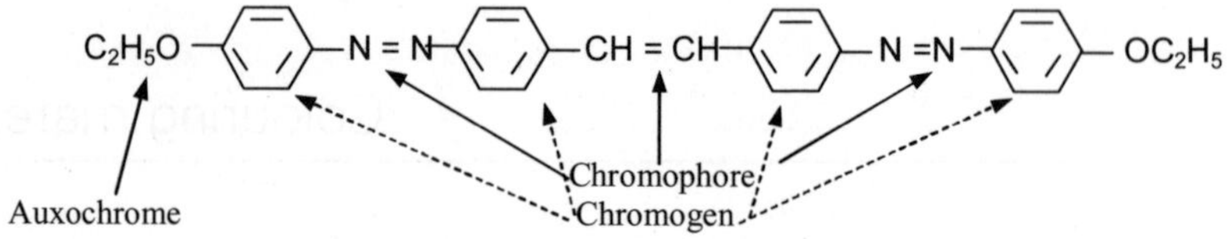

2.1 Various parts of chrysophenine G.

- Picric acid is an acid dye (Fig. 2.2), in which the three chromophores are nitro groups and auxochrome is the hydroxyl group (Colour Index, 1987).

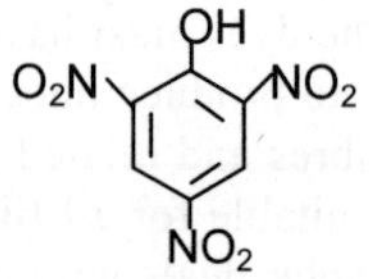

2.2 Chemical structure of picric acid.

- Green XBN, C I Vat Green 1, C I 59825 consists of C=O chromophores and –OCH$_3$ auxochromes and has the structure as shown in Fig. 2.3 (Colour Index, 1987).

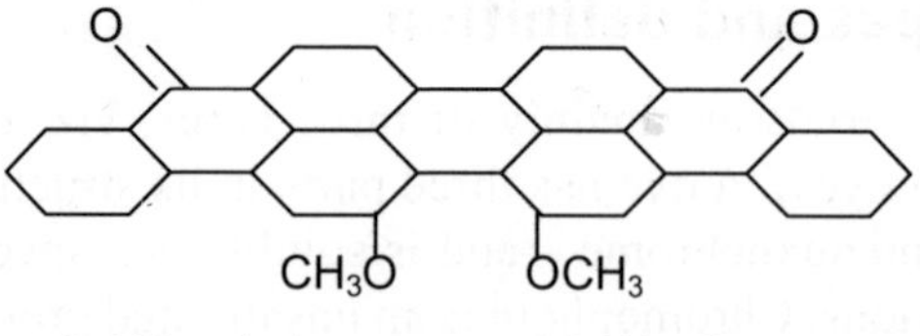

2.3 Chemical structure of C I Vat Green 1.

A pigment possesses typical structure: it may be a simple chemical, e.g. ferric oxide or insoluble colouring material with typical aromatic structure and is not soluble in any medium during application. Pigments formed *in situ* fibre are known as lakes or ingrain dyes, e.g. phthalocyanine blue formed through reaction between copper complex and alcian blue; insoluble azoic colours formed *in situ* through reaction of naphthol (coupling component) and a diazotized base; reaction product of metal oxides *in situ* fibres, etc. The basic difference between pigment and lake is that a pigment is produced in laboratory or manufacturing plant and is applied on textile along with a synthetic binder to retain it firmly on fibre surface through reaction between transparence binder and fibre when the pigment gets trapped in between fibre and binder; fastness of coloured textile depends on life of the binder; in contrast, lakes or ingrain dyes are formed when two chemicals are applied simultaneously or separately on fibre and the pigment is formed through reaction of these two chemicals *in situ* fibre which gets trapped inside fibre developing adequate fastness and binder is not applied for fixation. A textile coloured with lakes is superior to that with pigments, because lakes are formed throughout fibre due to thorough diffusion of chemicals inside textile. The coloured effect is visible on both side of fibre; wash fastness is superior and less stiff. In contrast, colouration with pigment makes fibre stiffer and mostly one-sided effect is produced if applied on one surface as pigment can not diffuse inside but remains on surface on which it has been applied.

2.2   Classification

Colouring materials are also classified in several other ways based on their source or origin, chemical structure, field of application, ionic nature and solubility in water, etc.

2.2.1   Source or origin

Natural colours

These are extracted from natural sources. Yellow natural dyes include turmeric, kamela, tesu, marigold, larkspur, harshingar, annato, berber's and dolu. In turmeric, the source is rhizomes of curcuma longa and the pigment is curcumin (diferuloyl methane derivative), produces brilliant shade, very fugitive to light and moderate wash fastness (Chakraborty et al, 2005; Gulrajani et al 1992a). Kamela (orange red powder occurring as a glandular pubescence on the pads of the tree *Mallotus philippinensis*) comprises several colouring matters, e.g. chalcones (rottlerin, isorottlerin), produced shade is fast to soap, alkalis and acids but has moderate fastness to light. A range of colours like bright golden yellow (with Sn) to mustard yellow (with Cu, Cr) and olive green (with Fe) are obtained on mordanting with metal salts. Tesu (flowers of the palm tree *Butea monosoperma*) retains colouring pigment butein and achalcone of orangish red colour. On adding mordant, pale yellow (with Al), yellow brown (with Cu, Cr), greenish grey (with Fe) and bright yellow (with Sn) colours can be obtained (Gulrajani et al 1992b). Marigold is the brightly coloured flowers of garden plant *Tagetes erecta*; it yields a shiny orange dye containing flavonol quarcetagetol (derivative of quercetol and some ellagic acid) which serves as natural mordant. It produces shades ranging from yellow to deep orange (with Sn) to brown (with Cu) and olive (with Fe) which are fast to light and wash. Yellow larkspur (distinguished by its yellow flowers, *Delphinium Zalil* is a plant) is rich in flavonol quercetol and isorhamnetol. Silk can be dyed to a sulphur yellow colour called gandhaki (with alum) and abisangur (with Cu). A bright golden yellow is developed with harshingar, (from the corolla tubes of flowers of *Nyctanthes arbortristis* or night flowering jasmine) in which colouring pigment is nictanthin (identical to crocin) – the pigment occurring in saffron and tun flowers. Annato, a yellow–orange dye is obtained from the seeds and pulp of fruits of the annatto-*Bixa orellana* tree pulp is rich in tannin, contains a mixture of eight colourants of carotenoid group of which the main two are nor-bixin and bixin. It acts as acid dye and dyes silk directly in orange–red shade. Berberis (the roots, bark and also the stems of *Berberis vulgaris* are the source of a fluorescent yellow–green colour) has colouring matter of *alkaloid-berberine* and its derivatives. Being the only known natural basic dye, it does not form complexes with metal ions, and mordants have no effect on either colour or fastness properties. Light fastness is very poor – the yellow colour changes to yellow–brown on exposure to light and wash fastness is moderate. Coloured pigment in 'dolu' comprising of a number of anthraquinone derivatives of emodin, chrysophanol, aloe-emodin, rhein, gallic acid, tennin and catechin (dye is extracted from the roots of the shrub *Rheum emodi*) produces light and washfast mustard yellow, also develop wide range of fast and deep colours in combination with mordant, e.g. brown (with Cu), olive green (with Fe), red–violet (with Cr) and light yellow (with Sn) (Gulrajani et al 1992c).

Red dyes include mainly safflower, manjit, patang and lac. Safflower (Florets of *Carthamus tinctorius*) contain two colouring matters – carthomin based on benzoquinone structure which is scarlet red, insoluble in water, and safflower yellow is water soluble. Red, pink, rose, crimson or scarlet shades are produced. Shades are fugitive to light, air and sensitive towards alkali, chlorine and sulphuric acid. Manjit (Roots of *Rubia Cordifolia*) contain a mixture of several anthraquinone pigments such as purpurin, munjistin, xanthopurpurin, and pseudo purpurin. Lively colours such as red (with Al + Sn), orange (with Sn) and violet (with Fe) are produced which are quite fast to light and washing. Patang (*Caesalpinia sappan* is a small thorny tree belonging to the class of soluble red woods) contents brazilin which when oxidized to brazilein yields a red colour. Wide range of colours having good fastness to light and washing can be produced like beige (with Al, Sn), brownish grey (with Fe), purple (with Cr) and brown (with Cu). Crude lac or stick lac (resinous protective secretion of tiny lac insect *Laccifer lacca* which is a pest on a number of plants) contains a water-soluble red dye, laccaic acid (hydroxyl-anthraquinone carboxylic acid) and an alkali soluble yellow dye – erythrolaccin; the dye yields lively scarlet, crimson and pink shades with good fastness to light and wash.

Indigo (leaves of the plant *Indigofera tinctoria*) is the only natural blue vat dye, which is synthesized too. The pigment is present in the form of a soluble, colourless glucoside indicant (a combination of glucose and indoxyl) which oxidizes to form insoluble pigment indigotin. It is fast to light, washing, milling, perspiration and peroxide bleach. Green shade is produced by over dyeing indigo shades with some natural dye, mostly turmeric. Another source of pale but fast green is *Gurhal* flowers.

Brown dyes are rich in tannins, moderately fast colours are obtained like henna (with Al, Fe), onion skins (with Cu), dolu (with Cu) and myrabolan (with Cu). Log wood (or campeachy wood is the product of a large tree of the leguminous family, botanically known as *Haematoxylon Campechianum*) retains coloured pigment as substituted di-hydro pyran, namely, haematin gives a distinct red colour but in combination with right mordant gives blue, violet and black. Some more attractive colours can be obtained by mixing different natural dyes in the same dye-bath and through top-dyeing.

Natural dyes are of two types: substantive or non-mordant dyes, which produce fast shades at boil and include walnuts and lichens; mordant dyes require an additional chemical to make the colour permanent. Dyeing with these dyes is basically a three step process – dye extraction, mordanting and dyeing (Gulrajani, 1993). Mordanting is taken into action before, after or concurrently with the dyeing process with efficient mordants – e.g. alum, chrome, iron and tin salts – brighter colours and better fixation are obtained only by pre-mordanting (Das, 1992).

Natural dyes are also extracted from unlimited sources of barks of trees and leaves of plants. In spite of all these, these are used to a limited extent owing to very high labour cost involved in the collection and preparation of dyes and difficulty in obtaining accurate reproduction of colour and shade.

Natural colours are easily biodegradable; fastness is moderate to good and dyeing technique is mostly through pre-mordanting. Shades lack brightness and shade range is limited.

Synthetic colours

A large amount of synthetic colours are obtained through reaction of dye intermediates, e.g., anthranilic acid, β–naphthol, benzidine, naphthionic acid, J-acid, metanilic acid, H-acid, hydroquinone, Schaffer's acid, G-acid, R-acid, etc., with other chemicals. These intermediates are first synthesized through various chemical reactions and are stored as starting material to produce various dyes. Well-established technology is available to apply these colours, shades are bright and fast except a few dye–fibre combinations; shade and hue range is unlimited for most classes of dyes.

Several alternate recipes are available to produce a hue and shade with desired fastness and cost. These colours are not easily biodegradable with exceptions, like reactive dyes which are self-hydrolysed in water.

2.2.2   Chemical structure

Colouring materials are also classified based on their chemical structure, e.g., azo, nitro, nitroso, stilbene, di- and tri-phenylmethane, xanthene, acridine, quinoline, methane, thiazole, azine, oxazine, thiazine, anthraquinone, indigoid and phthalocyanine, etc. Examples of dyes with these groups can be found separately in different chapters.

2.2.3   Field of application

Classification based on application is of commercial importance as a specific textile can be coloured with a single or a few classes of dyes with specific application technique (Table 2.1). In fact, a textile processing industry handles a few fibres which necessitate storage of only required dyes or dye classes irrespective of their chemical structure.

Table 2.1 Classification of dyes based on application

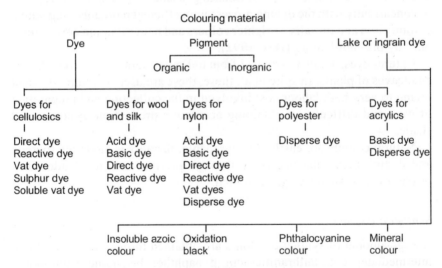

Pigments are suitable for all fibres as absence of affinity necessitates application with binder whereas lakes are developed generally *in situ* fibre if component chemicals possess affinity for fibre to diffuse inside. Metal complex and mordant dyes are derivatives of acid dyes; selective acid dyes with chelating sites react with metals to form these dyes. When the metal is reacted with dye followed by its application on fibre, known as pre-metallised dye; if the metal and dye are applied separately on fibre and are allowed to form the complex, it is called a mordant dye when the metal has affinity for fibre and act as mordant to facilitate dye–mordant–fibre combination. Before the application of a dye to produce a specific hue and shade on a fibre, related cost of dyeing, quality of textile, end use, infrastructure and fastness properties to be developed are to be judged.

Cellulosics are invariably dyed with any of the listed dyes depending on cost and fastness of dyeing; wool and silk are dyed with acid, metal complex or reactive dyes. Basic dyes do not produce light-fast shades on these fibres due to slow decomposition of dye against light. 'Acrylic fibres are dyed with only basic dyes to produce lightfast shades as incident light gets transmitted to fibre phase preventing dye decomposition because of poor moisture regain of fibre. Chemical bonding between dye and fibre ensures good wash fastness too'. Disperse dyes are not applied on acrylics as medium and deep shades can not be produced due to poor penetration of dye in highly crystalline phase. Acid, metal complex and reactive dyes are applied on nylon to develop fast shades; direct dyes are applied where the product is cheaper and will not face subsequent washing, but disperse dyes are never applied due to poor light fastness of dyeings. Polyester is non-ionic and is dyed with only disperse dye; anionic dyeable polyester can easily be dyed at or below boil with direct, reactive, vat dyes, etc., where cationic dyeable type can be dyed with basic dyes.

It is also important to mention here that dyes are also used to enhance whiteness of textiles. To counteract any deficiency in bleaching or to impart superior whiteness, white bleached textiles are often after-treated with optical whitening agents with a little of violet, red dye, etc, depending on type of textile to develop a tone, which increases brilliancy of whiteness. Some dyes are colourless and are used to improve performance of the dyeing process, e.g., colourless basic dyes are frequently used as cationic dye fixing agents to improve wash fastness of direct dyed cotton or as a retarder to produce levelled shades in dyeing acrylic with basic dye.

2.2.4   Ionic nature

Dyes can also be classified according to electrical charge that exists in its ionized coloured part before or during dyeing. Based on this, direct, acid, reactive, sulphur, metal-complex, vat and soluble vat dyes are anionic; basic dye is cationic and disperse dye is non-ionic in nature. Electrical nature of a dye ascertains strike rate and rate of dyeing; opposite electrical charges on dye and fibre enhances both the rates depending on force of attraction generated between these two. A very high rate may lead to unlevelled dyeing necessitating application of retarding or levelling agents. Similarly same electrical nature of dye and fibre retards both the rates requiring application of exhausting agents to promote dye uptake. A chemical bond is established between dye and fibre if both possess opposite electrical charge, otherwise a physical attachment occurs if only one of these is ionic.

2.2.5   Solubility in water

Water is the universal medium of dyeing. It is cheaper, non toxic, easily available and can be recovered and reused. That is why solubility of dye in water is an important factor in dyeing. Direct, reactive, solubilised vat and acid dyes are water-soluble dyes. Vat and sulphur dyes are water-insoluble but can be taken into solution by reduction and solubilisation. Basic dye is soluble in acidic warm bath while other dyes are insoluble in water under normal conditions.

2.3   Nomenclature of colouring materials

2.3.1   Colour Index specification

Nomenclature of a colouring material is of paramount importance and very interesting – a successful dyer must have some knowledge of it, which can also be had of manufacturer's literature. There is no symmetric nomenclature system due to which it is usual worldwide to have various names assigned to a single dye by different manufacturers, called commercial names. To avoid confusion in it, SDC and AATCC (Society of Dyers and Colourists and American Association of Textile Chemists and Colorists) have assigned CI specification (Colour Index generic name and constitution number) to a dye, which may have different commercial names but chemical structure and properties are identical and so the international name of a dye is expressed with its CI number, e.g., some acid red dye may have the name, CI Acid Red 1, etc., followed by its CI constitution number as CI 18050 (a five digit number). The CI specification assigned to a dye has following explanation as shown in Fig. 2.4.

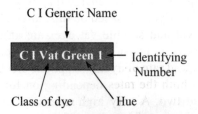

2.4 Description of C I specification of colouring materials.

2.3.2   Interpretation of commercial name of dye

Name of a commercial dye consists of three parts: brand name, hue and its specification. The name starts with a brand name assigned by a specific manufacturer, e.g., disperse dye produced by Indian Dyestuff Industries (IDI) is assigned a brand name 'Navilene', 'Dispersol' by ICI, etc.; naphthols as 'Amarthol' by Amar dye industries while that by Atul as 'Tulathol', etc. The second part is the hue, e.g., red, green, blue, yellow, etc., while the third part describes its specifications, e.g., Y for yellowish, G for greenish, R for reddish, B for bluish, etc.; this last part generally contain some more characters, e.g., 2G, 3G, etc., i.e., double greenish, triple greenish, just to express intensity of greenish tone and so on. Sometimes a letter in the third part indicates the dye class, e.g., Alizarin Blue D (D for direct dye); Alizarine light blue SE is an acid dye (S for sauer, i.e. acid), etc. Introduction of an 'F' in the third part indicates its good light fastness property (Finar, 1975).

To complete the description, let the name of a reactive dye is 'Procion Red M5B'. It means it is from ICI (ICI assigns its reactive dyes trade code procion), hue is red and M for cold brand reactive dye with 5B indicating that the red hue is with an intense blue tone.

2.4   References

SHENAI V A (1987) *Chemistry of dyes and Principles of dyeing*, 3rd Edition, Mumbai, Sevak Publications.

FINAR I L (1975) *Organic Chemistry*, Volume I, 6th Edition, London, Longman Group Limited.

Colour Index International (1987), 3rd Edition (3rd Revision), Society of Dyers & Colourists and AATCC, Bradford, UK.

CHAKRABORTY J N, GUPTA R, RANI V and DAS M (2005) 'Chemical processing of silk', *Indian Textile J.*, **115**, December, 39–51.

DAS S (1992) 'Application of natural dyes on silk', *Colourage*, **39**, 9, 52–54.

GULRAJANI M L (1993) *Chemical Processing of Silk*, Delhi, Indian Institute of Technology.

GULRAJANI M L, GUPTA D B, AGARWAL V and JAIN M (1992a) 'Some studies on natural yellow dyes: I. C 1 Natural Yellow 3: Turmeric', *Indian Textile J.*, **102**, 4, January, 50–56.

GULRAJANI M L, GUPTA D B, AGARWAL V and JAIN M (1992b) 'Some studies on natural yellow dyes: II. Flavonoids – Kapila / Onion / Tesu', *Indian Textile J.*, **102**, 5, February, 78–84.

GULRAJANI M L, GUPTA D B, AGARWAL V and JAIN M (1992c) 'Some studies on natural yellow dyes: III. Quinones: Henna / Dolu', *Indian Textile J.*, **102**, 6, March, 76–83.

3

Dye–Fibre interaction

Abstract: In dyeing, dye is attached with fibre by some sort of forces, may be physical or chemical in nature. This is essential keeping in view desired fastness of dyeings during domestic use. Various dyeings show wash, light, sublimation, rubbing fastness at varying extents based on dye–fibre system. This chapter discloses modes of dye–fibre attachment in order to forecast these fastness criteria.

Keywords: H-bond, van der Waals force, ionic bond, covalent bond, binder

3.1   Introduction

In a dyeing process, efficiency of dyeing is directly proportional to extent of diffusion of dye at the interior of fibre. A dye possessing higher affinity for fibre will significantly improve rate of surface deposition, moderate surface adsorption and slower diffusion. However, rate of diffusion can be improved by opening the fibre structure up to some extent by using swelling agent for natural fibres or by heating up man-made fibres to cause internal 'brownian movement of polymer chains', which breaks up few physical forces engaged to attach polymer chains and make the fibre less resistant to the dye. The dye, however, either reacts with fibre or retained by it with physical forces called fixation. Whatever may be the mode of attachment – physical or chemical, distribution of dye must be uniform to ensure a thorough shade. To achieve that, dye should migrate and diffuse throughout freely; this depends on size and shape of fibre pores which are to give adequate passage to adsorbed dye molecules. In fact, fibre pores may be bigger in size, but may not be compatible with size and shape of dye. In spite of having overall larger pore volume; it may not help in transporting the dye molecule through it (Bird and Boston, 1975).

Fastness as well as quality of dyed textile depends largely on the nature of dye–fibre attachment. Few basic concepts to explain dye–fibre interaction, called theories, those predict this attachment are divided into six categories, viz., physical theory, chemical theory, physico-chemical theory, fibre-complex theory, solid solution theory and mechanical or pigment theory. The concept of half-time dyeing, which ascertains affinity of dye for fibre and a viable test to assess compatibility of dye is a practical approach. During this time, half of dye molecules get exhausted on fibre and each requires to be established for separate dye–fibre systems alone or in combination.

3.1.1   Physical theory

Dyes after application are physically retained by fibre through hydrogen bond and van der Waals forces. Fastness of dyeings depend on size of dye molecule and its solubility in water – the larger the size and lesser the solubility, the better the fastness.

Examples are dyeing of cellulosics with direct, vat, sulphur and solubilised vat dyes; dyeing of man-mades with disperse dyes. Direct dyes are water-soluble causing poor wash fastness, whereas sulphur and vat dyes are water insoluble and larger in size showing good to excellent wash fastness.

3.1.2   Chemical theory

Dye and fibre, both possess required reactive groups to develop dye–fibre attachment through chemical bonding. Nature of bond is mostly ionic (electrovalent), though in some cases, covalent bonds are also formed. Fastness of dyeings depends upon number of reactive sites attached to the fibre – the greater the number of sites attached, the better the wash fastness. After half-time dyeing, electrolyte is added for further exhaustion of bath, e.g., dyeing of cotton with reactive dyes. Somewhere, rate of dyeing is so fast that retarding or levelling agents are applied at the start of dyeing to ensure level dyeing through reduction in rate of dyeing, e.g., dyeing of wool with acid dyes. Examples of ionic attachment are dyeing of silk with acid or basic dye, acrylic with basic dye, nylon with acid dye, cationic dyeable polyester with basic dye and anionic dyeable polyester with direct, vat, sulphur and reactive dyes, while dyeing of cotton with reactive dyes is a case of covalent bond formation between dye and fibre.

3.1.3   Physico-chemical theory

Dyes are applied to form physical bonds with fibre. To improve fastness of dyed textile, size of dye molecule is increased by reacting the dye (while on textile) chemically with some other external chemical or dye known as after-treatment. One component must be a dye and the second component is either a dye or a chemical, e.g., after-treatment of direct dyeings with either of metal salt, formaldehyde, cationic dye fixing agent, basic dye, resins or diazotized base. Other examples are back tanning of dyed protein fibres, mordanting of cotton for dyeing with basic dye and mordant dyeing of wool.

3.1.4   Fibre complex theory

Dye is incapable of entering into fibre matrix by itself due to larger structure and lack of affinity and is produced *in situ* by means of reacting two different chemicals under favourable conditions. The first component is applied to the textile by padding or exhaust method, followed by reacting previously treated textile with second component. In some cases, the two components are mixed and simultaneously applied – reaction takes place when the treated fabric is after-treated under favourable conditions. The desired colour is developed only after coupling. The coloured complex (lake or ingrain dye) is formed *in situ* and can not come out of fibre phase, e.g., dyeing of cotton with insoluble azoic, mineral, oxidation and phthalocyanine colours. Insoluble azoic colour is formed in textile when solubilised naphthol is applied on cotton followed by passing of the naphtholated cotton through a diazotised base. Mineral colours are formed *in situ*, when different metal salts are applied separately on cotton and converted to their oxides. Oxidation black is developed when aniline polymerises inside cotton. Phthalocyanine colours are formed when alcian blue is mixed with copper complex in solvent and applied; treated cotton is dried and polymerized to form phthalocyanine blue dye *in situ*.

3.1.5   Solid solution theory

Both the phases are solid – dyestuff and fibre, but when the dye is applied under suitable condition, it gets passage inside fibre resulting a solid–solid solution mechanism. Dyestuffs are retained within the fibre with physical forces. Fastnesses of dyeings are good in some cases (polyester) and not good in other cases (light fastness of disperse dyed nylon). Dyeing is done at higher temperature, when the fibre structure is opened up paving adequate passage to the dye. The dye gets mechanically deposited on fibre surface, diffuses inside and is trapped due to collapse in fibre structure during cooling, e.g., dyeing of man-mades with disperse dye. The term solid solution has been referred due to the fact that location of dye inside hydrophobic fibre can not be ascertained by electron microscopy or x-ray diffraction due to solubilisation of it in fibre matrix, looks like a solid vs. solid solution (Mittal and Trivedi, 1983).

3.1.6   Pigment or mechanical theory

The colouring material possesses no reactive site, has no affinity for fibre and is insoluble in water and other solvents. Dyeing is generally carried out in aqueous emulsion along with a binder. The latter on curing establish cross-link with textile and trap pigments on fibre surface. Though fastness properties are good, but it depends strictly on the longevity of the film. Fabrics become stiffer, e.g. dyeing of any fibre with pigment colours. The colour is retained by the fibre mechanically and no pigment–fibre attachment exists.

3.2   Chemical aspects related to theories of dyeing

3.2.1   Dyeing of cellulosics with direct dye

Direct dyes are water-soluble; split in dye anion and a cation in water. Cellulosic fibres, when dipped in water for dyeing, its outer surfaces acquire negative charge.

$$DSO_3H \longrightarrow DSO_3^-H^+$$
$$Cell\text{-}OH \longrightarrow Cell\text{-}O^-H^+$$

Due to presence of similar electrical charge on dye and fibre, stripping takes place and so, after half-time dyeing, electrolyte is added (NaCl or Na_2SO_4) to reduce this potential difference (zeta potential). Dye–fibre attachment takes place through hydrogen bonding and van der Waals force between –OH groups of cellulose and functional groups of the dye (NH_2 or OH).

3.2.2   Dyeing of cellulose with reactive dye

Reactive dyes are of two types, viz. hot (monochlorotriazine dyes) and cold brands (dichlorotriazine dyes). Due to presence of two chlorine atoms, cold brands are more reactive, even dyeing takes place at room temperature. Reactive dyes attach themselves with cotton through formation of covalent bonds with cotton, only when alkali is added to dyebath to neutralize acid formed during this bond formation.

Hot brand dyes possess only one chlorine atom and so dye–fibre attachment takes place at one junction.

3.2.3   Dyeing of cellulose with vat dye

Dyes are insoluble in water and possess C=O groups in their structure. These are reduced and solubilised to develop affinity for cellulosics.

$$D - C = O \xrightarrow{\text{Reducing Agent}} D - C - OH \xrightarrow{\text{Alkali}} D - CONa$$

After dyeing, these are recovered into their parent form by oxidizing the dyed fabric. Size of dye molecule is smaller than pore size of cotton fibres but form bigger stable aggregates after soaping to develop superior fastness. A physical theory of dyeing is followed, i.e. dyes are retained by fibre through hydrogen bonds and van der Walls forces.

3.2.4   Dyeing of cellulose with insoluble azoic colour

Naphthols are first solubilised with NaOH and applied on cotton. When a base is diazotised and the naphtholated textile is treated in this base, coupling takes place resulting the formation of coloured lake. The whole process passes through three different stages, viz.,

(a) solubilisation of naphthol

$$D - OH + NaOH \longrightarrow D - O - Na$$

(b) diazotisation of base

$$Ar - NH_2 \xrightarrow{\text{HCl + NaNO}_2, 5°C} Ar - N = N^+Cl^-$$

(c) reaction between naphthol and diazotized base to develop lake

$$Ar - N = N^+Cl^- + D - O - Na \longrightarrow Ar - N = N - O - D \text{ (lake)}$$

3.2.5   Dyeing of cellulose with sulphur colour

Dyes have sulphur linkages (S-S) and are insoluble in water. These are converted to thiols (-S-H-) and then to sodium salt (-S-Na-); the later shows affinity for cellulose. After dyeing, dyed fabric is oxidized to restore original dye structure. Physical theory of dyeing is followed here.

$$Dye - S - S - Dye \xrightarrow{\text{Reduction}} Dye - SH - HS - Dye \xrightarrow{\text{Solubilisation}}$$

$$Dye - SNa + NaS - Dye \xrightarrow{\text{Oxidation}} Dye - SH - HS - Dye$$

3.2.6   Dyeing of wool, silk and nylon

All these fibres have both acid and basic groups, but the basic groups mainly take part in dyeing. The simplest formula can be represented simply as

$$H_2N - F - COOH \quad (F = \text{fibre chain of wool, silk or nylon})$$

In acidic pH, the fibres are ionized as follows:

$$H_2N - F - COOH \xrightarrow{HX} X^{-+}H_3N - F - COOH$$

Acid dyes are water-soluble, ionizes in water:

$$RCOOH \longrightarrow RCOO^- + H^+$$

In acid medium, dyeing takes place forming electrovalent bond between dye and fibre

$$X^{-+}H_3N - F - COOH + RCOO^- + H^+ \longrightarrow H_3N^+ - F - COOH + HX$$
$$\underset{\text{(dyed fibre)}}{\overset{|}{{}^-OOC - R}}$$

3.2.7   Dyeing of acrylic with basic dye

Conventional acrylic fibres possess negative charge showing affinity for basic dye.

$$\left[\begin{array}{c} CH_2 = CH \\ | \\ CN \end{array} \right]_n$$

Basic dyes in slight acidic pH are also dissociated as

$$R_4NCl \longleftrightarrow R_4N^+Cl^-$$

Dyeing takes place by forming electrovalent bond between dye and fibre

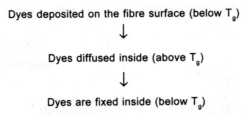

Rate of dyeing is very fast around and over T_g of acrylic fibre (~85°C) necessitating application of a cationic retarding agent for level dyeing.

3.2.8   Dyeing of man-made fibres with disperse dye

Polyester has no reactive sites and non-ionic dye also possesses no electrical charge; dyeing occurs in mild acidic medium. In dyeing, fibre structure is opened up at higher temperature causing deposition of dyes on fibre surface. Adsorbed dye diffuses inside and on cooling gets trapped. H-bonding and van der Walls attachments are developed between dye and fibre.

Dyes deposited on the fibre surface (below T_g)

↓

Dyes diffused inside (above T_g)

↓

Dyes are fixed inside (below T_g)

3.3   References

BIRD C L and BOSTON W S (1975) *The theory of coloration of textiles*, Bradford, England, Dyers company publication trust.

MITTAL R M and TRIVEDI S S (1983) *Chemical Processing of polyester/cellulosic blends*, Ahmedabad Textile Industries Research Association, India.

4

Dyeing with direct dye

Abstract: Direct dyes are water-soluble anionic dyes and retain sulphonic acid groups in structure. Though light fastness seems to be good in most of the cases, washfastness suffer due to its water solubility. Most of the direct dyes belong to azo class and are suitable for cellulosics, where the product faces no domestic washes, like cheaper carpets. Dyes are retained by the fibre through hydrogen bonds and van der Walls force. Washfastness may be improved by suitable after-treatment technique. Due to presence of acid groups, direct dyes react with protein fibres forming washfast shades but lack brilliancy.

Keywords: Direct dye, electrolyte, temperature, washfastness, after-treatment

4.1   Introduction

Direct dyes are mainly applied on cellulosics but are also suitable for nylon, wool and silk. Dyeings are good light fast but poor washfast due to smaller size of dye with good water solubility; which can be improved to some extent through after-treatment. Bright shades are not available in this series. Articles which are seldom washed, e.g., window coverings, upholstery and heavy bedding, or labeled as 'dry clean only' are dyed with this dye. Direct dyes are superior to other dye classes in terms of cost, better light fastness, ease in application, shorter dye cycle, low cost of auxiliaries, remarkably lesser use of water and much lower salt level in effluent (Esche, 2004). Developments in direct dyes for cellulosics have been discussed elsewhere in detail (Shore, 1991).

4.2   Properties of direct dyes

Dyes split up in water forming dye anion and sodium cation.

$$D - SO_3Na \longrightarrow DSO_3^- + Na^+$$

The free sulphonic acid form of dye is less soluble then corresponding

27

sodium salt, necessitating addition of little Na_2CO_3 for complete dissolution in bath.

$$2D - SO_3H + Na_2CO_3 \longrightarrow 2DSO_3Na + H_2O + CO_2$$

If hard water is used, a chelating agent like EDTA is to be used to remove hardness. Dye molecules exist as aggregates and during dyeing these aggregates are to be broken down to promote dye uptake and levelled shade.

Though dye retains negative charge while in bath, dye–fibre attachments are of various types: with cotton it is by physical means, i.e. H-bonds and van der Waals forces; with nylon, wool and silk it forms ionic bond. In the first case, presence of weak physical forces as well as solubility of dye in bath renders poor washfastness. As dyes are completely soluble in alkaline bath, domestic washing results severe stripping out of dye. In contrast, ionic attachment in case of wool, silk and nylon improves washfastness ratings but shades remains relatively dull and so are not viable for these fibres.

Presence of direct dye on cellulosics can be identified by boiling dyed material in bath along with scoured cotton when the latter is dyed with stripped-off colour.

Most of the dyes are sulphonated azo compounds with chemical groups, such as $-NH_2SO_3H$ and $N=N$. Here $-N=N-$ is the chromophore and others are auxochrome. Dyes are dischargeable with almost all reducing agents. Most of the commercial direct dyes conform to the general formula

$$R_1 - N = N - X - N = N - R_2$$

4.3   Classification

Dyes are classified based on their (i) chemical structure and (ii) method of application.

4.3.1   Chemical structure

Most of the direct dyes belong to azo dye class, viz. Monoazo dyes (Diamine scarlet B), Diazo dyes (Congo red), Triazo dyes (Direct brown) and Polyazo dyes (Chlorazol dyes). A few are stilbene derivatives, viz. Stilbene direct yellow and rest are thiazole derivatives, e.g., Primuline light yellow.

4.3.2   Method of application

Dyes are classified in three groups, viz. self-leveling (A), salt sensitive (B) and temperature sensitive (C) (Peters, 1975). Class A dyes show good levelling and exhaustion even in absence of electrolyte and heat, i.e. at room temperature. Class B dyes have poor levelling properties but with controlled addition of salt, level dyeing can be obtained even without application of heat while exhaustion of class C dyes is affected by heating up the dyebath and application of salt is not so effective. C dyes because of high exhaustion at 90°C with poor levelling, are best controlled by starting dyeing at room temperature for 30–45 min with slow heating up to 40–45°C, after which salt is added. The bath is further heated up slowly to 90–95°C and dyeing is carried out for required time. Rise of temperature increases rate of dye sorption, diffusion and migration, especially for class B and C dyes causing levelling while presence of electrolyte ensures formation of aggregates at the end of dyeing to improve some sort of fastness.

Direct dyes are marketed under brand names like Direct, Aizen, Atlantic, Atul, Benzanil, Benzo, Bitafast, Chlorazol, Coranil, Cotton, Cuprofix, Cuprophenyl, Diacotton, Diamine, Diazine, Diphenyl, Diazol, Duasyn, Enianil, Helion, Intralite, Dyrect, etc. (Colour Index, 1987).

4.4   Specification of some direct dyes

Structure and CI specification of some commonly used direct dyes are shown in scheme 4.1 (Colour Index, 1987).

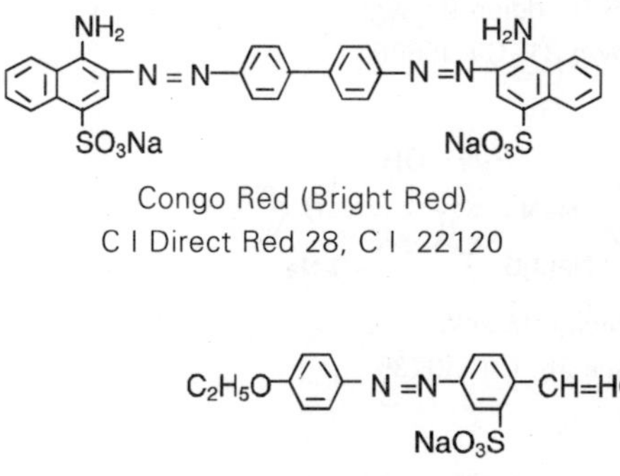

Congo Red (Bright Red)
C I Direct Red 28, C I 22120

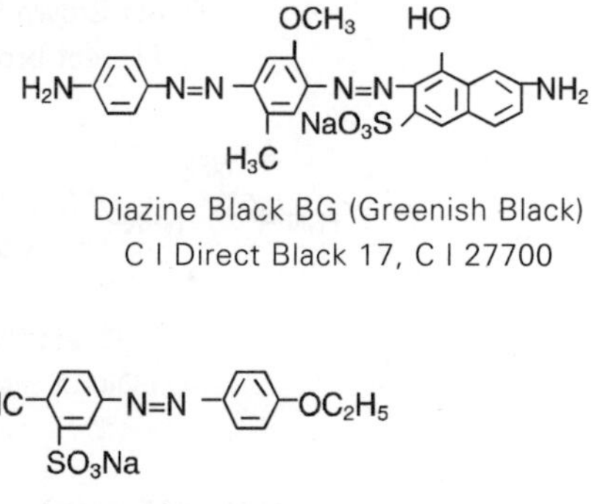

Diazine Black BG (Greenish Black)
C I Direct Black 17, C I 27700

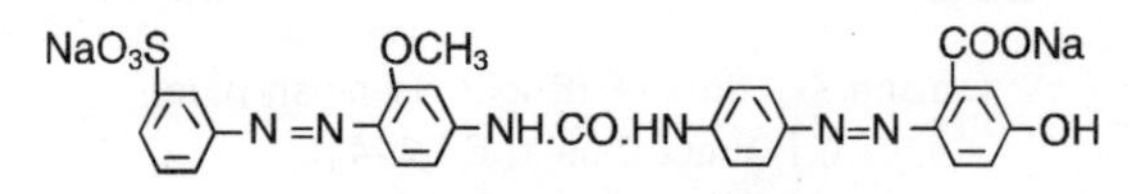

Chrysophenine (G / GX / R) (Reddish Yellow)
C I Direct Yellow 12, C I 24895

Diazine fast Yellow 4GL (Bright Yellow)
C I Direct Yellow 44, C I 29000

Scheme 4.1 Contd. ...

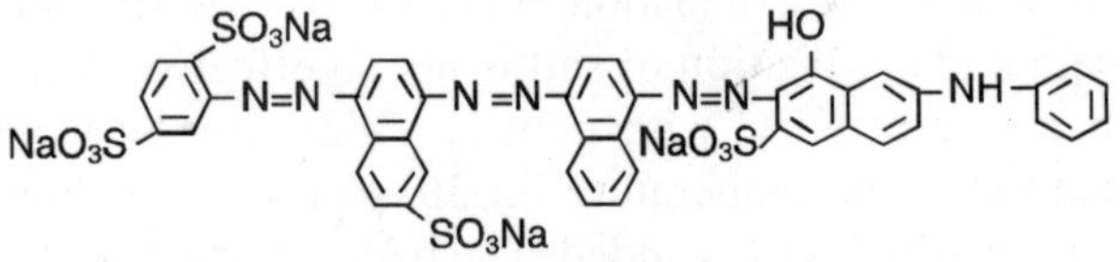

Diazine fast Blue 4GL (Blue)
C I Direct Blue 78, C I 34200

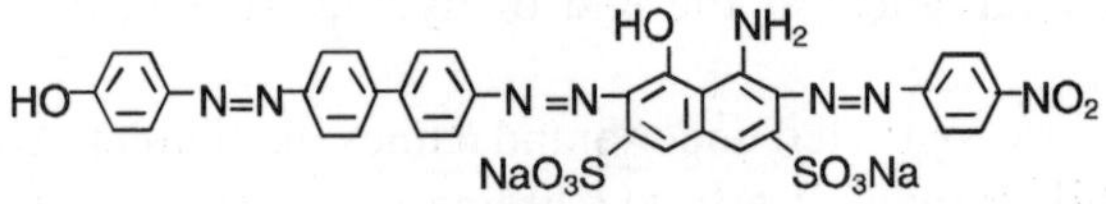

Direct Green A / BX / G (Dull Green)
C I Direct Green 6, C I 30295

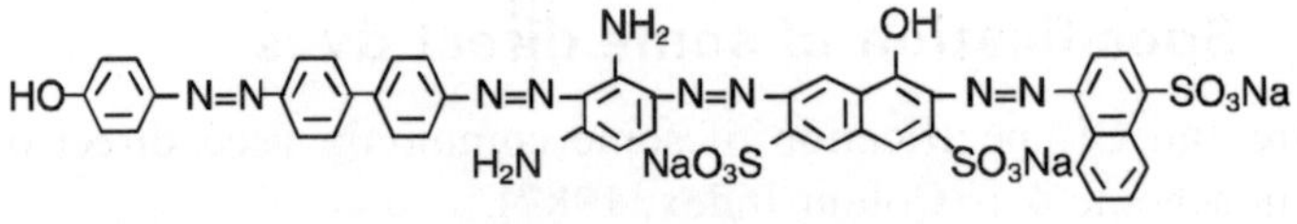

Direct Brown B (Reddish Brown)
C I Direct Brown 31, C I 35660

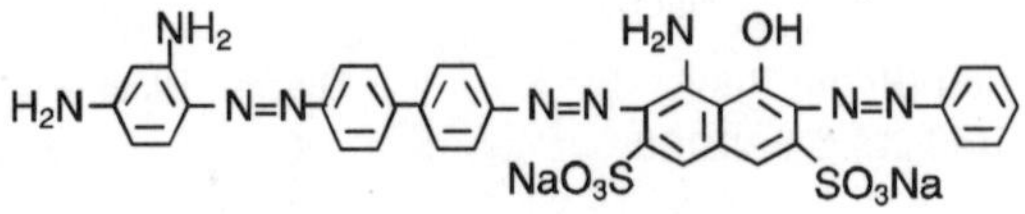

Direct Black E (Black)
C I Direct Black 38, C I 30235

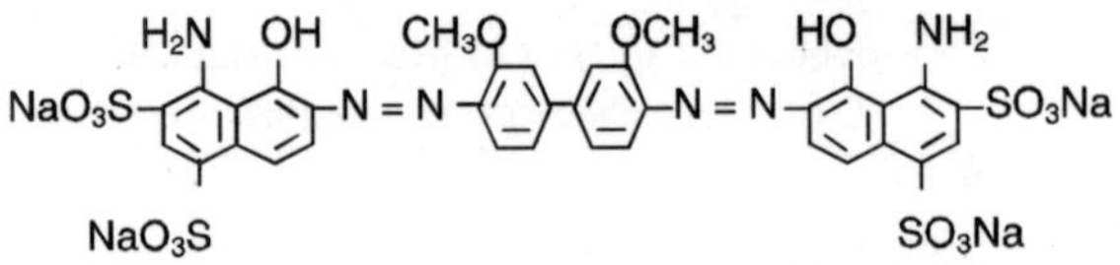

Diazine Sky Blue FF (Bright Greenish Blue)
C I Direct Blue 1, C I 24410

Scheme 4.1 Contd. ...

Diazine Orange 5 (Bright yellowish Red)
C I Direct Orange 26, C I 29150

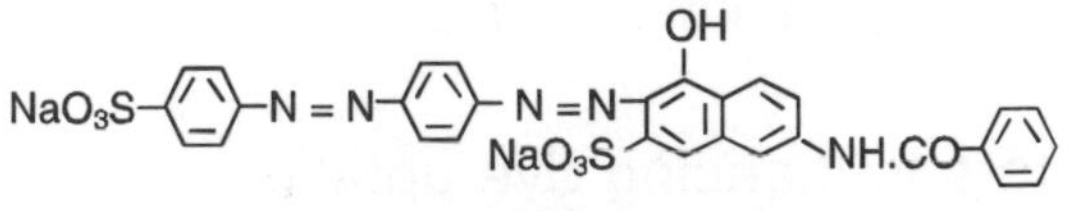

Durazol Red 2B (Bright Red)
C I Direct Red 81, C I 28160

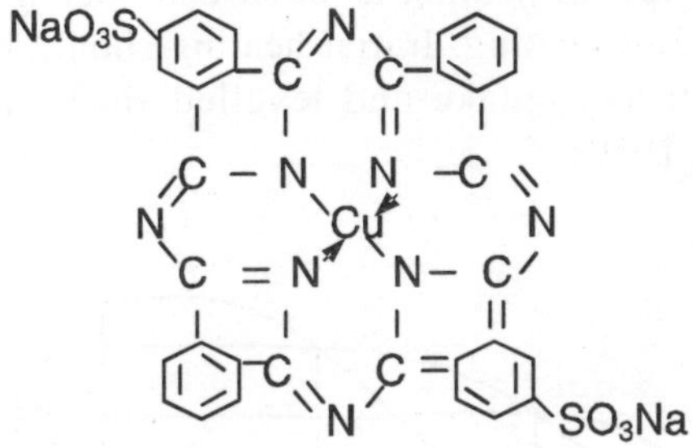

Durazol Blue 8G (Turquiose Blue)
C I Direct Blue 86, CI 74180

Scheme 4.1 Structure and C I specification of some important direct dyes

4.5   Dyeing of cellulosics

Though direct dyes belong to A, B and C classes having differential criteria for efficient dyeing, under practical conditions, either manufacturers do not provide adequate information on class of supplied dye or dyes from various classes are mixed to produce a specific hue necessitating to follow a common dyeing approach for all classes of dyes.

Required amount of dye is pasted with little Turkey red oil to prevent floating of dye in bath, required amount of water is added followed by addition of a trace of soda ash for complete solubilisation of dye. Cotton is wetted and placed in bath. Dyeing is started at room temperature with slow heating up to 45°C over a period of 15–30 min after which salt (50 g/l) is added; temperature is further raised slowly to boil for complete exhaustion of dye in 1–2 hrs depending on depth of shade. Soaping and washing are omitted; rather dyeings are directly dried and stored.

Typical recipe and conditions of dyeing

Dye   – $x\%$
T R Oil   – just little to make paste
Na_2CO_3 – a trace
Salt   – 50 g/l
Temperature   – room temperature to boil
Time   – 2–3 h

4.5.1   Factors influencing dye uptake

Influence of temperature

(a) Equilibrium sorption of dye (g dye/kg fibre) at higher temperature and at higher rate of heating is lesser than that at lower temperature at slower rate of heating. Initial heating should be at a slower rate for maximum dye uptake and levelled shade, shown in Fig. 4.1 (Vickerstaff, 1950).

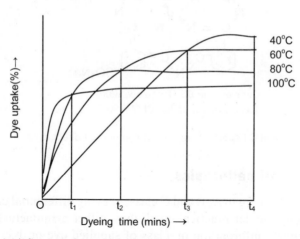

4.1 Influence of rate of heating on dye uptake.

(b) Cellulose acquires negative charge when dipped in water and dye acquires anion in bath resulting repulsion between dye and cellulose; the difference in potential of dye and fibre is called zeta potential. The potential barrier is overcome by introducing some extra heat energy into bath.

(c) The pores of fibres are very small; it only allows entering a small single dye molecule. Direct dyes exist in solution as aggregates. With rise in temperature, these aggregates are broken down progressively to single molecules causing better penetration and higher diffusion of dye. The net result is a levelled shade with maximum dye uptake.

For effective maximum dye uptake, rate of heating should be slower with more time for exhaustion; even a lower constant temperature for prolonged time is adequate for that.

Feasibility of dyeing of cotton with direct dyes at lower temperature in presence of triethanolamine was studied and it was observed that dye uptake varied with temperature and triethanolamine concentration; dyeing cost was considerably less along with substantially higher dye uptake even at 50°C with comparable fastness properties in comparison with those dyed in conventional method (Malik, 2003).

Effect of pH

Application of a trace of alkali is required to ensure complete solubilisation of dye; concentration must be too little otherwise dyebath exhaustion may be poor with a better levelled shade.

(a) Increase in alkalinity of bath lowers substantivity of dye. This is because the negative potential of dye and fibre both get increased causing more repulsion between dye and cellulose (higher zeta potential), showing lesser dye uptake. Concentration of Na_2CO_3 must be kept on too lower side.

(b) Addition of excess alkali raises the pH causing better solubility of dye but reduces affinity causing better diffusion as more and more anionic –OH groups are released in bath.

(c) Decrease in substantivity of dye at higher pH improves levelling of shade.

(d) Increase in pH counteracts hardness of water and so dyeing must be done in either neutral or slight alkaline medium.

Role of electrolyte

(a) Addition of electrolyte promotes dye uptake through reduction in zeta potential and promotes substantivity of dye (Fig. 4.2).

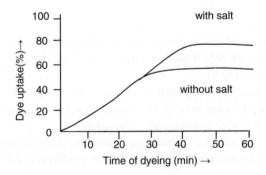

4.2 Effect of salt on dye uptake.

(b) During dyeing, single molecules of dye diffuse inside fibre through the pores. Addition of electrolyte promotes formation of dye aggregates inside the fibre and improves washfastness.

Concentration of electrolyte depends on: (a) number of SO_3H groups retained by a dye molecule – higher the number higher will be requirement of salt to suppress higher zeta potential and (b) percent of shade to be developed – deeper the shade, more will be requirement of salt.

All direct dyes require salt at varying concentrations for better exhaustion on cellulosics; even some exhaust well at low salt concentration (Herlant, 1993).

Influence of liquor ratio

Higher the effective concentration of dye in bath, higher will be dye uptake. This implies that a lower liquor ratio is more effective for efficient dyeing. Increase in liquor ratio will produce lighter shades and vice-versa, as shown in Fig. 4.3.

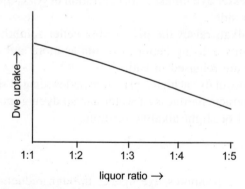

4.3 Dye uptake at varying liquor ratio.

Time of dyeing

(a) A slower rate of heating and increase in time of dyeing for a given dye concentration will affect higher dye uptake, as explained in Fig. 4.4 (Vickerstaff, 1950).
(b) The equilibrium absorption depends upon time and temperature. If rate of heating is high, lower will be the time to reach to equilibrium absorption with lower dye uptake and vice-versa.
(c) For mixtures of dyes, rate of exhaustion of each dye is important for level dyeing and desired shade which may be confirmed through compatibility test.

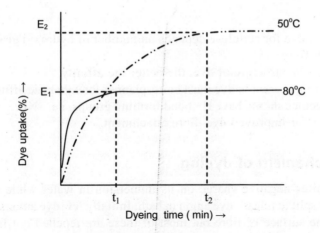

E_2 at 50°C for time t_2 > E_1 at 80°C for time t_1, when t_2 > t_1

4.4 Impact of time of dyeing on dye uptake.

The concept of half-time dyeing plays an important role to assess affinity and compatibility of dyes in mixture. It is the time required for transfer of half as much dye from bath to fibre in comparison to that is absorbed in fibre at equilibrium. Half-time dyeing of dyes A and B is t_3. Absorption of dye A ceases at t_4 while that for dye B at t_5; t_4 and t_5 are remarkably higher than t_3, even double of that as shown in Fig. 4.5 (Vickerstaff, 1950). It shows that about half of dye is exhausted from bath during first few minutes of dyeing and the rest is exhausted throughout a long time thus conforming levelling of shades.

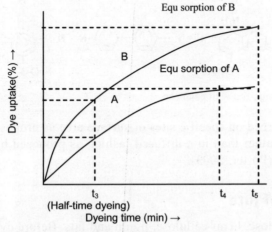

4.5 Concept of half-time dyeing.

Affinity of dye

(a) Affinity of dye for cellulose depends on number of hydroxyl groups present in fibre.
(b) The linear the structure of dye, the better the affinity.
(c) Benzene ring present in dye must be co-planner for enhanced affinity.
(d) Dye molecules should have H– bond forming groups, i.e. $-NH_2$, $-OH$ or $-N=N-$ for improved dye–fibre attachment.

4.6   Mechanism of dyeing

Cellulose acquires negative charge on its immersion in water while dye molecules also split to release dye anion in bath. Initially few dye anions are absorbed on the surface of fibre but most of these are repelled out from surface due to huge negative potential difference (zeta potential) and further dyeing ceases. Addition of electrolyte reduces the zeta potential and promotes absorption by releasing sodium cation which gets attached with dye, carries it to the surface and reduce extent of osmotic work required to transfer the accompanying metal ions. Sodium salt of dye molecules deposited on fibre surface slowly diffuse inside water swollen cellulose matrix and places itself alongside the polymer chain with H-bond and van der Waals forces in the way as shown in Fig. 4.6 (Vickerstaff, 1950). Application of heat facilitates breaking of dye aggregates as well as reduces zeta potential.

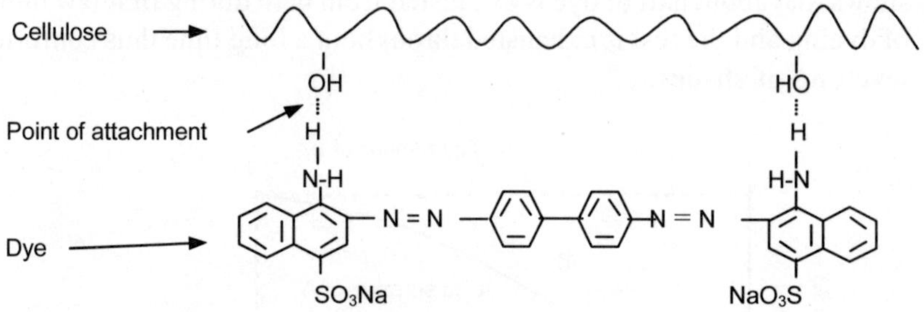

4.6 Attachment of Congo red with cellulose.

Direct dyes are sorbed on specific sites or surface area as proposed in Langmuir equation rather than in a diffused fashion as proposed by the Freundlich equation (Porter, 2003).

4.7   Dyeing of jute

Jute consists of cellulose, hemi-cellulose, lignin and fats. Before dyeing, pre-treatment is imparted with mild alkali. Application of strong alkali attacks hemi-cellulose. Jute is a cheap fibre, has more amorphous region than cotton and it contains hemi-cellulose, so –OH groups show enhanced affinity for direct dye. Method of application is same as that with cotton. Dye uptake is also more than that on cotton for a given dye concentration requiring less dye for a specific shade.

4.8   Dyeing of protein fibres

Protein fibres like wool and silk, even nylon have –NH$_2$ and –COOH groups on either sides of polymer chain. Direct dyes can react with –NH$_2$ groups of fibre to form ionic bonds. When wool or silk is dipped in acidic bath, –NH$_2$ groups are protonated to form –NH$_3^+$ and dye anion (DSO$_3^-$) react with –NH$_3^+$ groups to establish ionic bonds. Dyeing rate is so high that electrolyte is added at the start of dyeing to retard dye uptake to form levelled shades. Direct dyes produce fast but dull shades on wool and silk.

Protein fibre $\longrightarrow$ –NH$_2$–P–COOH $\longrightarrow$ NH$_3^+$–P–COO$^-$ $\longrightarrow$ NH$_3^+$–P– COOH

Dye $\longrightarrow$ DSO$_3$Na $\longrightarrow$ DSO$_3^-$ + Na$^+$

Fixation $\longrightarrow$ NH$_3^+$–P– COOH + DSO$_3^-$ $\longrightarrow$ DSO$_3^-$ NH$_3^+$–P–COOH

Concentration of electrolyte obviously depends on shade to be developed, e.g. for light shade – 5 g/l, medium shade – (10–20) g/l and deep shade (>=3%) – (30–50) g/l.

4.9   After-treatment of dyeings

Shades produced on cellulose with direct dyes are not washfast as dyes are water-soluble and molecular size of dye is smaller than pore size of cellulose in most cases. Washfastness can be improved by increasing molecular size of dye through its reaction with other chemicals or dyes after dyeing, while the dye will remain attached with fibre with H– bond and van der Waals force, i.e. a physicochemical theory of dyeing is followed. Various post-treatment methods are (i) treatment with metal salts, e.g. CuSO$_4$, K$_2$Cr$_2$O$_7$ etc or combination of these, (ii) treatment with HCHO, (iii) diazotization and development, (iv) coupling with diazotized base, (v) treatment with dye cationic fixing agents, (vi) topping with basic dye, and (vii) treatment with resin (Peters, 1975).

Selection of a specific after-treatment process is based on the chemical structure of dye that has been used in dyeing; random selection of any method will not show any result as chemicals or dyes to be applied on dyeing only show reactivity for a specific dye structure. Exceptions are treatment with cationic dye fixing agents, topping with basic dye and treatment with resin – these after-treatments are based on anionic charge on dye and not on the chemical structure.

4.9.1   Treatment with metal salt

Metal salts get attached with dye increasing size of dye molecule. Dyes possessing chelating sites in their structure, e.g., O-O'hydroxy, O-O'carboxyl, O-hydroxy O'carboxyl, only show response to this after-treatment process through chelation of metal atom in their structure. The chelation site may be on the same benzene ring too.

(a) O–O'dihydroxy azo

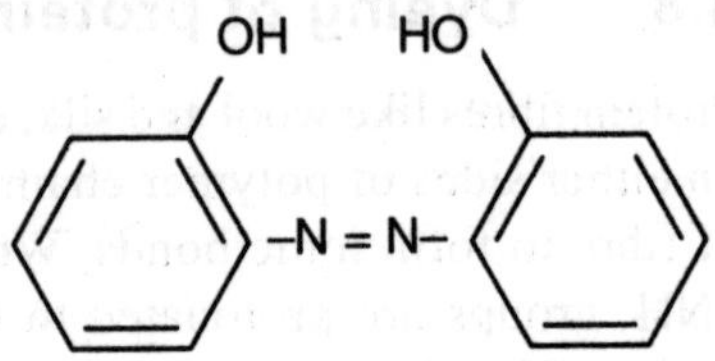

(b) O- hydroxy O'-carboxyazo

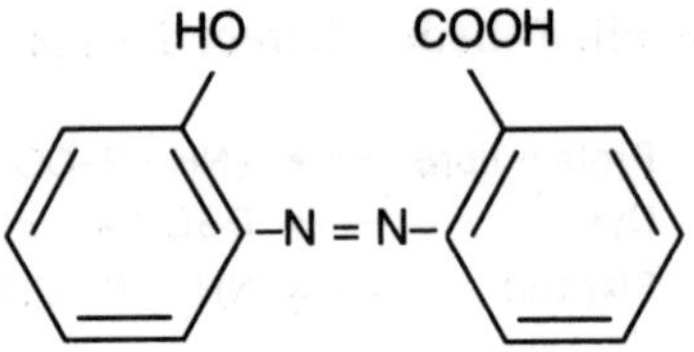

(c) Salicylic acid

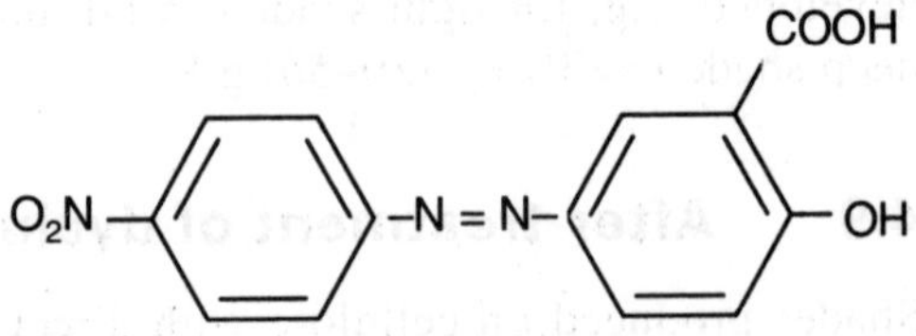

If a direct dye possesses O-O'dihydroxy azo-based structure, its attachment with metal takes place as follows:

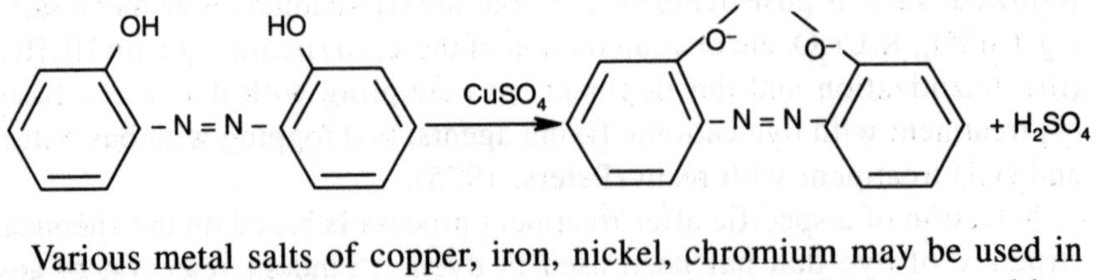

Various metal salts of copper, iron, nickel, chromium may be used in this after-treatment either alone or in combination based on cost and hue to be developed but generally salts of copper and chromium are preferred, i.e. $CuSO_4$ and $K_2Cr_2O_7$.

Dyeings are treated in a bath containing CH₃COOH (1–3%) and metal salt (1–3%) based on depth of shade at 50–60°C for 30 min followed by washing and drying. However, presence of metal in dyed textile makes it acid sensitive, stiffer and change in hue occurs. Reproducibility of exact hue is troublesome.

4.9.2   Treatment with formaldehyde

Dye must have 2-hydroxy or 2-amine groups in m-position to each other around its structure to react with formaldehyde which forms methylene bridges (D-CH₂-D-CH₂) between two successive dye molecules to double the size suppressing its solubility.

The process sequence and parameters are same as those are used for metal salts with exception to formaldehyde which is used in place of metal salt. Shades are difficult to reproduce, polluted environment due to use of formaldehyde, release of same at higher temperature and no such improvement in light fastness though washfastness is improved.

4.9.3   Diazotisation and development

Dye possessing –NH₂ groups in p-position w.r.t azo group undergoes this treatment. When dyed cotton is passed through a diazotized bath, the dye in it is diazotized. It is then followed by coupling with a developer, e.g., phenol, α and β-naphthol, resorcinol, m-phenylene diamine, phenyl methyl pyrazolone, amino diphenyl amine, toluene, etc., (chemicals containing –OH groups). The final shade depends on coupling agent – developer combination.

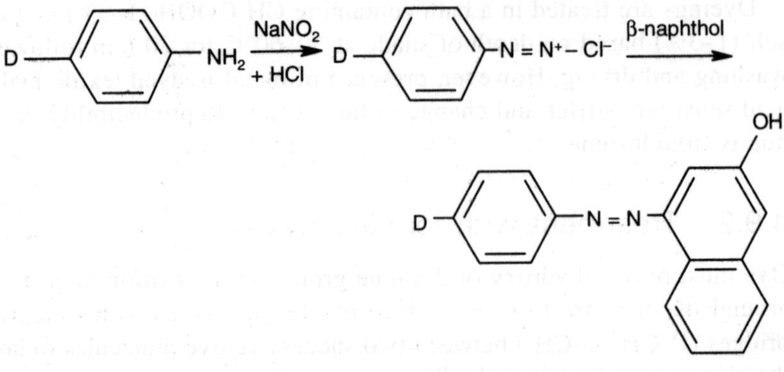

Dyed fabric is treated with $NaNO_2$ (1–3%) and HCl (6–7%) at 5°C for 15 min, when $-NH_2$ groups in dye form diazonium salt, which is coupled with a developer (coupling agent). A coupling bath is prepared with a developer (1%), solubilising agent (NaOH – 1% for phenolic compounds, Na_2CO_3 – 1% for amines, HCl – 0.5% for diphenylamine compounds) in 10 times water. In this process, though all the fastness properties are improved, change in shade makes it commercially impracticable.

4.9.4   Coupling with diazotized base

Dyes possessing –OH or $-NH_2$ groups are only after-treated in this method. Dyed cotton, when treated with diazotized base, e.g., with benzo–p–deep brown G, the reaction finally becomes

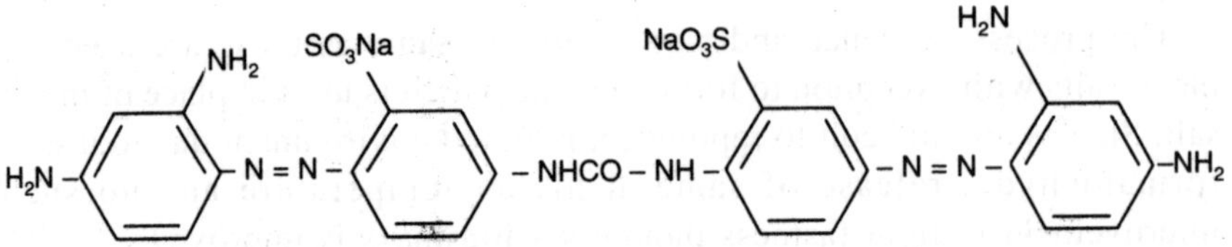

The stepwise reaction occurs as follows

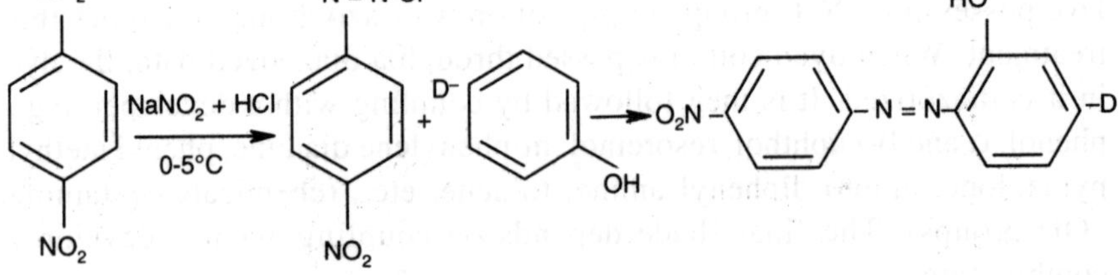

The only base used is p-nitro aniline as it is cheaper and has very high coupling energy. Base is treated with $NaNO_2$ (1–3%) and HCl (3–7%) at 5°C. Once the diazotization is over, pH of bath is adjusted to 4–5 with

CH_3COOH and CH_3COONa and the dyed cotton is treated in it at room temperature for 10–15 min followed by soaping and washing.

4.9.5   Treatment with cationic dye fixing agents

Cationic dye fixing agents are colourless basic dyes and can react with anionic dye to increase its size – wet fastness is improved but light fastness is adversely affected. Cellulose dyed with any direct dye, irrespective of its structure, can be after-treated in this method. No colour change takes place and reproducibility is excellent with a fear of change in tone in certain cases. Cationic dye fixing agent splits up into long cation and short anions when direct dye also splits up into long anion and short cation; subsequent reaction forms the complex.

Dye fixing agent   – $R_4NCl \longleftrightarrow R_4N^+ + Cl^-$

Direct dye   – $DSO_3Na \longleftrightarrow DSO_3^- + Na^+$

$(R_4N^+ + Cl^-) + (DSO_3^- + Na^+) \longrightarrow (DSO_3^- + R_4N^+) + NaCl$

Cationic dye fixing agent (0.3%) is dissolved in water along with CH_3COOH (1–3%). Dyeings are treated in this bath for 15 min at 50–60°C. Exact quantity of fixing agent depends on depth of shade. Some common dye fixing agents used are fixanol C (cetyl pyridinium bromide, ICI), sandofix WE (sandoz), etc.

4.9.6   Topping with basic dye

Basic dyes have the same chemical nature as that is possessed by cationic dye fixing agents, only difference is that basic dyes are coloured molecules. When direct-dyed cellulose is treated in a basic dyebath under identical conditions, reaction occurs to form direct dye–basic dye complex. Bright washfast shades are produced with some deterioration in light fastness. Colour of both basic and direct dye must be same to avoid any change in final colour.

Basic dye – $R_4NCl \longleftrightarrow R_4N^+ + Cl^-$, Direct dye – $RSO_3Na \longleftrightarrow RSO_3^- + Na^+$

$(R_4N^+ + Cl^-) + (RSO_3^- + Na^+) \longrightarrow (RSO_3^- + R_4N^+) + NaCl$

4.9.7   Treatment with resin

Resins are cationic surfactants, dissolved in water to develop cation which in turn reacts with dye anion to form dye–resin complex.

$$
\text{Resin —} \quad
\begin{array}{c}
H_2N \\ \\ H_2N
\end{array}\!\!\!\diagdown\!\!\diagup\; C = NH
\quad\xrightarrow{\;H_2O\;}\quad
\left[\;
\begin{array}{c}
H_2N \\ \\ H_2N
\end{array}\!\!\!\diagdown\!\!\diagup\; C = NH_2
\;\right]^{+}
$$

$$
\text{Dye —} \quad DSO_3Na \longrightarrow DSO_3^- + Na^+
$$

$$
\text{Complex} \quad\longrightarrow\quad
\left[\;
\begin{array}{c}
H_2N \\ \\ H_2N
\end{array}\!\!\!\diagdown\!\!\diagup\; C = NH_2
\;\right]^{+} DSO_3^-
$$

Dyed cotton is padded with resin, catalyst and acid at room temperature, dried at 70–80°C and cured at 140°C for 5–10 min followed by soaping, washing and drying. Some commonly used resins are Fibrofix (cynamide–formaldehyde), lyofix SB (melamine–formaldehyde), etc.

4.10   Stripping of colour

It becomes essential to strip out dye to rectify unleveled shades. Boiling in water for 10–15 min causes partial stripping, addition of little NH_4OH in bath strips out most of colour; complete stripping may be done with strong reducing or oxidizing agents. To strip out dye from after-treated cotton effectively, the chemical used in after-treatment is to be removed prior to stripping out the dye. Sequestering agent like EDTA in presence of mild acid can remove metal salts; $Na_2S_2O_4$ can destroy colour developed through diazotization–coupling and shades developed with diazotized base. HCHO can be removed in 3% HCl for 30 min; HCOOH can remove dye fixing agents. Removal of after-treatments is succeeded by general methods of stripping out of dye.

4.11   References

Colour Index International (1987), 3rd Edition (3rd Revision), Bradfor, UK, Society of Dyers & Colourists and AATCC.

ESCHE S P (2004) 'Direct dyes -Their application and uses', AATCC Rev., 4, 10, 14–17.

HERLANT M (1993) 'Low salt dyeing of cellulosics', Am. Dyestuff Rep., 82, 4, 19–25.

MALIK S K, BHAUMIK S and MUKHERJEE R N (2003) 'Low-temperature dyeing of cotton by direct dyes', Indian J. Fibre Textile Res., 28, 4, 462–465.

PETERS R H (1975) Textile Chemistry, Volume-III, New York, Elsevier Scientific Publishing Company.

PORTER J J (2003) 'Understanding the sorption of direct dyes on cellulose substrates', AATCC Rev., 3, 6, 20–24.

SHORE J (1991) 'Developments in direct dyes for cellulosic materials', Rev. Prog. Color Related Topics, 21, 23–42.

VICKERSTAFF T (1950) The physical chemistry of dyeing, London, Oliver and Boyd.

5

Dyeing with sulphur dye

Abstract: Sulphur dyes are exclusively used in dyeing cotton with deep black, blue and a few other shades at remarkably cheaper price alongwith good light and washfastness. Dyes are non-ionic and are applied in reduced and solubilised anionic state succeeded by oxidation to restore parent non-ionic dye structure when sulphur dye–cellulose physical attachment is established alongwith aggregation in situ. Produced shades lack tinctorial brilliance which may be modified by topping while brilliant red, orange and yellow are the limitations. Excess free sulphur is liable to damage cotton in moist atmosphere forming tiny holes. Sodium sulphide, the most effective reducing cum solubilising agent pollutes waste water and corrodes concrete discharge pipes.

Keywords: Sulphur dye, sulphur, reduction, sodium sulphide, tendering

5.1   Introduction

Sulphur dyes are chiefly used for dyeing cotton with medium to heavy black, navy blue, brown and olive green shades primarily because of cheaper cost in expense of good light and washfast shades. Dyes are non-ionic and insoluble in water; the desired anion is developed on reducing and solubilising at higher temperature when it shows affinity for cellulose. Sodium sulphide (Na_2S) performs reduction and solubilisation both severing sulphur linkages into thiols followed by formation of sodium salt of thiol or mercaptides or thiolates, which are soluble in water and substantive towards cellulose. Higher rate of exhaustion occurs at 90–95°C in presence of electrolyte.

$$
\begin{array}{ccccc}
 & Na_2S & & Na_2S & \\
\text{Dye} - \text{S} - \text{Dye} & \longrightarrow & \text{Dye.SH} + \text{HS.Dye} & \longrightarrow & \text{Dye.SNa} + \text{NaS.Dye} \\
\text{(Sulphur dye)} & \text{Reduction} & \text{(thiols)} & \text{Solubilisation} & \text{(Na–salt of thiols or} \\
 & & & & \text{thiolates)}
\end{array}
$$

Dyed cotton is oxidized to restore parent dye structure in situ cotton and dye is retained by cellulose as aggregates with H-bonds and van der Waals forces.

$$\text{Dye.SH + HS.Dye} \xrightarrow{[O_2]} \text{Dye.S – S.Dye + H}_2\text{O}$$

Sulphur dyes are marketed under various brand names viz. Sulphur, Nissen, Asathio, Diresul, Pirocard / Pirosol, Coranil, Kayaku, Mitsui, Sodyeco / Sodyesul, Sulphol / Sulphosol, etc., and have been assigned C I constitution numbers ranging from 53000 to 53830 (Colour Index, 1987). Being a cheaper class of dye, regional manufacturers across the globe produce these dyes to meet local demand.

5.2   Features of sulphur dyes

Sulphur dyes are of variable composition, generally available in paste or powder form, insoluble in water, cheaper and easy to apply; posses good to excellent wash fastness (~4–5), good light fastness (~5–6), except black 1 (~7) (Wood, 1976; Guest and Wood, 1989) . Shades are relatively dull and incomplete in hue range; not fast to chlorinating agents and hence not applied on swimming costumes on fear of decolourisation, These dyes are important in producing cheap black, navy blue, khaki and olive green hues only which is a dream with other classes of cotton dyes at such a lower cost with good fastness. Brilliant red, orange and violet are not available in this class of dye. Light shades lack brilliance and solidity and hence are not produced; dyed cotton is occasionally topped with basic dye to brighten shade. Higher alkalinity of dyebath restricts use in dyeing protein fibres due to fear of hydrolysis.

Sulphur dyes are important in dyeing garments and denim in rope form for high quality and performance; a part of costlier indigo may be effectively replaced with blue sulphur dye through topping or bottoming to cut cost without affecting fastness properties as such (Dixon, 1988; Stromberg, 1994; Etters and Hurwitz, 1985). Lack in imposing right dyeing conditions causes bronziness. H_2S liberated during dyeing, which is the characteristic feature of all sulphur dyes, forms corrosive metal sulphide and this restricts use of metal vessels except stainless steel.

$$\text{Fe + H}_2\text{S} \longrightarrow \text{FeS + H}_2$$

Sulphur dyes are identified by boiling a piece of dyed cotton with $SnCl_2$ (alternately Zn) and HCl in a test tube and covering mouth of test tube with a piece of filter paper soaked in lead acetate. H_2S formed reacts with lead acetate to form PbS to convert the paper black.

$$\text{H}_2\text{S + (CH}_3\text{COO)}_2\text{ Pb} \longrightarrow \text{PbS + 2CH}_3\text{COOH}$$

Free sulphur, if present, gets oxidized to SO_2 during storage in moist atmosphere to form H_2SO_3; the later develops tiny holes on cellulose through local hydrolysis.

5.3   Chemistry of sulphur dyes

Sulphur dyes possess neither well defined chemical structures nor consistent in composition; are just specified with raw material and process used to manufacture. Indeed, excess sulphur is isolated from dye after synthesis. Sulphur linkages are the integral part of chromophore and are basically complex mixture of polymeric molecular species comprising of large proportion of sulphur in the form of sulphide (–S–), disulphide (–S–S–) and polysulphide (–S$_n$–) links in heterocyclic rings. Chromophoric systems are based on thiazole, thiazone, thianthrenes and phenothiazonethioanthrone, illustrated in Fig 5.1 (Allen, 1971; Christie, 2001).

Synthesis of sulphur dyes, in general, involves sulphurisation, in which sulphur, polysulphide or both in mixture is heated at around 180–350°C alongwith aromatic amines, phenols or aminophenols or refluxed in solvents under pressure. Sulphurised vat dyes are synthesized in the same way too but their reduction requires a strong reducing agent like sodium hydrosulphide. Numerous literatures describe synthesis of sulphur dyes in detail (Allen, 1971; Colour Index, 1987; Shankarling et al, 1996).

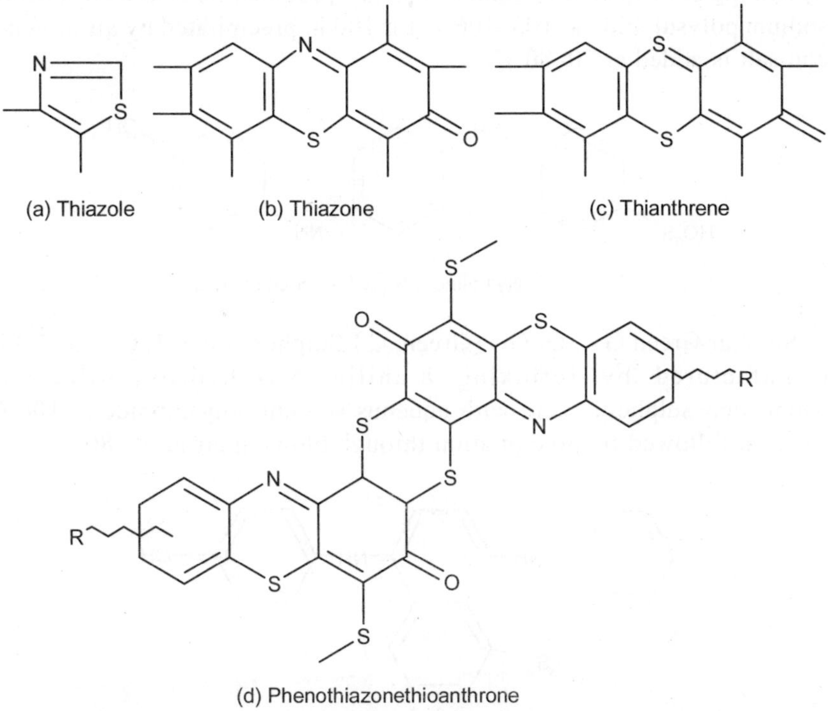

(a) Thiazole   (b) Thiazone   (c) Thianthrene

(d) Phenothiazonethioanthrone

5.1 Various chromophoric systems in sulphur dyes
   [Figure (d) has been reproduced by the permission of the
   Royal Society of Chemistry]

5.3.1   Synthesis of sulphur dyes

Sulphur Black BG / GXE / GXR (Bluish Black, C I Sulphur Black 1, C I 53185) is the most important member having empirical formula $C_{24}H_{16}N_6O_8S_7$ or $C_{24}H_{16}N_6O_8S_8$ and is produced by heating 2,4-dinitrophenol with sodium polysulphide (in proper ratio) at 110–120°C for 48–72 h at 130–140°C under pressure. The melt is diluted and black dye is precipitated by means of addition of acid or air-oxidation.

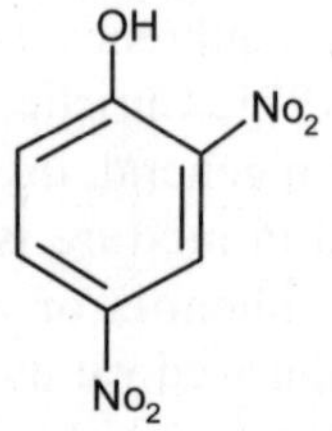

(2, 4-dinitrophenol)

Sulphur Blue FBL (Blue, C I Sulphur Blue 10, C I 53470) is produced by heating N-[p-(p-hydroxyanilino) phenyl] sulphanilic acid with aqueous sodium polysulphide at 105–106°C for 100 h, precipitated by air blowing and salt is added at 70–80°C.

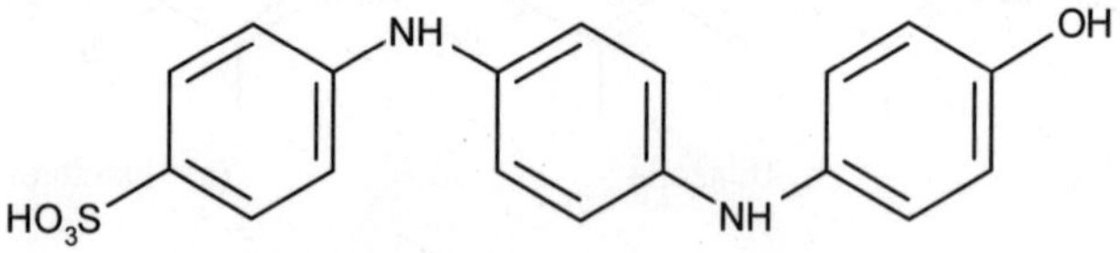

N–[p-hydroxyanilino) phenyl] sulphanilic acid

Sulphur Green G / 3G/ GG (Green, C I Sulphur Green 3, C I 53570) is manufactured by refluxing 8-anilino-5-(p-hydroxyanilino)-1-naphthalenesulphonic acid with aqueous sodium polysulphide at 106°C for 24 h followed by precipitation through blowing air at 60–80°C.

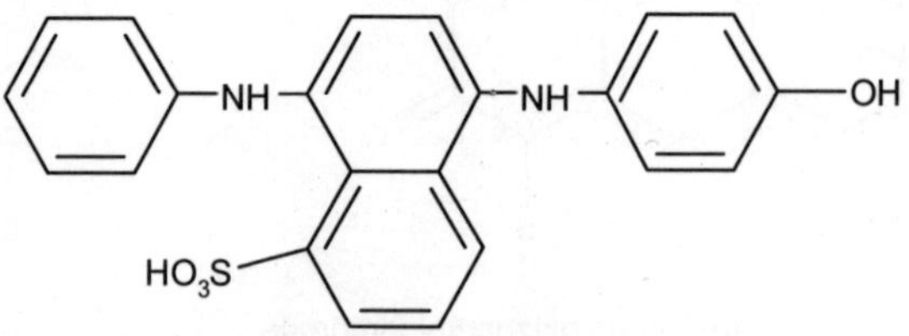

8-anilino-5-(p-hydroxyanilino)-1-naphthalenesulphonic acid

Sulphur Brown GS (Reddish Brown, C I Sulphur Brown 7, C I 53275) is obtained by reducing 1,8-dinitronaphthalene with aqueous Na_2S at 90°C followed by baking at 270°C with sodium polysulphide, $CuSO_4$ and NaCl.

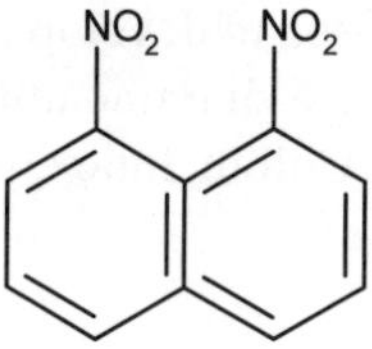

1,8-dinitronaphthalene

Sulphur Yellow G (Yellow, C I Sulphur Yellow 9, C I 53010) is produced by mixing 133 parts of 2',4'-dinitroacetanilide, 93 parts of phthalic anhydride, 280 parts of Na_2S and 94 parts of previously molten sulphur at 95°C; salt is added and the mixture is heated up to 130°C and continued for 10–11 h beyond which it is heated up to 180°C and maintained for 3–4 h.

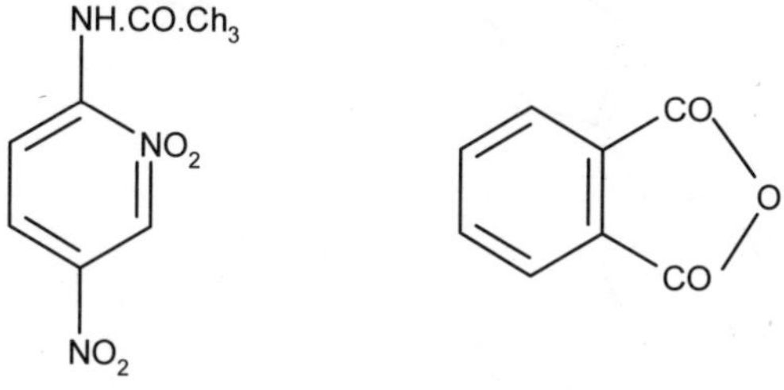

2',4'–dinitroacetanilide

phthalic anhydride

5.3.2 Synthesis of sulphurised vat dyes

Sulphurised vat dyes are produced through sulphurisation process that is used for other sulphur dyes but are reduced with $Na_2S_2O_4$ and applied like vat dyes (Bernhardt, 1984; Heid, 1978). One such commonly used dye is Hydron Blue R (C I Vat Blue 43, C I 53630). Indophenol produced from condensation of p-nitosophenol and carbazole in presence concentrated H_2SO_4 at -20°C is reflxed at 107°C for 24 h with sodium polysulphide in butanol to produce this dye.

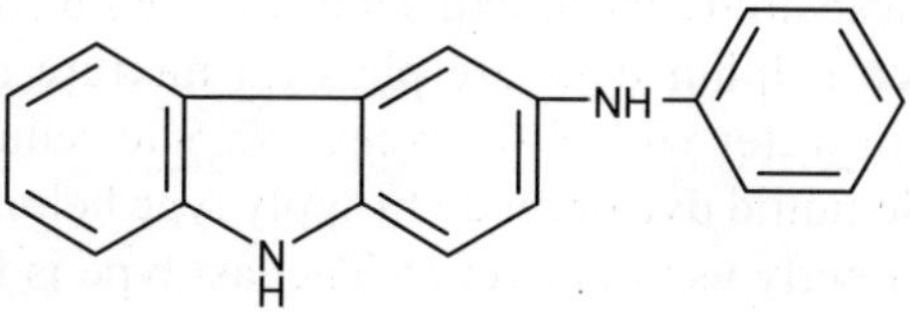

Indophenol

Hydron Blue R is an inexpensive substitute for indigo and can not be completely reduced by Na_2S, so it is reduced invariably by $Na_2S_2O_4$ in combination with NaOH.

Another example of sulphurised vat dye is Novatic Yellow 5G, C I Vat Yellow 2 (Atic). It is widely used to develop greenish yellow hue on cotton and is prepared by refluxing 2,6-diaminoanthraquinone, benzoin chloride and sulphur in naphthalene (Colour Index, 1987).

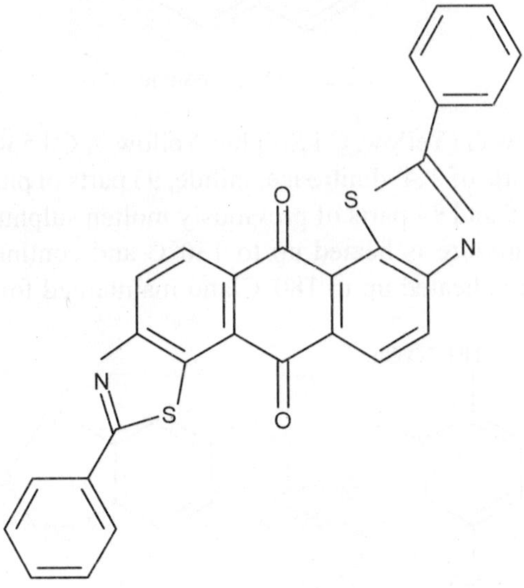

Novatic Yellow 5G, C I Vat Yellow 2, C I 67300

5.4   Classification

Sulphur dyes are classified based on chemical structure and application. Chemical classification includes two types, viz. sulphur dyes and sulphurised vat dyes. The first type includes only sulphur linkages and are reduced with Na_2S at boil. Sulphurised vat dyes retain both sulphur linkage as well as carbonyl group as chromophore and are reduced with $Na_2S_2O_4$ and NaOH combination at specific temperature but can not be reduced with Na_2S.

From application point of view, dyes are available in three main forms, viz. powder, reduced and water soluble. Powder form is the conventional insoluble type; these are reduced and solubilised with Na_2S at boil before application. Leuco sulphur dyes are physical mixture of dye and Na_2S; requires heating in water with little excess Na_2S to reduce and solubilise dye. Water soluble liquid dye or ready to apply type belongs to sulphurised vat class and is directly used in dyeing. The last type is free from fly-out, suitable for continuous dyeing but too costly.

5.5   Batch application

Exhaust method is extensively followed due to lesser problems associated and ease in rectifying uneven dyeing. Dyeing includes a few stages, viz. reduction, dyeing, washing, oxidation, soaping and final washing.

5.5.1   Reduction

Reduction involves conversion of sulphide, disulphide or polysulphide into leuco thiols. Dye is pasted with little turkey red oil to improve wetting and prevent floating of dye molecules. Water is added to it followed by a little Na_2CO_3 to counteract any H_2SO_3 that might has been formed during storage of dye. Na_2S is added gradually with constant stirring the mixture at 90–95°C for 10–15 min for complete reduction and solubilisation of dye which is marked by formation of stable foam on surface of liquor. Reduction occurs through formation of NaHS (sodium hydrosulphide) and NaOH during hydrolysis of Na_2S; NaHS reduces dye while NaOH solubilises leuco dye to thiolates (Teli et al, 2001).

$$Na_2S + H_2O \quad NaHS + NaOH$$

$$D\text{-}S\text{-}S\text{-}D \xleftarrow{\quad Na_2S \quad}{[NaHS + NaOH]} D\text{-}S\text{-}Na + Na\text{-}S\text{-}D$$

Stable reduced dyebath is highly alkaline; addition of little Na_2CO_3 neutralizes this acid prior to reduction, avoids wastage of Na_2S and stabilizes reduced form of dye.

$$Free\ sulphur + O_2 \longrightarrow SO_2$$
$$SO_2 + moisture\ (H_2O) \longrightarrow H_2SO_3$$
$$H_2SO_3 + Na_2CO_3 \longrightarrow Na_2SO_4 + H_2O + CO_2\uparrow$$

Concentration of Na_2S required is based on its purity. Crude and flake grades have a purity of 30–35% and 60% respectively and so almost double amount is used for reduction and solubilisation; increase in purity will result in lesser dosing proportionately.

Over-reduction of dye with excess Na_2S must be avoided as it leads to product having lower affinity for cellulose. A chelating agent, like EDTA counteracts interference of metals in hard water.

5.5.2   Dyeing

A commercial recipe for dyeing cotton with sulphur dye consists of dye ($x\%$), turkey red oil (a little), Na_2CO_3 (2 g/l), Na_2S ($2x\%$ for crude and

1.5x% for flakes) and NaCl (50 g/l). Once reduction and solubilisation is over, wet cotton is dyed at 90–95°C for 30–60 min after which salt is added and dyeing is continued for further 1.5–2 h; the bath is drained out and dyed cellulose is washed thoroughly. Due to presence of large amount of dye required to produce deep shades and its moderate affinity for cellulose, dyeing time before and after salt addition should be kept on higher side to promote exhaustion of bath. Reduction as well as dyeing of cotton is preferably carried out in jigger or winch. At least four turns are imparted in dyebath after which salt is added and further eight turns are imparted at boil. The bath is dropped, dyed cotton is cold washed for 2 turns, oxidized for 2–4 turns followed by soaping, washing and padding with CH_3COONa. Detailed description of exhaust dyeing in jigger and winch are also disclosed elsewhere (Krauzpaul, 1987; Aspland, 1992a, 1992b; Anon, 1987).

5.5.3   Oxidation

Dyeings are oxidized in $K_2Cr_2O_7$ (1–2 g/l) and CH_3COOH (1–2 ml/l) at 50–60°C for 30 min, when the dye restores its parent oxidized form. Reaction of dichromate and acid produces nascent oxygen required for oxidation.

$$K_2Cr_2O_7 + 8CH_3COOH \longrightarrow 2CH_3COOK + 2\,(CH_3COO)_3Cr + 4H_2O + 3[O]$$

Dichromate oxidation is cheaper but stiffens dyed cotton through deposition of chromium and changes tones of blue dyeings. Stiffness results more needle cuttings of sewing threads forcing to oxidize these with H_2O_2.

5.5.4   After-treatment

After oxidation and soaping, dyeings are treated with CH_3COONa at low liquor ratio and dried to counteract inorganic corrosive acid, if formed in post-dyeing stage.

However, deposition of salt ($NaHSO_3$) makes dyeings marginally stiffer. To improve light fastness, dyeings may be treated with metal salts in presence of acid and light fastness may be improved by treatment with dye-fixing agents. Dullness of shades may be improved to some extent by topping with little basic dye.

5.6   Continuous dyeing

Conventional sulphur dyes are not suitable for continuous application due to inconsistency in penetration of large amount of dye to produce deep shades within reasonably short time. Micro-dispersed powdered sulphur dyes or solubilised sulphur dyes, viz. hydron stabilisol dyes are most suitable and may be applied on mercerized cotton in pad–steam process with intermediate drying using caustic lye and hydrosulphite. Polyester–cotton blends can be dyed in thermosol / pad-steam in single or two stage process. The blend is padded with samaron dyestuff, followed by drying and thermosoling; repadded with hydron stabilisol dyes with 40–60% liquor pick up and steamed. Washing, oxidation and soaping are carried out as usual (Muller and Steinbach, 1986). However, cost of dyeing and chances to develop well levelled shade poses problem in continuous dyeing with sulphur dyes.

5.7   Reducing systems

Sulphur dyes require reduction potential around (-550 to -650) mV for proper reduction and to retain its reduced form throughout dyeing (Bechtold et al, 1991). Various effective reducing systems are available; few are cheaper but hazardous to ecological balance.

5.7.1   Conventional reducing agents

Either Na_2S alone or glucose (4 g/l) along with NaOH (2 g/l) are two effective formulations to reduce and solubilise sulphur dyes at 90–95°C and produced thiolates are very much stable for prolonged time (Kaushik et al, 1993). Sodium hydrosulphide (NaHS) is another important reducing agent and the dosing is equal to that of Na_2S flakes but require addition of either of Na_2CO_3 (10 g/l) or NaOH (5 g/l). In fact, Na_2S produces NaHS and NaOH when dissolved in water. Sodium hydrosulphite or dithionite ($Na_2S_2O_4$, 3 g/l) with NaOH (1.5 g/l) is more powerful reducing agent than Na_2S with reduction potential around -700 mV. Reduction with hydrosulphite may cause over-reduction of dye leading to a product having poor affinity for cellulose. A few red-brown, green and olive dyes decompose in presence of hydrosulphite, while a few dyes give better yield but have not been used for dyeing sulphur blacks because of reduced colour yield and poor reproducibility. Solubilized sulphur blacks are applied using hydrosulphite only at higher cost. Leuco sulphur dyes and sulphurised vat dyes are often reduced with $Na_2S_2O_4$ with a consequence of inadequate stable bath and dyeings possess lower wash fastness.

Other reducing agents, viz. thioglycolic acid and thiosalicylic acid though are effective in certain cases but the shades are changed to some extent (Klein, 1982). Sodium bisulphide reduces scopes of bronziness by keeping dyebath in reduced condition consistently and prevents surface oxidation. Sodium sulphide, the cheaper and most efficient reducing agent is associated with huge waste water load.

5.7.2   Sulphur-free reducing agents

Glucose supresses dye uptake in jiggers and winches due to inconsistency in maintaining temperature beyond 90°C. It is a strong hydrocarbon and so though readily degradable, produces higher BOD_5 and COD. β-mercaptoethanol alongwith NaOH results efficient dye uptake in dyeing with leuco sulphur dyes in exhaust and pad-dry-steam processes without releasing any odour but costly (Chavan, 2001).

Iron salts along with alkali may act as efficient reducing agent at room temperature, but the extent of reduction of dye remains proportional to solubility of iron. Iron (II) salts on reaction with NaOH produces $Fe(OH)_2$ generating a reduction potential around $-(700–750)$ mV; the potential is well above that required for reduction of conventional sulphur dyes. Negligible solubilisation of iron can not reduce dye upto any remarkable extent till $Fe(OH)_2$ is not complexed with weaker ligands, viz. tartaric acid, citric acid, triethanolamine, etc. However, the reduction potential becomes too higher to cause over reduction of dye (Semet, 1995; Chavan and Chakraborty, 2001).

Hydroxyacetone generates a reduction potential as high as -810 mV, biocompatible and is applied alongwith NaOH, but chances of over-reduction of dye exists and is costlier (Marte, 1989).

Molasses –NaOH combination produce well stable dye bath with higher dye uptake (Chavan and Vhanbatte, 2002; Shukla and Pai, 2004). In-house preparation of reducing sugar by hydrolyzing edible cane sugar at pH 3.5 for 16 hours at room temperature shows best yield in form of clear viscous solution with 80% conversion and clarity of product with stability till 3 months. Dyeing results in terms of colour yield, Lab values and wash fastness properties were found to be parallel with those obtained form conventional dyeing using sodium sulphide except tonal change which could be overcome by dyeing with reducing sugar alongwith NaOH at 80°C with intermittent addition of both throughout exhaustion (Chavan and Vhanbatte, 2004).

5.8   Oxidizing agents

Two most commonly used oxidizing agents are $Na_2Cr_2O_7$ or $K_2Cr_2O_7$ in combination with an acid.

$$K_2Cr_2O_7 + 4H_2SO_4 = K_2SO_4 + Cr_2(SO_4)_3 + 4H_2O + 3[O]\uparrow$$

Chances of over-oxidation exists as oxygen is formed rapidly; H_2SO_4 may be replaced with a weak acid like CH_3COOH to retard the process and temperature is maintained at 50–60°C for 15–30 min. Before oxidation, a thorough wash is imparted to ensure absence of alkali; presence of little alkali in dyeings may resist oxidation as $K_2Cr_2O_7$ produces CO_2 and not oxygen when reacts with alkali.

$$K_2Cr_2O_7 + Na_2CO_3 = 2KNa\ CrO_4 + CO_2$$

Both $Na_2Cr_2O_7$ and $K_2Cr_2O_7$ are cheaper, precipitate chromium on cellulose and impair handle that is why sulphur dyed sewing thread is invariably oxidized with H_2O_2 to reduce needle cutting and to preserve suppleness of yarn. Sulphur blue shades are oxidized with H_2O_2 to avoid reddish impression imposed with dichromates.

H_2O_2 is generally not used on cost ground and chances of over-oxidation exists if used alone at pH 7; but H_2O_2 (1 ml/l, 35% or 130V) may be used at pH little over 7 and 50–60°C to release required nascent oxygen. Acidified KIO_3 is costlier but does not change tone of dyeings. Both H_2O_2 and KIO_3 are eco-friendly oxidizing agents; in fact KIO_3 was first established as the ideal oxidizing agent for sulphur dyes (Klein, 1982).

5.9   Bronzing of shade

Sulphur dyes are liable to develop bronze like look mainly on black and blue shades. There may be several reasons behind this problem, viz. (i) excessive heavy dyeing, due to which most of dye molecules occupy surface areas only reflecting back more incident light, (ii) exposure of goods to air during dyeing, when a layer of reduced dye is formed on surface of cotton and the same gets dried and oxidized; the same is repeated during next passing through dyebath and so on, (iii) use of excess dye in dyeing, and (iv) inadequate Na_2S in dye bath causing incomplete reduction of dye. The root of this problem lies on insufficient diffusion of dye with more surface deposition and to rectify it, dye molecules are to be distributed homogeneously which can be achieved by redyeing in a fresh bath with Na_2S at 90–95°C for 30–60 min at lower liquor ratio adding 10% of total dye earlier used.

5.10   Fastness of dyeings

Light and wash fastness of sulphur dyed cotton, in general, are in the range of 5–6 and 4–5 respectively. Light fastness may be enhanced by topping with either of vat dye or aniline black which helps to improve wash fastness too. Various washing cycles discharge a part of dye progressively beyond 50°C due to presence of bleach formulation in modern detergents, but decolourised dye does not stain adjacent apparels because of its non-ionic nature. Treatment with polymeric or conventional cationic dye-fixing agents facilitates to improve washfastness in both batch and continuous methods

(Burkinshaw and Collins, 1998). After-treatment of dyeings in unoxidised thiolate state with alkylating agents based on polyhalogenohydrins enhances washdown property and in such a case, anionic dye molecules form complex with alkylating agent requiring no post-oxidation as required for conventional sulphur dyes (Cook, 1982; Wood, 1976; Colin, 1995). However, change in shade, decrease in light fastness and trouble in repairing faulty dyeings restrict use of this technique. Various techniques to improve wash fastness have been explained in numerous literatures (Perkins and Crew, 1975; Burkinshaw et al, 1996; Burkinshaw and Collins, 1995).

Fastness to chlorine is mostly poor; hypochlorite bleach decolourises sulphur dye permanently with exceptions of sulphurised vat dyes and few red, green and black. However, in latter cases, a remarkable part of shade is discharged. Peroxide bleach too decolourises sulphur dyed shades but the action is not so severe like that with hypochlorites (Colin, 1995).

Dry crock fastness is reasonably good (4–5) while wet crock fastness is inferior (2–3). Dyeing of cotton in jet dyeing machine improves wet crock fastness considerably due to hydrodynamic shearing action produced during passing of fabric through ventury and steel tubes of the jet (Tobin, 1979). Perspiration fastness grades are generally very good.

5.11   Tendering of cellulose

Free sulphur, if remains present in sulphur dyed cotton, is oxidized to SO_2 which in turn absorbs moisture to form H_2SO_3 in humid atmosphere and develops tiny holes on cellulose due to local hydrolytic action. The problem can be handled by passing soaped and washed dyeings through CH_3COONa bath at too lower liquor ratio so that H_2SO_3 as and when produced will be converted to harmless CH_3COOH.

$$H_2SO_3 + CH_3COONa \longrightarrow CH_3COOH + Na_2SO_3$$

To assess extent of tendering, dyed sample is covered on both sides with two white pieces of cotton fabric, stitched, treated at 140°C for 1 min, is kept in open air when loose sulphur is oxidized and in contact with moisture form H_2SO_3. Stitches are removed, three pieces are separated out and is tested for presence of loose sulphur.

5.12   References

ALLEN R L M (1971) 'Colour Chemistry', 1st Edition, London Thomas Nelson & Sons Ltd
ANON (1987) 'Sulphur dyes are important', *J. Soc. Dyers Color.,* **103**, 9, 297–298.
ASPLAND J R.(1992a) 'A series on dyeing: Chapter 4: sulphur dyes and their application', *Textile Chem. Color.*, **24**, 3, 21–24.

ASPLAND J R (1992b) 'A series on dyeing: Chapter 4 /Part 2: practical application of sulphur dyes, *Textile Chem. Color.*, **24**, 4, 27–31.

BECHTOLD T, BURTSCHER E, GMEINER D and BOBLETER O (1991) 'Investigations into the electrochemical reduction of dyestuffs', *Melliand Textilber.*, **72**, 1, 50–54.

BERNHARDT H (1984) 'Sulphur / sulphur vat dyes: Which commercial form is the right one for my plant', *Melliand Textilber.*, **65**, 12, 833–835.

BURKINSHAW S M, COLLINS G W (1995) 'Improvement of the wash fastness of sulphur dyeings on cotton', Proceedings: International conference and exhibition, AATCC, Atlanta, 169–183.

BURKINSHAW S M, COLLINS G W and GORDON R (1996) 'Continuous dyeing with sulphur dyes: aftertreatments to improve the wash fastness', Proceedings: International conference and exhibition, AATCC, Nashville, September, 296–303.

BURKINSHAW S M and COLLINS G W (1998) 'Aftertreatment to reduce the washdown of leuco sulphur dyes on cotton during repeated washing', *J. Soc. Dyers Color.*, **114**, 5/6, 165–168.

CHAVAN R B (2001) 'Environment-friendly dyeing processes for cotton', *Indian J. Fibre Textile Res.*, **26**, 2, 93–100.

CHAVAN R B and CHAKRABORTY J N (2001) 'Dyeing of Cotton with Indigo using Iron(II) salt Complexes', *Coloration Tech.*, **117**, 2, 88–94.

CHAVAN R B and VHANBATTE S (2002) 'Alternative reducing system for dyeing of cotton with sulphur dyes', *Indian J. Fibre Textile Res.*, **27**, 2, 179–183.

CHAVAN R B and VHANBATTE S (2004) 'Studies on use of reducing sugar as an alternative of sodium sulphide for dyeing of cotton with sulphur dye', *Ph D Thesis*, IIT, Delhi.

Colour Index International (1987), 3rd Edition (3rd Revision), Bradford, UK, Society of Dyers & Colourists and AATCC.

CHRISTIE R M (2001) *Colour Chemistry*, Great Britain, Royal Society of Chemistry.

COLIN S (1995) *Cellulosics dyeing*, Ed. Shore J, Society of Dyers & Colourists, Bradford.

COOK C C (1982) 'Aftertreatments for improving the fastness of dyes on textile', *Rev. Prog. Color Related Topics*, **12**, 1, 73–89.

DIXON M (1988) 'Role of sulphur dyes in garment dyeing, *Am. Dyestuff Rep.*, **77**, 5, 52–55.

ETTERS J N and HURWITZ M D (1985) 'Determining indigo and sulphur dye contributions to denim shade depths; *Am. Dyestuff Rep.*, **74**, 10, 20–21.

GUEST R A and WOOD W E (1989) 'Sulphur dyes', *Rev. Prog. Color Related Topics*, **19**, 63–71.

HEID C (1978) 'Hydron Blue: Indigo of the 20th Century', *Textil Praxis Int.*, **33**, 3, 285–287.

KAUSHIK R C D, SHARMA J K and CHAKRABORTY J N (1993) 'Investigation of exact reducer and oxidizer to control effluent load in sulphur dyeing', *Asian Textile J.*, **2**, 1, 47–50.

KLEIN R (1982) 'Sulphur dyes: today and tomorrow', *J. Soc. Dyers Color.*, **98**, 4, 106–113.

KRAUZPAUL G (1987) 'Sulphur and sulphur vat dyes in exhaust dyeing: a review from the dyer's standpoint', *Textil Praxis Int.*, **42**, 2, 140–142 + IX–XI.

MARTE E (1989) 'Dyeing with sulphur, indigo and vat dyes using the new RD process: Hydroxyacetone makes it possible', *Textil Praxis Int.*, **44**, 7, 737–738.

MULLER J and STEINBACH J (1986) 'Hydron Stabilosol dyes extend the applicability of sulphur dyestuffs', *Textil Praxis Int.*, **41**, 10, 1102–1104 + X–XII.

PERKINS W S and CREW B J (1975) 'Improving wash fastness of sulphur dyes', *Textile Chem.Color.*, **7**, 6, 108–111.

SEMET B, SACKINGEN B and GURNINGER G E (1995) 'Iron(II) salt complexes as alternatives to hydrosulphite in vat dyeing', *Melliand Textilber.*,**76**, 3, 161–164.

SHANKARLING G S, PAUL R and JAYESH M (1996) 'Sulfur dyes – constitution, synthesis and application', **43**, 8, 47–54; **43**, 9, 57–62.

SHUKLA S R and PAI R S (2004) 'Sulphur dyeing using non-sulphide reducing agents, *Indian J. Fibre Textile Res.*, **29**, 4, 454–461.

STROMBERG W (1994) 'Indigo-rope dyeing for high quality and performance', *Int. Textile Bulletin*, **40**, 1, 19–22.

TELI M D, PAUL R and PARDESHI P D (2001) 'Liquid dyes: Preparation, properties and Applications', *Asian Textile J.*, November, 72–79.

TOBIN H M (1979) 'Jet dyeing with sulphur and vat colours', *Am. Dyestuff Reporter*, **68**, 9, 26–28.

WOOD W E (1976) 'Sulphur dyes: 1966–1976', *Rev. Prog. Color Related Topics*, **7**, 80–84.

6

Dyeing with reactive dye

Abstract: As the name implies, these dyes react with cellulose to establish covalent bond, are water soluble. Dichlorotriazinyl reactive dyes are highly reactive, suitable for exhaust method; cellulose dyed with these develop poor wash fastness due to partial hydrolysis of dye, which is very difficult to remove even in repeated washings while monochlorotriazinyl and vinyl sulphone dyes are less reactive, suitable for application by padding or printing and possess no partial hydrolytic property. Lack of overall affinity for cellulose in exhaust dyeing necessitates application of huge salt, which increases dissolved solid content of waster water. Dyes are eco-friendly; full hue as well as shade range exists. When applied to silk, wool and nylon, is attached with chemical bond with good fastness.

Keywords: Reactive dye, covalent bond, hydrolysis, electrolyte, washing

6.1   Introduction

Reactive dyes form covalent bond with cotton through neucleophilic substitution or neucleophilic addition mechanism and the dyes are familiar as substitutive and additive dyes respectively. Due to presence of strong dye–fibre interaction, fastness properties are remarkably good except wash fastness which is poor to moderate due to hydrolytic nature of dyes. Pre-dissolved dye is applied on cotton; salt is added for better exhaustion followed by fixation with alkali. Subsequent soaping and washing remove all superficial and hydrolysed dyes.

A reactive dye does not show any change in colour during dyeing as that occurs for vat and sulphur dyes, as the change takes place at the site of reaction and not at the chromophore, as shown in Fig. 6.1 (Gokhale and Shah, 1981). In general, structure of all reactive dyes can be represented as C-B-A.

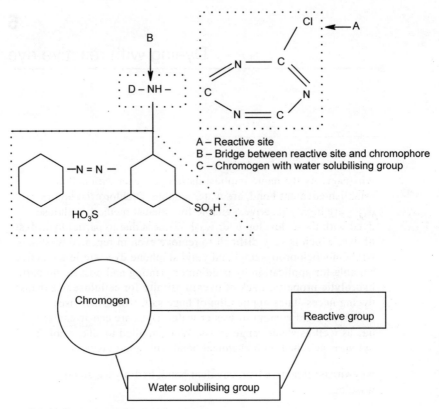

A – Reactive site
B – Bridge between reactive site and chromophore
C – Chromogen with water solubilising group

6.1 Various parts of a reactive dye.

6.2 Classification

Reactive dyes basically belong to two classes, viz. (i) substitutive and (ii) additive. Substitutive dyes include mono and dichlorotriazines [Procion MX, (ICI), Reactofix (Jay synth), Cibacron (Ciba) and Procion H(ICI)], etc.; 2,4,5 trihalogenopyrimidines [Reactone (Geigy), Drimarene (Sandoz)], 2-methylsulphonyl-4-methyl-5-chloropyrimidine [Levafix P, Bayer], 2,3 dichloroquinoxaline [Levafix E, Bayer] etc., out of these the mono and dichlorotriazine dyes shares the most. Additive types include sulphuric acid esters of β-hydroxyethylsulphones [Ramazol (Hoechst), Navictive (IDI)], N-substituted - β-amino ethylsulphones [Hostalan and Remazolan dyes (Hoechst)], sulphuric acid esters of β-hydroxypropionamides and β-chloropropionamides [Primazin (BASF)] etc.; ramazol dyes holds the importance in this type (Bird and Boston, 1975).

Dichlorotriazine dyes possess two reactive chlorine atoms and are applied at room temperature, called cold brand or M dyes while monochlorotriazines, possessing only one chlorine atom are applied under application of heat and is known as hot brand or H dyes.

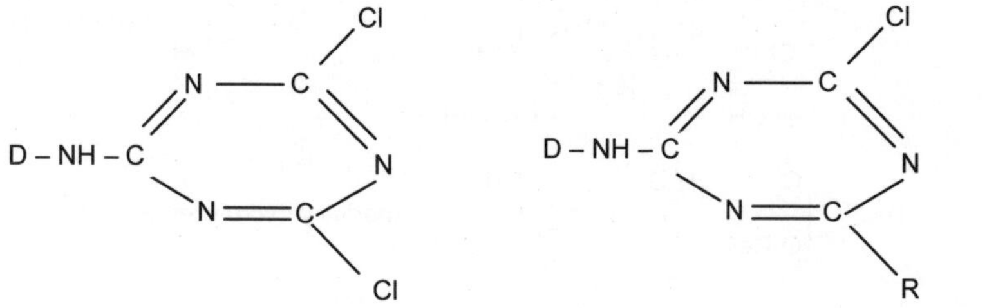

Dichlorotriazine dyes (cold, M)   Monocholorotriazine dyes (hot, H)

Simply represented as

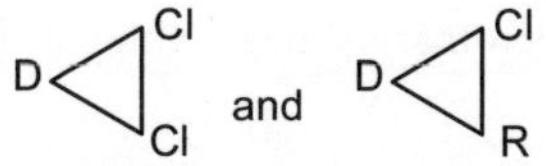

Additive type includes mainly vinyl sulphone dyes (ramazol dyes) possessing a general formula $DSO_2 CH=CH_2$. These dyes do not give any by-product during reaction with cotton or water, follows a neucleophilic addition mechanism and are applied through application of heat. Chemical attachment takes place in alkaline pH.

6.3   Properties

Reactive dyes, in general, are water soluble – solubility is enhanced by addition of little urea. Overall affinity for cotton is poor, necessitating dyeing for prolonged time along with excess of salt. Dyeings show excellent wash fastness except with cold brands which lack required wash fastness due to chances of simultaneous reaction with cotton and water. Full shade range is available; dyes are eco-friendly – resulting in self-decolouration of discharged liquor through hydrolysis under natural conditions; the reaction is same as that happens in dyebath. Dichlorotriazine dyes are applied at room temperature or slightly beyond that requiring no installation of boilers.

Cold brand dyes possess better affinity for cotton and are suitable for exhaust dyeing only; colour value of dyeings may be partially lost if dyed in pad-batch method. These are not suitable for localized dyeing (tie and dye, batik and other printing, etc.) as hydrolysis of dye during dyeing and subsequent domestic washing may stain adjacent garments spoiling look and aesthetics (Fig. 6.2).

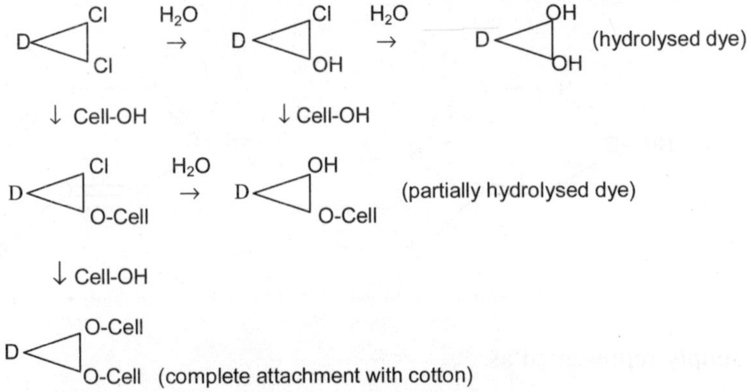

6.2 Reaction of dichlorotriazinyl dyes with cellulose and water.

Being highly reactive due to presence of two chlorine atoms, cold brands are more susceptible to hydrolysis and fixation rate is also high; dye molecules engage one chlorine atom to react with cotton while another with water causing poor wash fastness as partially hydrolysed dye can not be taken out from dyed cotton effectively.

In contrast, hot brand and ramazol types are suitable for padding or printing only due to their poor affinity for cotton. Due to presence of only one chlorine atom or reactive site, these are either hydrolysed or react with cotton but do not stain adjacent ground or garments during washing and are extensively used in padding and printing (Fig. 6.3). In practice, dyeing rate is quite higher (rate of dyeing / rate of hydrolysis ~1.8×10^4) under standard dyeing conditions lowering chances of hydrolysis, though partial hydrolysis is evident to some extent.

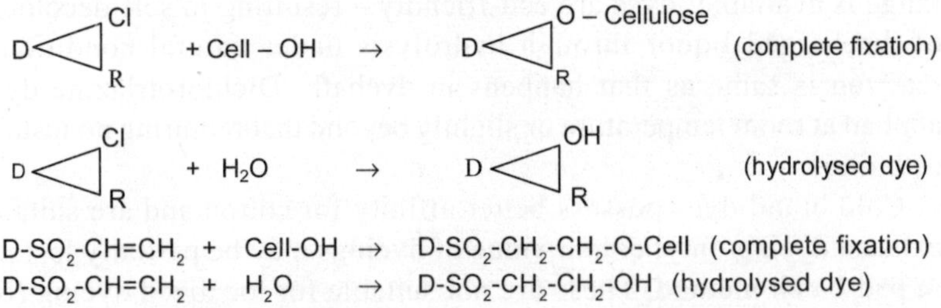

6.3 Reaction of mono-chlorotriazinyl and ramazol dyes with cellulose and water.

6.4   Specification of some reactive dyes

Chemical structure and colour index specification of some commonly used reactive dyes are depicted in scheme 6.1 (Colour Index, 1987).

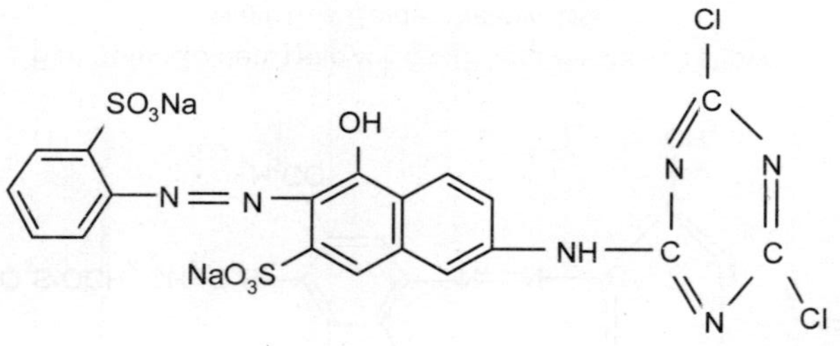

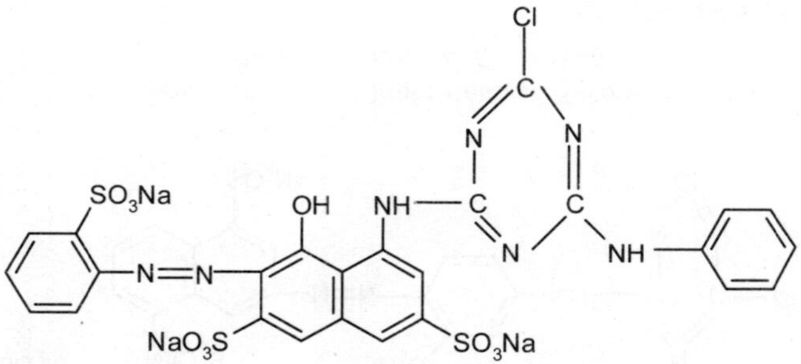

Amaryl Red 3BX
(Bright Bluish Red)
C I Reactive Red 3, C I 18159

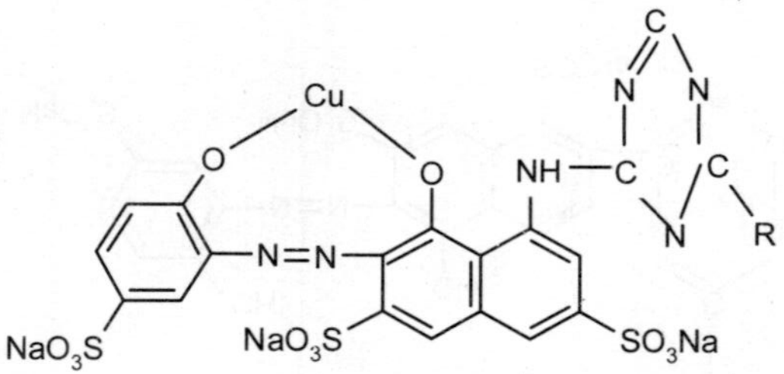

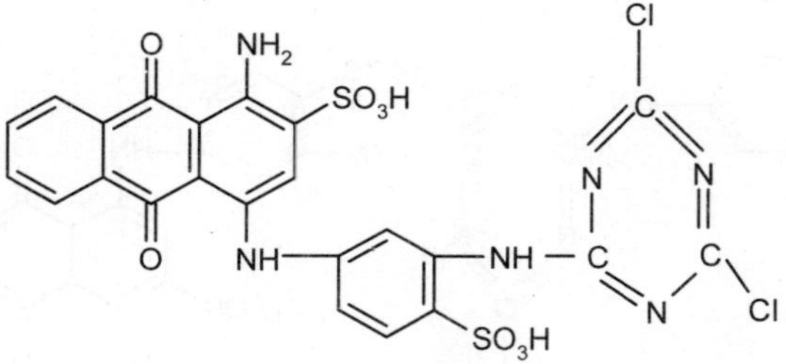

Reactofix Blue MR (Bright blue)
C I Reactive Blue 4, C I 61205

Reactofix Orange MGR / Amaryl Orange 2GX
(Bright Orange)
C I Reactive Orange 1, C I 17907

Cibacron Violet 2R (Reddish Violet)
C I Reactive Violet 2, C I 18157

Scheme 6.1 Contd. ....

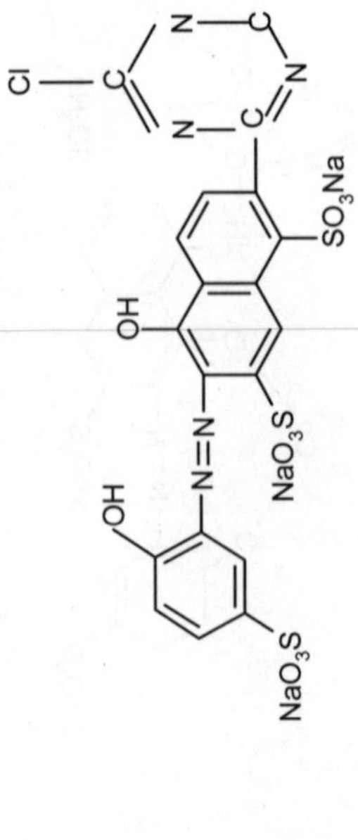

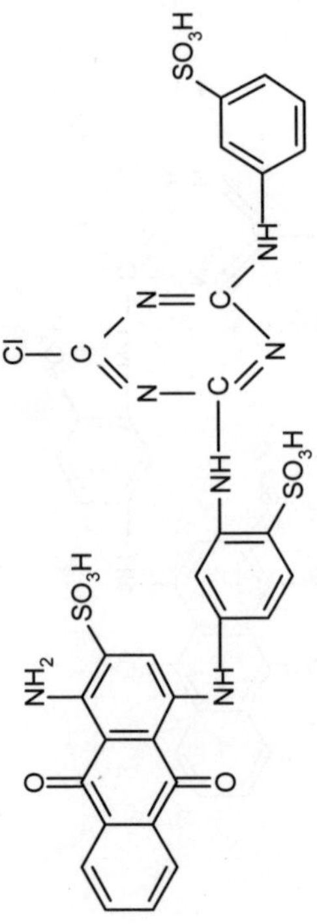

Reactofix Blue HGR (Bright Blue)
C I Reactive Blue 5, C I 61205 : 1

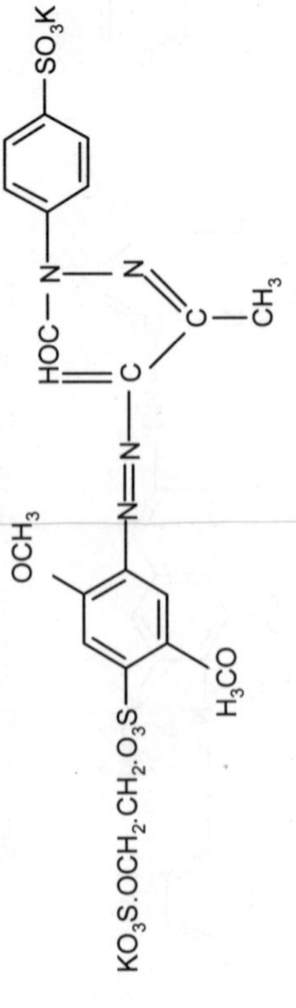

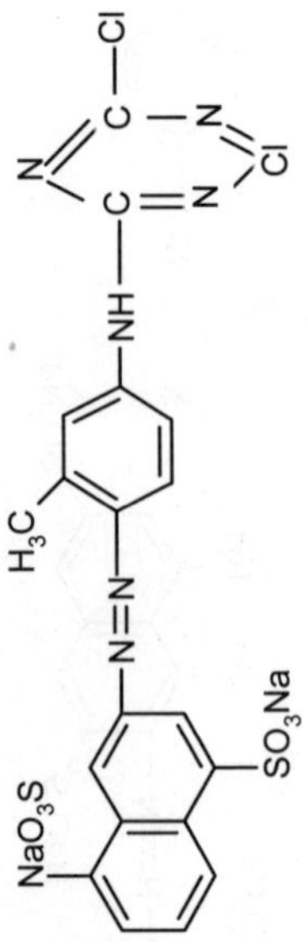

Amaryl Yellow R(Bright Reddish Yellow)
C I Reactive Yellow 4, C I 13190

Procion Rubine MXB (Dull Bluish Red)
C I Reactive Red 6, C I 17965

Reactofix Golden Yellow HGL (Bright Reddish Yellow)
Navictive Golden Yellow GG
C I Reactive Yellow 17, C I 18852

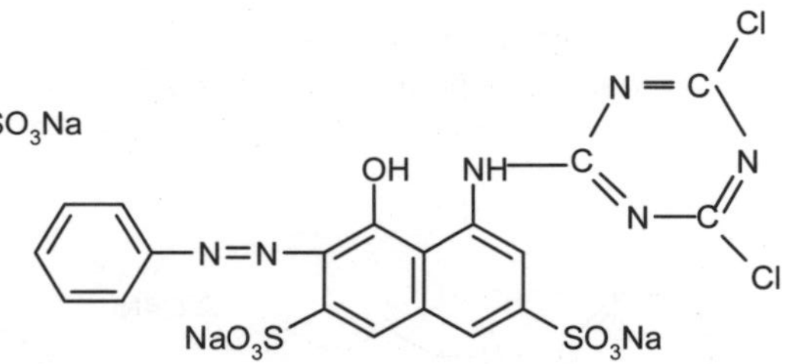

Reactofix Red M5B (Bluish Red)
C I Reactive Red 2, C I 18200

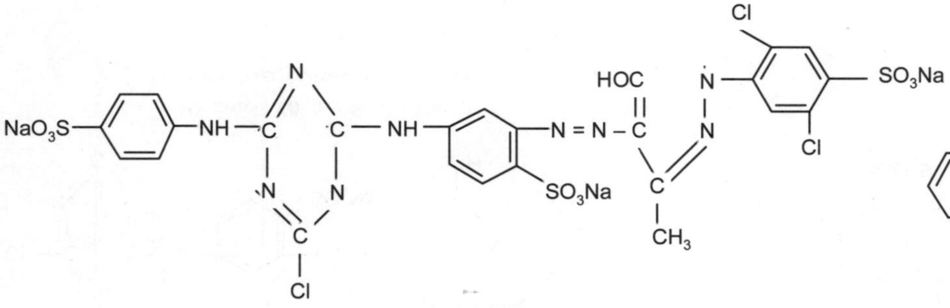

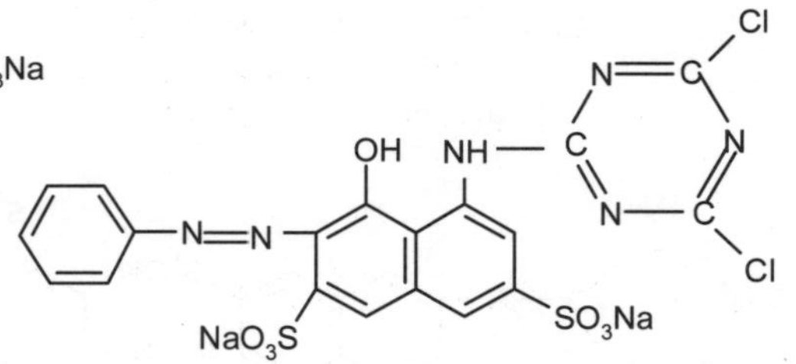

Reactofix Orange M2R / Amaryl Orange 2R
(Reddish Orange)
C I Reactive Orange 4, C I 18260

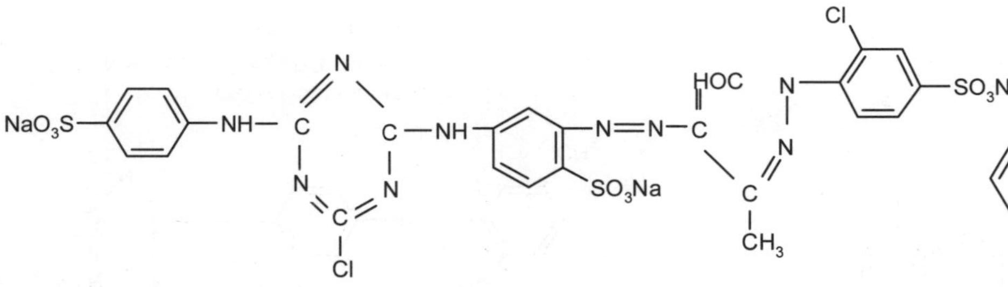

Cibacron Yellow 3G (Greenish Yellow)
C I Reactive Yellow 2, C I 18972

Helaktyn Yellow F5G / Ostazin Yellow S6G
(Bright Greenish Yellow)
C I Reactive Yellow 1, C I 18971

Scheme 6.1 Contd. ....

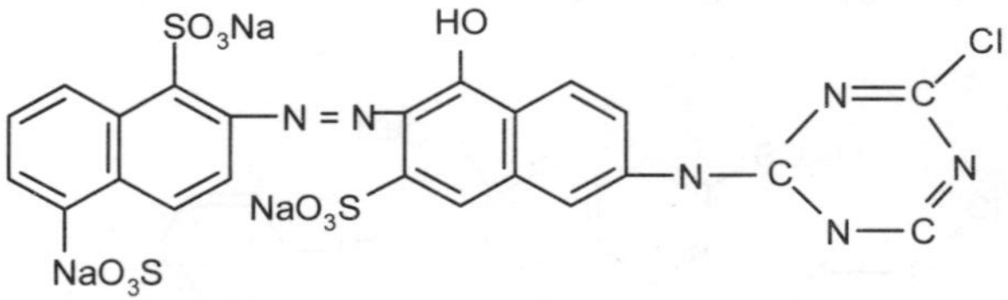

Reactofix Orange H2R
(Reddish Orange)
C I Reactive Orange 13, C I 18270

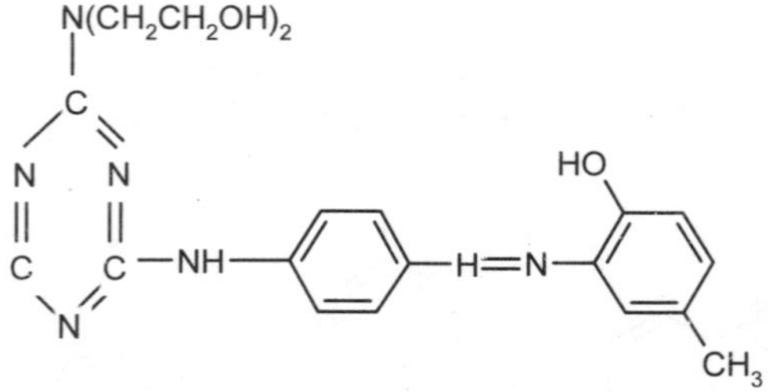

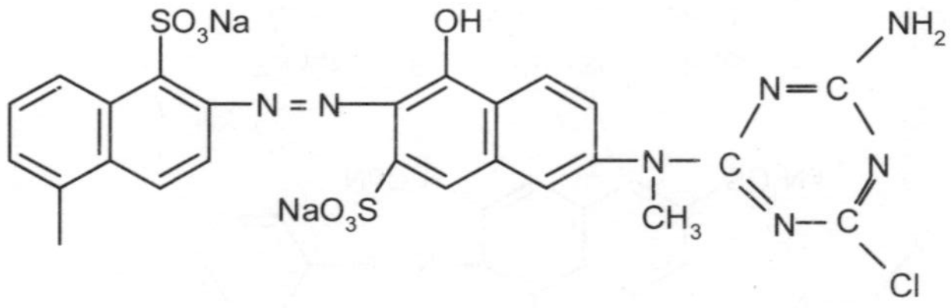

Reactofix Supra Red Violet HRL /
Navictive Red Violet R (Reddish Violet)
C I Reactive Violet 4, C I 18096

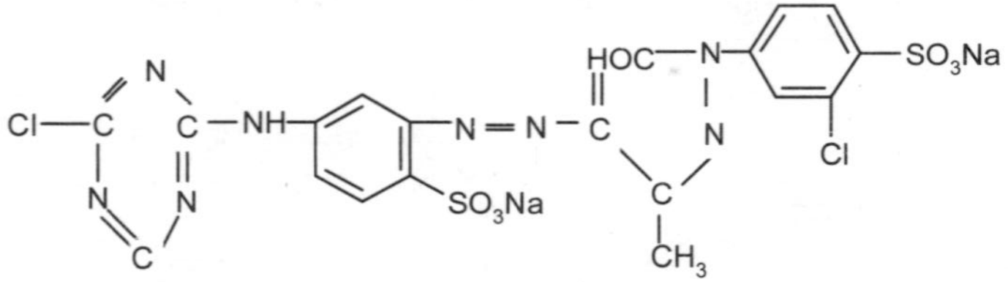

Procinyl Yellow G (Yellow)
C I Reactive Yellow 5, C I 11859

Reactofix Supra Orange H3RL/
Navictive Orange 3R (Bright Yellowish Red)
C I Reactive Orange 16, C I 17757

Scheme 6.1 Contd. ....

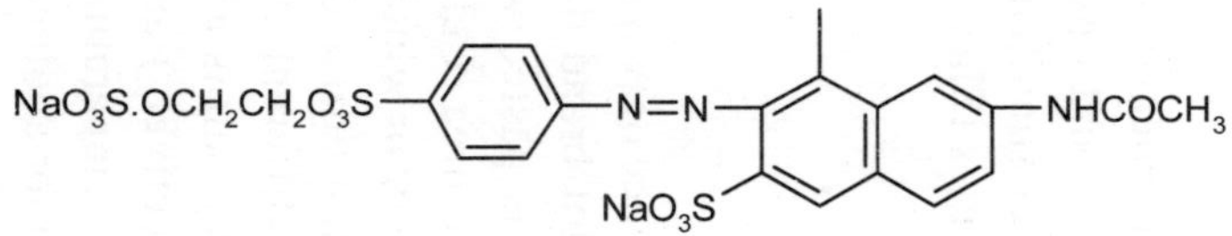

Reactofix Supra Orange H3RL/
Navictive Orange 3R (Bright Yellowish Red)
C I Reactive Orange 16, C I 17757

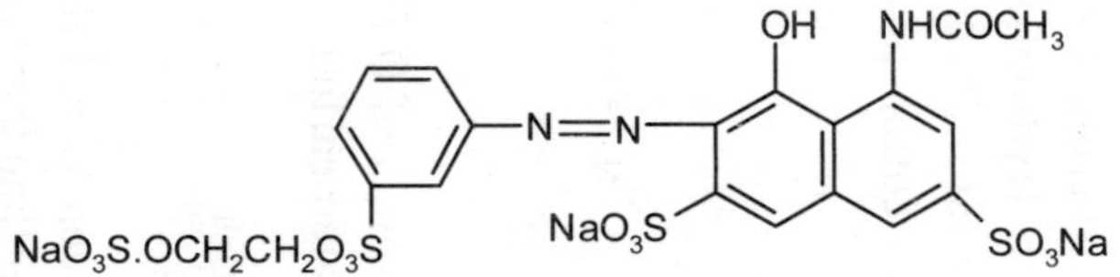

Reactofix Supra Red Violet HRL/
Navictive Red Violet R (Reddish Violet)
C I Reactive Violet 4, C I 18096

(2:1 copper complex of
C I Reactive Violet 4)

Reactofix Supra Violet H5RL/
Navictive Violet 5RN (Violet)
C I Reactive Violet 5, C I 18097

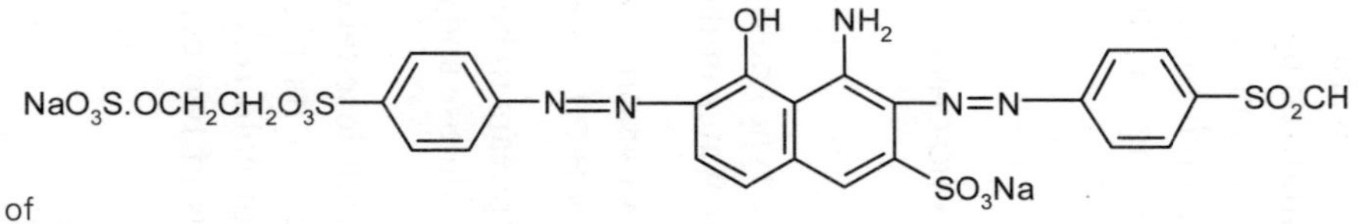

Amaryl Navy Blue GX /
Reactofix Supra Black HBL (Black)
C I Reactive Black 5, C I 20505

Scheme 6.1 Contd. ....

Scheme 6.1 Structure and C I specifications of some important reactive dyes.

6.5   Reaction with cellulose and water

Cellulose gets attached with substitutable reactive dyes producing acid. Even at neutral pH, some sort of reaction is expected but with liberation of more and more acid, forward reaction ceases. To favour fixation, liberated acid is to be neutralized by adding alkali in bath. Hydrolysis also occurs in water in the same way producing HCl. However, rate of hydrolysis is not so prominent compared to reaction of dye with cotton provided parameters of efficient dyeing are strictly maintained.

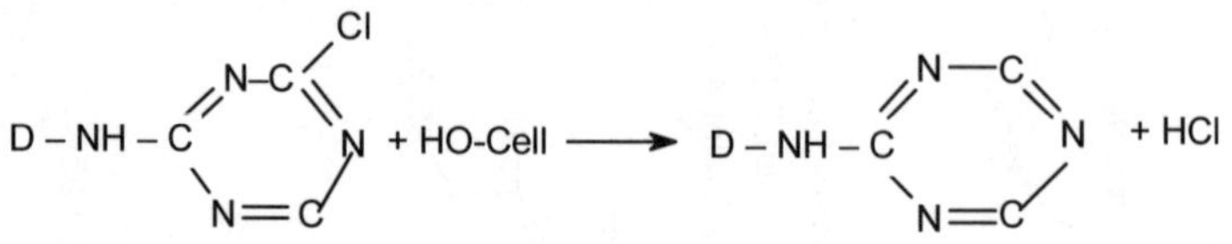

Ramazol dyes (additive type) possessing general formula of $D\ SO_2CH = CH_2$ react with cotton or water in the same way as stated earlier:

$$D\ SO_2CH = CH_2 + \text{Cell-OH} \rightarrow D\ SO_2CH_2CH_2\text{-O-Cell}$$
$$D\ SO_2CH = CH_2 + \text{H-OH} \rightarrow D\ SO_2CH_2CH_2\text{-OH}$$

The reactivity of dye appears to be somewhere between M and H dyes; rate of reaction between cellulose and dye is much higher than that between dye and water.

Reaction of dye with cellulose takes place with primary –OH group at C_6 position (CH_2OH) due to its highest reactivity and obviously, two reactive chlorine atoms of cold reactive dyes react simultaneously with two different –OH groups of two cellulose molecules (Shenai, 1987).

6.6   Application

6.6.1   Exhaust method

Cold-brand dyes possess better affinity due to presence of two substitutable chlorine atoms in structure and are preferred. Hot brand dyes are also applied in exhaust technique occasionally. Dye is pasted with T R oil followed by dissolution in water along with little urea. Cellulose to be dyed is wetted and treated in bath or loaded in jigger or winch. Dyeing is carried out for 30 min with half dye solution in first turn and rest half in second turn of jigger at required temperature for different classes (30°C and 40–45°C for cold and hot brands respectively), salt is added (30–50 g/l and 75–100 g/l for cold and hot brands respectively) and dyeing is continued with slow heating of bath to required temperature (40–45°C and 60–80°C for cold and hot brands respectively) for another 1–2 h after which alkali is added (5–10 g/l Na_2CO_3 for cold brands, 6–10 g/l Na_2CO_3 along with 2 g/l NaOH for hot brands) and fixation is carried out over a period of 45 min at this temperature. Dyebath is discharged; goods are thoroughly washed, soaped at boil and hot washed for removal of partially hydrolysed / unreacted / fully hydrolysed dye. Addition of dye, if required for shade matching must be done in salting stage and prior to addition of alkali, otherwise chances of formation of unleveled shades exist. Dyeing conditions as well recipe formulation is explained in Table 6.1.

Dyeing with ramazol dyes is started at 40–45°C with half dye solution along with half salt and half alkali for the first turn followed by rest half of dye, salt and alkali during the next turn. Dyeing is continued for further 1 h and temperature is raised to 80°C after which 45 min is further allowed for fixation. Washing, soaping and rest are as usual.

Cold brands are best suitable for exhaust or batch method where others classes are suitable for semi-continuous and continuous methods but on requirement may be used for exhaust dyeing too in expense of more use of dye for a given shade due to lesser affinity of dye for fibre. Turquoise blue shades are frequently developed with hot brand dyes only to achieve brilliant shades.

6.6.2   Continuous method

Due to overall low affinity of reactive dyes for fibre, production of levelled shades is comparatively easy. Cold brand dyes are applied through a pad–dry or pad–dry–steam technique, the latter is suitable and popular for all brands of reactive dyes. In pad–dry method, fabric is padded with dye, $NaHCO_3$ (15–20 g/l), urea (50–100 g/l) and wetting agent (2–5 g/l), followed by batching for 2–12 h or drying at 100–110°C when water evaporates and transports dye at the interior of fibre for fixation. $NaHCO_3$ turns to Na_2CO_3 at higher temperature in presence of moisture raising dye bath pH to facilitate fixation. Urea acts as potential solvent, activates moisture to retain soluble form of dye and is essential for deeper shades. Pad-dry-steam method gives excellent result, steaming may be done at 105°C for 5 min. To avoid oxidation during steaming, resist salt L, e.g. ludigol (5–7 g/l) may be incorporated in padding liquor. Reactive dyes, being water soluble, migrate during drying throughout cotton from a point at lower temperature to a point at higher one as evaporation of water from latter point is fast causing movement of water molecules along with dye. Drying must be uniform at each point of fabric to avoid development of unlevelled shades. Float dryers are the ideal drying machines; otherwise hot flue or cylinder drying machines with proper thermostatic control may be engaged. The efficiency of the method however decreases as shades become deeper. A significant improvement in dye pick up can be obtained

Table 6.1 Dyeing recipe and stepwise conditions

	M-Brands (% shade)				H-Brands (% shade)			
	≤ 0.5	0.5-2.0	2.0-4.0	≥4.0	≤0.5	0.5-2.0	2.0-4.0	≥4.0
Starting Temperature(°C)			30				40	
Dyeing before salt addition (min)	15	20	25	30	15	20	25	30
Salt (g/l)	20	30	40	50	50	60	75	100
Time of salt treatment (min)	30	30	45	60	30	45	60	60
Temperature of salt treatment (°C)			40-45				75-85	
Na_2CO_3 (g/l)	5	6	8	10	6	6	8	10
and NaOH (g/l)	–	–	–	—	2	2	2	2
Temperature for fixation (°C)			40-45				70-80	
Time of fixation (min)			45				45	
Washing		Thorough and repeated				Moderate		

by adding 100–120 g/l of urea to the padding liquor; this pick up is more apparent with cotton then with viscose. Addition of urea is expensive, so fixation can be enhanced by passing the goods through a streaming chamber for 5–15 s after dyeing.

A double padding makes it possible to work with a very stable stock solution of the dye containing no alkali. Cloth is first padded with dye solution and then dried, after which it is impregnated with alkali and salt, passes through streaming chamber for fixation of dye.

The molten metal bath is well adopted for continuous dyeing with reactive dyes. The sequence involves padding with dye solution, passing through a solution of sodium meta silicate and salt resting on the surface of the metal in the entry limb of the U-tube, applying heat while immersed in the molten metal followed by rinsing with water when the cloth emerges from the exit limb.

In padding with ramazol dyes, dye is dissolved in presence of urea (50 g/l). Na_2SiO_3 (120 g/l) and NaOH (10 g/l) are added to the dye liquor just before padding. Cotton fabric is padded at 70–80% expression, batched and covered with PE sheet, rotated overnight to avoid by-passing, debatched and neutralized with CH_3COOH (1 ml/l) at 40°C, soaped with non-ionic detergent at boil for 10–15 min and thoroughly washed.

For cottage industries, where infra-structure for heating up of bath is not available, cotton may be padded in a liquor containing H-dyes along with urea (5 g/l), NaOH (0.5 g/l) and Na_2SiO_3 (100–120 g/l), batched and covered with a polyethylene sheet. The batch is rotated overnight or 8–10 h followed by washing.

Continuous dyeing of cotton with all reactive dyes may be best done by padding with a liquor consisting of dye, urea (50 g/l), $NaHCO_3$ (20 g/l), resist salt L (5 g/l) and sodium alginate (2 g/l) followed by drying, steaming, soaping and washing.

6.6.3   Function of chemicals

(a) Salt

Reactive dyes possess negligible affinity for cotton. Cotton and dye both acquire negative charge when dipped in bath and repel each other reducing substantivity of dye. Salt is applied to reduce potential difference between the two resulting better exhaustion. It acts as exhausting agent and helps in migration of dye molecules from bath on to fibre.

(b) Alkali

During fixation of dye, HCl is liberated in bath. Addition of alkali neutralises the HCl formed and maintains the rate of forward reaction. Addition of alkali increases availability of maximum –OH groups in the dye bath promoting rate of neutralization.

$$HCl + NaOH \rightarrow NaCl + H_2O$$

If produced HCl is not removed, backward reaction that is stripping of dyes from fibre occurs. In case of hot brand dyes, rate of fixation is too slow; produced NaCl from reaction between HCl and NaOH promotes exhaustion during fixation. Some part of Na^+ ($NaOH \leftrightarrow Na^+ OH^-$) reacts with cellulose, swelling of fibre takes place and hence better dye penetration occurs.

Type of alkali to be applied for fixation depends on reactivity of dye. For dyes of higher reactivity, a mild alkali e.g. Na_2CO_3 may be best used to control rate of fixation reducing rate of hydrolysis. Reactitivity of hot and ramazol brand dyes are too less; application of combination of Na_2CO_3–NaOH show best result while $NaHCO_3$ is applied in continuous dyeing through padding or in printing with reactive dyes as the latter due to its lower reactivity does not react with dye prematurely before steaming. Na_2CO_3 or $NaHCO_3$, when heated up produces NaOH.

$$2NaHCO_3 \xrightarrow{H_2O} Na_2CO_3 \xrightarrow{H_2O} 2NaOH + CO_2\uparrow$$

pH of the bath should be kept on lower side to reduce rate of hydrolysis of dye and higher rate of fixation. NH_4OH can produce maximum –OH groups at a very low pH to give promising results.

6.7   Factors affecting dye uptake

Various activities in dyeing, viz. affinity, exhaustion, extent of reaction, fixation, hydrolysis, etc., depend on rate of reactivity of dye, liquor ratio, temperature, electrolyte, pH, nature of fibre and time of dying.

6.7.1   Affinity and reactivity of dye

Affinity of reactive dyes is directly proportional to their molecular structure – dyes of lower molecular weight have lower affinity, lower will be its hydrolysis and vice-versa. Fastness of dyeings will depend on the extent up to which fully and partially hydrolysed dyes have been removed. Though removal of fully hydrolysed dye is relatively simple, but the same for partially hydrolysed one is rather difficult; the larger the structure of dye the higher the energy requirement for its removal as bond energy directly varies with structure of dye. Selection of a dye with higher affinity results in higher dye uptake at shorter time indeed but with an inferior fastness –

partially hydrolysed dyes are not completely removed during soaping and washing, rather slow removal occurs during domestic washing cycles causing poor wash fastness.

Fixation of dye and rate of hydrolysis depend on presence of reactive groups in dye molecule. Even with the same brand, reactivity differs from one dye to another.

Rate of diffusion depends on temperature of bath and size of dye molecules. Higher the temperature, higher is the rate of diffusion for dye molecules of lower size.

Low affinity dyes (H- and ramazol brands) are mainly selected for printing as partial hydrolysis of dye is absent implying excellent fastness and no staining of ground pattern / colour during washing. In dyeing, dyes possessing higher affinity (M-brands) give better result as far as yield is concerned but fastness is to be ensured through repeated washing and soaping.

6.7.2   Liquor ratio

Hydrolysis of dye increases with increase in liquor in bath; reactivity of dye, temperature, etc., also enhance hydrolysis. A lower liquor ratio is the preferred one keeping in mind solubilisation of huge amount of dye, salt and alkali. During dyeing, free flow-down of dye solution must occur after passing out of textile through bath.

Liquor ratio is also governed by the machine at disposal. A jigger or winch can work up to a liquor ratio of 1:6 (assuming jigger capacity as 300 l and minimum cotton material loaded is 500 m ~50 kg) and as low as 1: (1–2). But too lower a ratio may result uneven dyeing due to higher effective dye concentration in bath and higher strike ratio.

6.7.3   Temperature

Rise in temperature enhances hydrolysis. During dissolution of dye and dyeing, temperature of dye bath should not go beyond 40°C for M-brands and 70°C for H-brands. In dyeing with cold brand dyes, dyes of high molecular weight exhaust more at higher temperature, while those of lower molecular weight at lower temperature. M-dyes show highest rate of absorption and diffusion at room temperature at a pH of 10.5–11.5. H-dyes show reactivity when alkali is added above 70°C. Phthalocyanine based H- reactive blue dyes require a dyeing temperature of 10–15°C higher, i.e. 85°C, e.g. brilliant blue H5G, H7G, turquoise blue 2GFL, 5GFL, etc. Under no circumstances, temperature of bath should be raised beyond the threshold value especially when alkali is in bath or liquor ratio is high.

6.7.4   Electrolyte

In general, all reactive dyes possess very little affinity for cotton; affinity varies among dyes in a specific brand and also within brands. Enhancement in dye uptake is best affected by applying salt at varying concentration for various brands based on affinity of dyes in a specific brand. Higher the affinity, lesser the liquor ratio and lesser the shade depth, lower is the dosing of salt. Cold and hot brand dyes are applied at a salt concentration of 50 and 75 g/l respectively. Quick exhaustion is very essential to resist loss of colour through hydrolysis. A few reactive dyes require application of salt in excess due to lack in affinity, e.g., yellow 6G, H5G, orange brown HG, scarlet HR, blue C4GP, etc.

6.7.5   Alkalinity or pH of bath

Dye bath should be free of alkali till exhaustion is completed to resist premature fixation of dye and corresponding uneven dyeing. Rise in pH up to 11 increases exhaustion and reactivity, but beyond this, exhaustion decreases. Excess alkali in bath promotes hydrolysis of dye. A pH around 10.5–11 with only Na_2CO_3 and 11–12.5 with combination of NaOH and Na_2CO_3 are suitable for cold and hot brand dyes respectively. Due to higher reactivity of M dyes, a mild alkali alone is sufficient for fixation, but a combination of NaOH and Na_2CO_3 (2 and 6 g/l respectively) promotes fixation of H and ramazol dyes through common ion effect due to their lesser reactivity with cotton.

6.7.6   Nature of fibre

Dye uptake is also influenced by accessible free volume in fibre – higher the volume, higher will be dye uptake. Scoured cotton has the least free volume, bleached cotton has little higher and mercerized cotton has the highest free volume due to removal of more impurity or immature fibre. Viscose has more free volume than mercerized cotton and so dye uptake will be according to these free volume data.

6.7.7   Time of dyeing

Reactive dye solution must not be stored for long time; otherwise hydrolysis will occur with loss of colour value. Time of dyeing must be reduced keeping exhaustion of bath on higher side. Dyeing time must be shorter, especially when the process is carried out at higher temperature or with more alkali or more water to reduce hydrolysis.

Time for both exhaustion and fixation are to be optimized as running the process beyond calculated time is meaningless. When no further exhaustion is possible or the reaction between dye and fibre has ceased, dyeing for more time unnecessarily lengthens the process.

6.8   Washing of dyed goods

Complete removal of hydrolysed dye from dyed cotton is a pre-requisite factor for better fastness of dyeings. H and ramazol dyed cotton show better response as dyes are either hydrolysed or react with cotton only, but problem is encountered with cold brand dyes due to partial fixation of few dye molecules on fibre with one hydrolysed end. Soaping is done in open-soaper or in rotary beam washing machine.

In beam washing, the beam is rotated slowly and hot water at 80°C is passed through one end of the beam when hydrolysed dye comes out completely. However, absence of open-soaper or beam washing unit may be compensated with hot jigger washing.

Removal of hydrolysed dyes from surface of fibre is not as typical as the case for hydrolysed dyes inside fibre. M-dyes, if partially hydrolysed inside, pose severe trouble for complete removal even after several washings.

Efficiency of washing may be assessed by sandwiching dyed cotton between two pieces of white cotton followed by ironing. Extent of straining of white fabric indirectly varies with efficiency of washing process.

All reactive dyed goods reveal true hue and tone in absence of alkali and so neutralization with CH_3COOH (1 ml/l) at 40–50°C is essential after fixation and prior to soaping.

6.9   Reactive dye on silk

Reactive dyes develop bright shades with good wash, light and perspiration fastness properties due to its reaction with $-NH_2$ group of silk. Silk should be thoroughly degummed before dyeing to remove sericin completely which may otherwise react with dye to show poor wash fastness. Silk can be dyed in either of alkaline or acid methods.

In alkaline method, dye bath is made up with dissolved cold brand dye and NaCl (10%). Dyeing is carried out at room temperature for 30 min after which temperature is slowly raised to 50°C and dyeing is continued for further 30 min; fixation is done by adding Na_2CO_3 (6–8 g/l) at 50°C for 40 min. Shading, if required, can be affected by addition of dye to the alkaline bath. A final wash-off with soap or a synthetic detergent at 90–95°C completes the process. Hot brand dyes can also be applied in the same way as with cold brand dyes, only difference is in temperature of fixation which is 70°C. Building-up of shade is not satisfactory.

Acid method is preferred when silk / cotton blend is dyed to produce reserved shades in acidic pH when reactive dyes do not react with cotton. Dye bath is prepared with dye and formic acid (0.5%), temperature is raised to 40°C and silk is added to bath. Dyebath is further slowly heated up to 85°C, when further addition of formic acid (3.5%) is made. Dyeing is continued for further 30 min followed by rinsing and finally washing-off. All cold brand dyes can be used in this method though alkali method is easier to control. Hot brand dyes are not used in acid dyeing of silk due to poor build up of shade.

6.10   Reactive dye on wool

Dye–fibre attachment resembles the same with that of acid dye on wool, i.e. chlorine of dye forms ionic bond with amino group of fibre with relatively higher bond energy. Reactive dyes produce very bright shades on wool with light fastness 5–7 and wash fastness 4–5. Affinity of reactive dye for wool is very high posing problem on production of level shades. Non-uniform initial absorption of dye on fibre must be controlled with non ionic surfactants. Dye-fixing agents are only applied during last wash to avoid staining and to improve wash fastness.

Bath is prepared with dye, 30% CH_3COOH (2–5%) and non ionic dispersing agent (2%). CH_3COOH may be replaced with CH_3COONH_4 or $(NH_4)_2SO_4$ (1–3%) when producing light shades. Wool is entered in bath and temperature is raised to 90–95°C over a period of 45 min. Dye fixing agents may be applied in the last rinsing bath. The process gives excellent result on woollen yarn in package form as well as knitwears.

6.11   Stripping

Partial stripping is possible with hot dilute CH_3COOH or HCOOH or with glacial CH_3COOH (5–10 g/l) at 90°C followed by washing and soaping. Full stripping is not possible so easily once the dye has chemically interacted with fibre. Build-up of shade at the initial phases of dyeing is to be strictly monitored and any correction required is to be made before fixation of dye.

Most of the reactive dyes are stripped out with NaOH + $Na_2S_2O_4$ (5 and 2 g/l respectively) at 50–60°C followed by NaOCl (0.5–1 g/l available chlorine) treatment, souring and washing.

Dyeings finished with resin is to be treated with HCl (10%) twice at 40–50°C separately to remove resin, followed by washing and removal of dye.

However, stripping process varies from dye to dye and unsatisfactory results are obtained in most cases.

6.12   Fastness properties

Efficient soaping and washing of reactive dyeings produce very good fastness due to attachment of dye with cotton through covalent bond. Inefficient washing may be compensated with cationic dye-fixing agents, e.g. fixanol PN up to some extent. Fastness to NaOCl bleach is poor in most cases. Treatment with hot CH_3COOH strips out most of dyes. A final soaping with ethylene diamine (1 g/l) reduces chance of acid hydrolysis with a very little fall in light fastness. Fast to dry ironing; exposure to light under high humidity reduces light fastness of pale shades. In contact with gas fumes, fastness properties are very much improved. Dyes are sensitive to metal ions, especially to copper. In presence of such metals, suitable metal sequestering agents should be used but with care, as excess use results in loss of colour.

Primary –OH groups of cellulose take part in dye–fibre attachment. It is interesting that M-brand dyes possessing two chlorine groups react with two different cellulose molecules using one chlorine group for each molecule of cellulose.

6.13   References

BIRD C L and BOSTON W S (1975) *The Theory of Colouration of Textiles*, Bradford, UK, Dyers Company Publication Trust.

Colour Index International (1987), 3rd Edition (3rd Revision), Bradford, UK, Society of Dyers & Colourists and AATCC.

GOKHALE S V and SHAH R C (1981) *Cotton Piece Dyeing*, India, Ahmedabad Textile Industries Research Association.

SHENAI V A (1987) *Chemistry of dyes and Principles of dyeing*, 3rd Edition, Mumbai, Sevak Publications.

7

Dyeing with reactive H-E dye

Abstract: Reactive HE (highly exhaustive) dyes are bi-functional and the reactive groups are located throughout the structure compared to conventional reactive dyes where two reactive groups are located side by side. This, in turn, impairs hydrolysis and imparts stability enabling to carry out dyeing even at high temperature in jet dyeing machine for dyeing of polyester–cellulosic blends. Dyes possess good fastness with full hue and shade range. Leveling is excellent with higher exhaustion rate as well as tinctorial value.

Keywords: Reactive HE dye, exhaustion, substantivity, reactivity

7.1   Superiority over conventional reactive dyes

Reactive highly exhaustive (H-E) dyes belong to substitution type and are suitable for exhaust dyeing of all cellulosic fibres in single or in blend with higher fixation at and above 85°C. Dyes exhibit exceptionally high tolerance to variations in dyeing conditions, well stable at higher temperature and are extensively used in dyeing of cellulosics part of polyester/cellulosic blends in one bath 2-step method in jet or HTHP beam dyeing machines. These dyes offer unique dyeing profiles including controlled rate of primary exhaustion in neutral alkali, high migration and diffusion properties, and controlled rate of secondary exhaustion after addition of alkali, and are the preferred choice for package dyeing (Khatri, 2004); levelness in shades and reproducibility are excellent even in little variation of dyeing parameters (Anon, 1982). These dyes differ from M and H brand dyes in respect to the location of reactive groups in dye structure – when one or two reactive sites are available on the same nucleus of H and M dyes, the reactive groups in H-E dyes are oriented throughout the whole structure of dye molecule (Betrabet et al, 1977); the advantage is that if one reactive group is hydrolysed, another is left for reaction with cellulose from another location (Peters, 1975). Their tendency for hydrolysis lies in between that of H and M dyes. These dyes are also useful in dyeing silk like that with cold brand reactive dyes (Gulrajani et al, 1995).

7.2   Features of H-E dyes

These dyes are superior to other reactive dyes as these are: (i) suitable for exhaust dyeing with higher exhaustion rate, (ii) higher substantivity, (iii) higher tinctorial value, (iv) possess excellent compatibility within group, even with H dyes, (v) good build-up of shade, (vi) insensitivity to liquor ratio, (vii) freedom from listing' and 'ending' in jig dyeing, (viii) economy in use because of high tinctorial yield, (ix) consistently high fixation, (x) reliability – excellent reproducibility owing to insensitivity to probable variations in liquor ratio, dyeing time, salt concentration, temperature, etc., (xi) complete shade range, (xii) higher molecular weight to provide good light and excellent wash fastness, (xiii) very good fastness-resistant to acid and alkaline hydrolysis and to attack by detergents containing mild oxidising agents and (xiv) too low waste-water load (Atul, 2009) Ltd. Commercial and generic names of reactive HE dyes are listed in Table 7.1. Only few dyes are adequate to reproduce any hue with desired depth and tone. The turquoise blue lacks wash fastness and inclusion of the same from H dyes completes the range.

Table 7.1 Commercial and C I generic names of reactive HE dyes

Commercial name	C I Generic name	C I Number	CAS Number
Reactive Yellow-HE4R	Reactive Yellow 84	—	61951-85-7
Reactive Yellow-HE4G	Reactive Yellow 81 / 105	—	59112-78-6
Reactive Yellow-HE3G	Reactive Yellow 105	—	176023-34-0
Reactive Yellow-HE6G	Reactive Yellow 135	—	77907-38-1
Reactive Golden Yellow-HER	Reactive Yellow 84A	—	61951-85-7
Reactive Red-HE3B	Reactive Red 120	25810	61951-82-4
Reactive Red-HE7B	Reactive Red 141	—	61931-52-0
Reactive Red-HE8B	Reactive Red 152	—	71872-80-5
Reactive Orange-HER	Reactive Orange 84	—	91261-29-9
Reactive Orange-HE2R	Reactive Orange 94	—	129651-47-4
Reactive Green-HE4BD	Reactive Green 19	—	—
Reactive Green-HE4B	Reactive Green 19 A	—	61931-49-5
Reactive Turquoise Blue-HA	Reactive Blue 71	—	12677-15-5
Reactive Navy Blue-HER	Reactive Blue 171	—	77907-32-5
Reactive Navy Blue-HE2R	Reactive Blue 172	—	85782-76-9
Reactive Blue-HERD	Reactive Blue 160	—	71872-76-9
Reactive Blue-HEGN	Reactive Blue 198	—	—
Reactive Black HEBL	—	—	—

7.3   Individual dye characteristics

Dyeing is started at 50°C in general, but good exhaustion and levelling depend on right temperature. Yellow dyes show moderate exhaustion beyond 60–65°C, setting of dyebath from the very beginning at such a temperature provides successful results – adequate salt is to be applied for good exhaustion. Yellow H-ER is well exhausted at room temperature and so starting temperature can be varied. Red, black and navy blue dyes are sensitive to temperature, random heating will cause unlevel dyeing and deep selvedge – especially in producing light shades; any addition to hot dye bath will show tailing effect. It is suggested to start dyeing at room temperature for two ends in jigger with dye in two halves followed by addition of salt; temperature is also slowly raised to desired level (Atul Ltd, 2009).

Green H-E4B and turquoise blue dyes are applied above 50°C. Their exhaustion property is good; turquoise blue is exhausted more in presence of excess salt. Temperature for fixation with alkali is 80–85°C for all these dyes. Thorough washing and soaping are essential for good fastness. Other dyeing criteria, e.g. liquor ratio, pH, time, etc., must be maintained – though liquor ratio has no such influence on dyebath stability and good exhaustion.

Reactive H-E dyes are most suitable for exhaust dyeing. Solubility in water is quite good, addition of common salt in bath reduces the extent of solubility and so it is desired to exhaust as much dye as possible before addition of salt in bath. Dyeing in vessels made of copper or iron turns the hues duller with change in tone, stainless steel vessels are recommended to produce brilliant shades. Dischargability varies from dye to dye. Light fastness ranges from 3 to 5, while wash fastness is outstanding provided efficient washing has been imparted to dyeings.

7.4   Structure of H-E dyes

The chromogen of the colour is supported through a stillbene in the molecule. Though reactive HE dyes were developed during 1970's and were put on commercial application since 1980's, structure and C I constitution number of most of the members are not in public domain except a very few as described in scheme 7.1 (Colour Index online, 2009). For the sake of understanding, a few structures are also mentioned here without colour Index specification and commercial names (Shawki et al, 1989).

7.5   Dyeing of cotton

7.5.1   Exhaust dyeing in jigger / winch

Dyeing technique somewhat lies in between those for cold and hot brand dyes. Pre-treatment of fabric with acetic acid (0.5–1.0 g/l) at room temperature for 1 turn at a pH ~5.5–6.5 results in levelling of shade (Jaysynth Dyechem, 2009). Dyeing is started at room temperature with

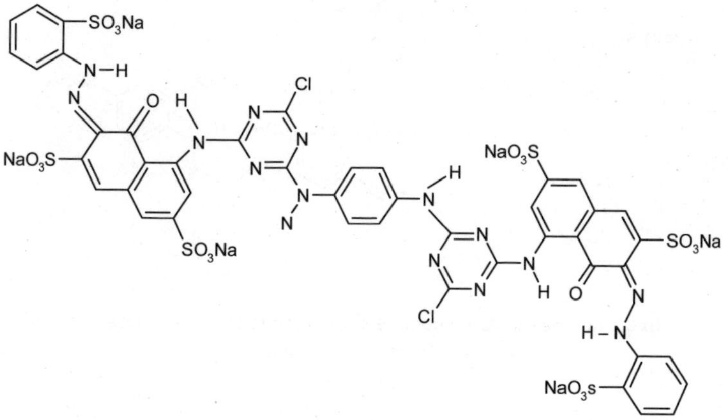

Reactive Red-HE3B (C I Reactive Red 120, C I 25810)

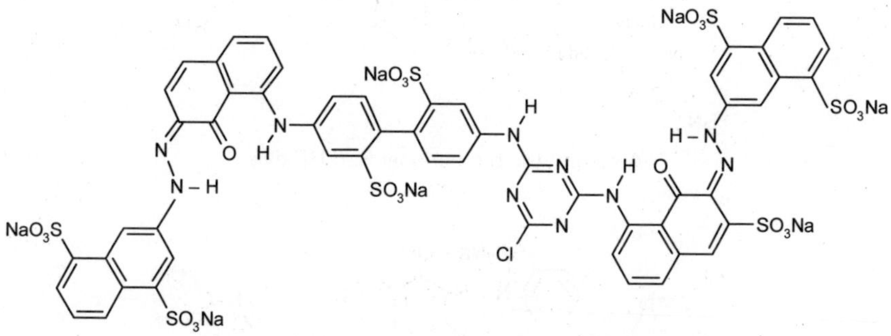

Reactive Orange-HER (C I Reactive Orange 84)

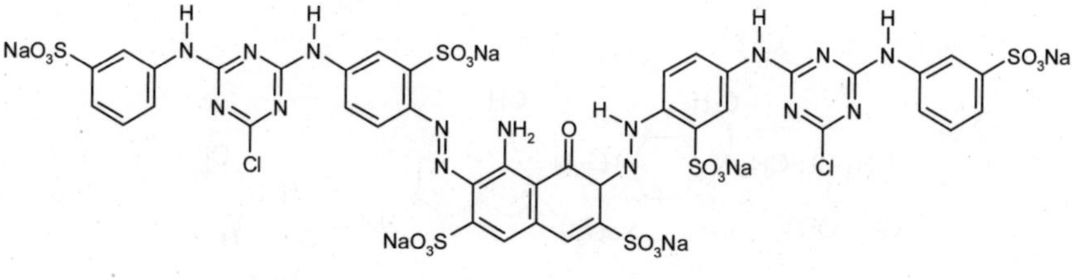

Reactive Green-HE4BD (C I Reactive Green 19)

Scheme 7.1 Contd.

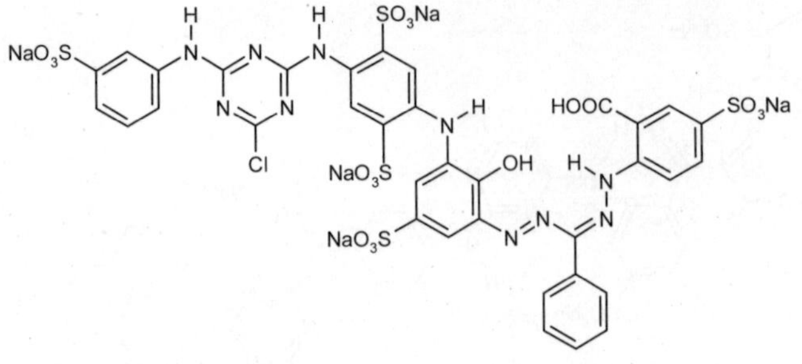

Reactive Blue-HERD (C I Reactive Blue 160) is the 1:1 copper complex of above structure

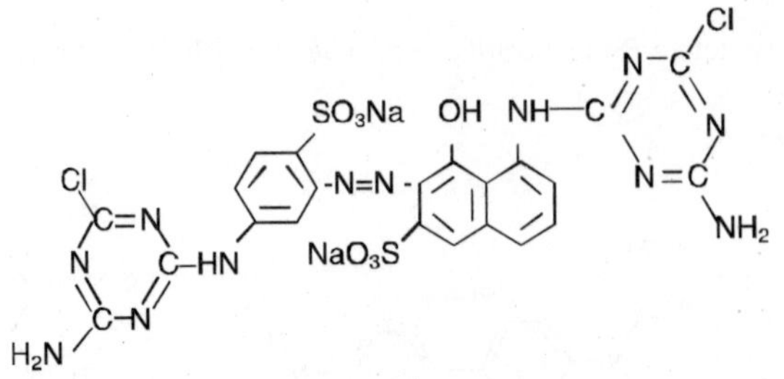

A symmetric bifunctional Red HE dye

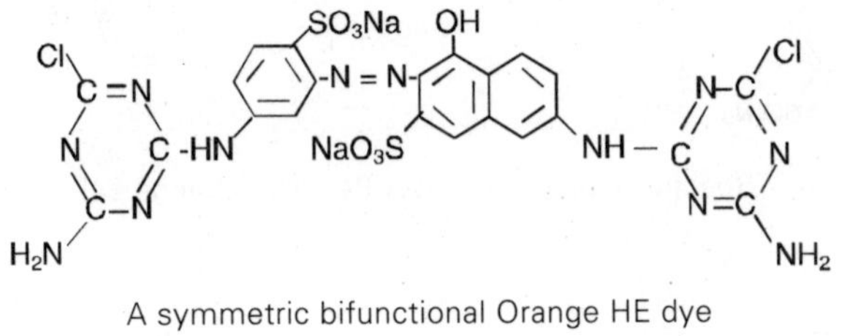

A symmetric bifunctional Orange HE dye

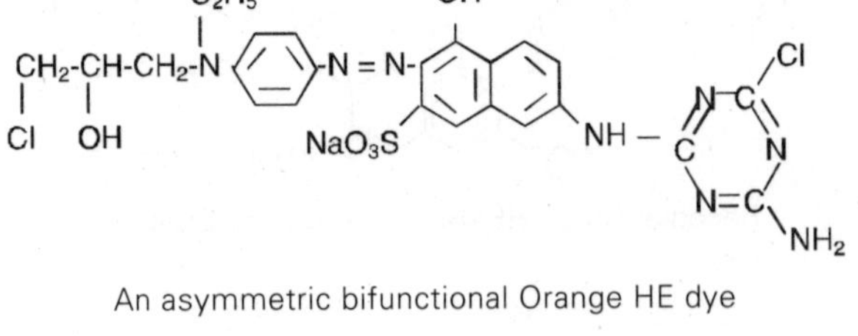

An asymmetric bifunctional Orange HE dye

Scheme 7.1 Chemical structure and C I specification of reactive H-E dyes.

dye in two parts and temperature is raised slowly after addition of salt, especially where light composite shades are to be produced with red, navy blue and / or black dyes in combination, and where a lot of batches are to be dyed in identical hue, tone and shade in exhaust method thud preventing variation in build-up of shades. Jiggers used in dye houses mostly do not have indirect heating system; constant injection of steam causes temperature difference between two ends resulting in tailing effect. It is advisable to heat up before each turn and keep the steam valve open slightly so that an overall constant temperature of dyebath for each turn is ensured. During third turn in jigger in salt, temperature is raised to 80–85°C, sampling is done after fourth turn followed by addition of alkali. If addition of dye becomes essential, dyebath is cooled down for further addition. In dyeing with deep shades, addition can be made at higher temperatures if amount of colour to be added is less. Dyebath pH should be in the range of 10.8–11.3 throughout period of fixation at 85°C; salt and alkali required for these dyes and their fixation times are listed in Table 7.2 (Betrabet et al, 1977; Chakraborty, 1990; Dawson, 1981).

Table 7.2 Salt and alkali requirement and fixation times (Atul, Tulactiv XLE dyes)

Depth of shade (%)	NaCl or anhydrous Na_2SO_4 (g/l)		Alkali (g/l)			Fixation Time (min)
	Unmercd. cotton	Mercd. cotton/ Viscose rayon	Na_2CO_3 only	Mixed alkali Na_2CO_3 + NaOH		
Upto 0.10	10	5	10	5	0.2	30
0.11–0.30	20	10	10	5	0.2	30
0.31–0.50	30	20	10	5	0.2	45
0.51–1.00	45	30	15	5	0.2	45
1.01–2.00	60	40	15	5	0.5	45
2.01–4.00	70	55	20	5	0.5	60
4.01 and above	90	65	20	5	0.5	60

Only Na_2CO_3 is sufficient for fixation, but a mixed alkali system works better, especially when hot brand dyes are in combination. Thorough soaping and washing are imparted to ensure good wash fastness. Further improvement in fastness can be done by using Fixanol PN (1–2 g/l) at 40–50°C or ethylene diamine and lissapol ND combination (1 g/l each) at boil for 15–20 min. Final treatment in mild acidic solution will result in a fabric of around neutral pH and improved tensile strength for further finishes at higher temperature. Dyed fabric must not be kept in open air for long time without drying. Elsewhere five different dyeing methods are described in the manufacturer's literature and are detailed below (Atul, 2009; Jaysynth Dyechem, 2009).

Method 1: Addition of salt by parts

This method is suitable where circulation of liquor is absent and electrolyte is manually added to control rate of exhaustion; suitable for all depths of shade. For better levelling and penetration, temperature may be raised to 95°C for 20 min during salt treatment and is to be lowered down to 80°C before alkali is added. In dyeing with turquoise H-EG, glauber's salt (Na_2SO_4, $10H_2O$) is used. Dyeing is started at 50°C along with sequestering agents. Pre-dissolved dye is added over 20 min at 50°C, temperature is raised to 95°C and electrolyte is added in portions of 10%, 30% and 60% with 15 min intervals. Dyeing is continued at 95°C for further 30 min, followed by cooling down to 80°C. Alkali is added (Na_2CO_3 –5 g/l + NaOH –0.2 g/l) and dyeing is continued for further 60 min at 80°C (Fig. 7.1).

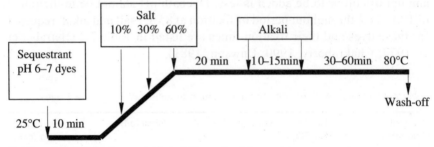

7.1 Reactive H-E dyeing cycle in 'addition of salt by parts'.

Method 2: Salt at start

To start dyeing, salt is added and is suitable for machines with liquor circulation and for dyeing of all medium to heavy shades. In dyeing pale shades (<0.5%) on mercerised yarns or with high density packages, temperature may be raised to 95°C and maintained for 20 min, then cooled down to 80°C and circulated at 80°C for 5 min before addition of alkali (Fig. 7.2).

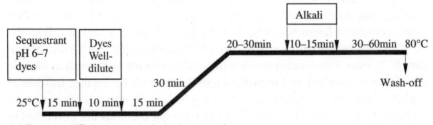

7.2 Reactive HE dyeing cycle in 'salt at start'.

Method 3: All in

In this case, alkali and salt are added at the start of dyeing and is suitable for machines with liquor circulation. Combined alkali system is preferred. Green H-E4BD, green H-E3B and navy blue H-ER should not be used by this 'all-in' method (Fig. 7.3).

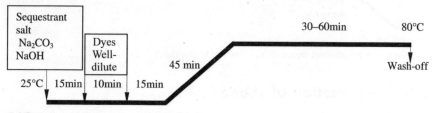

7.3 Reactive H-E dyeing cycle in 'all in'.

Method 4: Migration technique

Useful for machines with microprocessor controlled addition systems for dyeing pale shades ($< 0.5\%$) and for all shades on difficult substrates such as mercerised cotton and viscose packages (Fig. 7.4).

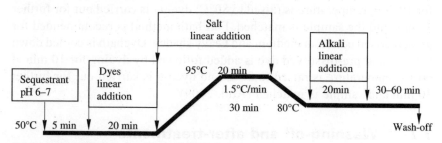

7.4 Reactive H-E dyeing cycle in 'migration technique'.

Method 5: Isothermal technique

This is for machines with microprocessor controlled addition systems for medium to heavy depths ($>0.5\%$) on unmercerised cotton where exposed selvedge is protected from being cooled down by (i) using closed jigger, (ii) controlled batching to avoid slippage during dyeing and using all pieces of fabric having equal width, (iii) controlling dye bath temperature at 90°C during salting stage ; better penetration of difficult fabrics is achieved by raising the bath temperature to 100°C during this stage, (iv) adjustment of dyebath temperature to 85–90°C to ensure that the fabric is maintained at a minimum of 80°C during the whole period of dye fixation (the alkaline stage) and (v) dyeing time given was adequate to fix dye on the selvedges (Fig. 7.5).

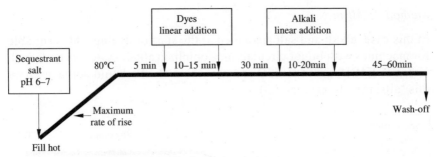

7.5 Reactive HE dyeing cycle in 'isothermal technique' .

7.6   Correction of shade

Dyed sample is checked at the end of salt treatment stage and before fixation with alkali. A fresh bath should be set for additions of over 25% of the original recipe. If addition for correction is less than 25% of original recipe, a half or a full bath shading method may be adopted. Half bath method is suitable for all shades on mercerised cotton, viscose and pale shades on unmercerised cotton. In this method, half of bath is dropped and refilled with water, pre-dissolved dye is added and dyeing is continued for 10 min, temperature is raised to 80°C; dyeing is carried out for further 30 min and the sample is matched. Full bath method is recommended for unmercerised cotton in medium and heavy shades. Dyebath is cooled down to 60°C and pre-dissolved dye is added followed by dyeing for 10 min at 60°C, temperature is raised to 80°C and dyeing is carried out for further 30 min and sample is matched (Atul, 2009).

7.7   Washing-off and after-treatments

Before soaping, salt and alkali are to be removed completely. Hot rinsing is imparted thrice at 70°C, each for 10 min followed by soaping at boil for 30 min with Lissapol D (1 g/l) and hot rinse at 70°C for 10 min. Efficiency of washing may be assessed by staining of white cotton.

Reactive H-E dyes possess excellent wash fastness and to improve fastness further Fixanol PN is applied to fabric for 20 min at 40°C at pH 5.0. If softening is also to be carried out, softener may be added to the exhausted after-treatment bath and continued for a further 20 min at 40°C.

7.8   Stripping

H-E dyes may be stripped-off at boil with Na_2CO_3 (or NaOH) + $Na_2S_2O_4$ (3–6 g/l each) for 1 hr followed by hot and cold rinses succeeded by NaOCl bleach to get good whiteness. Alternately, stripping with only NaOCl (10–

15% available Cl_2) at room temperature for 20 min succeeded by anti-chloring with $NaHSO_4$ (2.5 g/l) for 15 min at 50–60°C will remove residual chlorine. A few H-E dyes cannot be stripped completely with Na_2CO_3 (alternately NaOH) + $Na_2S_2O_4$ combinations, e.g. blue H-EGN but can be bleached with NaOCl to good white. H_2O_2 bleaching should not be used on fabric dyed with copper-containing dyes as cotton may get disintegrated.

7.9   Disperse-reactive H-E combination for polyester–cotton blends

H-E dyes are stable at higher temperature like 130°C and are applied in one bath-one step and one bath-two step methods in dyeing polyester–cotton or polyester–viscose blends. The first method is used in producing light shades and the second one to produce heavier shades – mainly, where reduction clearing after disperse dyeing is essential.

In the first method, i.e. one bath one step method, H-E dyes are mixed with disperse dyes. After dyeing of polyester part with disperse dye as usual, bath is cooled down to 95°C, salt is added followed by alkali at 85°C. Dye, salt and alkali – all should be increased by 10–15% over dyeing in jigger, as the liquor ratio in this case is higher. A part of alkali neutralizes acetic acid used in dyeing of polyester. In second method, i.e. one bath-two step method, rather than using H-E dyes along with disperse dyes, polyester part is dyed, followed by reduction clearing, washing and mild acid treatment to the fabric, which is then dyed with H-E dyes.

Application of other dyes is restricted on cotton fabric in rope form due to fear of only surface dyeing, H-E dyes produce level dyeing on rope form too. Maximum temperature of dyeing in alkali must be kept below or at 85°C. Further rise in temperature affects polyester, results loss in tensile strength, weight, etc. (Shenai and Nayak, 1981).

7.10   References

ANON (1982) 'New findings in the discontinuous treatment of cellulosic–fibre knitted fabrics with hot-dyeing reactive dyes', *Textil Praxis Int.*, **37**, 11, 1188–1193

ATUL LTD (2009) '*Tulactiv XLE dyes for excellent performance in exhaust dyeing of cellulose*', Reactive HE dyeing manual, Gujarat, India

BETRABET S M, BAGWE V B and DARUWALLA E H (1977) 'Behaviour of bifunctional and polyfunctional reactive dyes applied to cotton cellulose', *J. Soc. Dyers Color.*, **93**, 9, 338–345

CHAKRABORTY J N (1990) 'Reactive dyes: chemistry techniques and effluent problems involved', *Textile Dyer Print.*, **23** , 26, December, 31–33

Colour Index International (2009), on line, Bradford, UK, Society of Dyers & Colourists and AATCC

DAWSON T L (1981) 'pH and its importance in textile coloration', *J. Soc. Dyers Color.*, **97**, 3, 115–125

GULRAJANI M L, JAIN A and AGARWAL D (1995) 'Dyeing of silk with Procion HE dyes', *Indian J. Fibre Textile Res.*, **20**, 3, 150–155

JAYSYNTH DYECHEM LTD (2009) '*Reactive HE dyeing Manual*', Mumbai, India,

KHATRI A R (2004) 'Exhausting options', *Textile Month*, **2**, 13–14

PETERS R H (1975) *Textile Chemistry*, Vol-III, New York, Elsevier Scientific Publishing Company

SHAWKI A , MOHAMED A and KHARABAD (1989), Synthesis of some mono/ bifunctional reactive dyes, *Indian Textile J.*, **99**, 4, 164–167

SHENAI V A and NAYAK N K (1981) 'Action of alkali on polyester fibres', *Textile Dyer Print.*, **14**, 23, November 1, 25

8

Dyeing with vat dye

Abstract: Vat dyes retain C=O chromophore in structure, which renders water insolubility. Dyes are reduced and solubilised for application from aqueous media, when the dye shows affinity and substantivity for cellulosics. Rate of exhaustion is excellent with higher strike rate raising chances of unlevel dyeing. Shades possess all round fastness properties and are brilliant. All hues are available with high tinctorial power. Reduction process is little complicated; any mishandling may over-reduce dye changing hue and shade. Sodium hydrosulphite used for reduction of vat dyes increases waste-water load abruptly forcing to search for alternate reducing systems.

Keywords: vat dye, hydrosulphite, over-reduction, fastness, alternate reducing systems

8.1   Basic principle for application

Vat dyes are insoluble in water and nonionic; these are converted to leuco compounds on reduction followed by solubilisation with alkali in which state these show substantivity towards cellulose. After dyeing, parent dye structure is recovered by oxidising it within fibre, when the dye molecule get trapped *in situ* and establish linkage with fibre through H-bonds and van der Waals forces (Fig. 8.1). In ancient times, dyeing was often carried out in wooden vat assigning the name 'vat dye'. Carbonyl groups are the chromophores which are changed to >C=OH groups on reduction and then to soluble sodium derivatives (>C=ONa) in presence of alkali.

8.1 Basic principle of dyeing cotton with vat dye.

The dyeing process generally involves three basic steps:

(i)   Preparation of the vat, i.e. reduction of dye to its leuco form followed by its conversion to sodium salt.
(ii) Dyeing
(iii) Oxidation to parent dye

Dyes are non-ionic – develop anion on solubilisation; possess excellent affinity for cellulosics and are retained by fibre with H-bonds and van der walls forces on oxidation. Though molecular size of dye is too small compared to pore size of cellulosics, after diffusion and oxidation, these crystallize to form big stable aggregates during soaping and can not come out providing superior fastness properties.

8.2   Physical form of dye

Vat dyes are available in various commercial forms. Powder form has large particle size, (ultra-concentrated or uc grade) and is designed for exhaust dyeing. Fine micro-form (small particle size, fm), fine ultra dispersed form (fd, ud) and superfine (sf) qualities – all are suitable for continuous dyeing while paste form is suitable for printing.

8.3   Properties of vat dyes

All vat dyes are water insoluble, anthraquinones are mostly soluble in hot DMSO (dimethyl sulphoxide) and can be extracted from dyed sample at boil without any change in λ_{max}; C=O groups in anthraquinone dye act as chromophore and groups such as NH_2, OH, alkylamino (NHR, NR_2), benzamide (NH–CO–RH), alkoxy (–OR) are auxochromes. Sublimation temperature is very high and possesses excellent fastness properties. One characteristic feature of anthraquinone dyes is that on reduction these show typical change in colour and original colour is restored on oxidation. Dyes are costlier, full shade range is available; some of yellows, browns and oranges have marginal poor light fastness (Fox, 1948).

Table 8.1 Vatting and dyeing conditions as well as relative concentration of chemicals

Dye class	Vatting temp (°C)	Dyeing Temp (°C)	$Na_2S_2O_4$	NaOH	Electrolyte
IK	35–50	30	less	less	higher
IW	45–50	40–45	moderate	moderate	moderate
IN	55–60	50–55	higher	higher	less
IN special	≥ 60	≥ 60	very high	very high	not required

Indigoid dyes contain

$$\underset{-C-\,C-\,C-C-}{\overset{\overset{O}{\|}\quad\overset{O}{\|}}{}}$$

as chromophore and –NH or –S– groups as auxochrome. Indigoids are soluble in DMSO at room temperature, boiling glacial CH_3COOH, boiling $C_6H_5NH_2$ or C_6H_5N, partially soluble in $CHCl_3$, C_6H_5OH and H_2SO_4. On reduction, indigoids develop a nearly colourless solution and original colour is restored on oxidation. At 170°C, it sublimes-off and condenses into blue crystals of pure indigotin. Reduced and solubilised sodium compound of indigo has limited affinity for cellulose and very readily oxidized in the fibre to insoluble blue indigotin by atmospheric oxygen.

8.4   Classification

Vat dyes are classified based on chemical structure and method of application.

8.4.1   Chemical structure

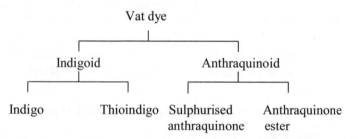

8.4.2   Method of application

This classification is based on conditions and amount of chemicals required for reduction, solubilisation and dyeing. Based on these facts, all vat dyes can be applied in any of these four methods (Fox, 1948), viz.

   (i) *IK dyes (Indanthrene cold dyes)*: Dyes belonging to this class require less alkali, less $Na_2S_2O_4$, low vatting (35–50°C) and dyeing temperatures (30°C) but lacks affinity. To promote exhaustion, large quantities of electrolyte is applied, e.g. C. I. Vat Blue 1 (indigo).

   (ii) *IW dyes (Indanthrene warm dyes)*: These dyes are applied in presence of moderate amount of alkali, moderate $Na_2S_2O_4$, lower vatting (45–50°C) and dyeing temperatures (40–45°C). Dyes possess moderate affinity and require moderate amount of salt to promote exhaustion, e.g. C.I. Vat Blue

6 (Blue BC), Blue 4 (Blue RSN), Brown 3 (Brown R), Brown 1 (Brown BR), Orange 15 (Orange 3G), Red 10 (Red 6B) etc.

(iii) *IN dyes* (*Indanthrene normal dyes*): Dyes from this class require relatively high alkali concentration, higher amount of $Na_2S_2O_4$, high vatting (55–60°C) and dyeing temperatures (50–55°C). Dyes show very good affinity for cellulosics and application of salt at very lower concentration completes exhaustion, e.g. C.I. Vat Violet 1(Purple 2R), Blue 20 (Dark Blue BO), Blue 17 (Dark Blue 2R), Green 1(Green XBN), Yellow 2 (Yellow GCN), Black 25 (Olive D), Green 3 (Olive green B), Green 8 (Khaki 2G), Direct Black 8 (Grey 2B) etc. (iv) IN special dyes: Dyes belonging to this group require exceptionally high alkali as well as $Na_2S_2O_4$ concentration. Vatting and dyeing are done at or above 60°C, no electrolyte is used due to excellent affinity of solubilised dye for fibre, e.g. C. I. Vat Brown 5 (Brown G), Green 9 (Black 2B) etc.

Dosing of $Na_2S_2O_4$ (sodium hydrosulphite or sodium dithionite) is proportional to number of C=O groups to be reduced for a specific dye and this reduced state is to be ensured till end of dyeing. The same is true for NaOH too. It implies that an IK dye possesses less C=O groups, dye structure are simple favouring reduction at lower temperature with less reducing and solubilising agents. Also, a vat dye may retain a few numbers of C=O groups in structure, but only a very few are to be reduced and solubilised for efficient dyeing, reduction of all C=O groups may cause over-reduction.

Indigo comes under IK class but follows completely different technique of application. Sulphurised vat dyes are applied either as vat dye or sulphur dye but reduction or dyeing is done as that used for a vat dye.

8.5   Structure of some important vat dyes

Vat dyes are marketed under various brand names, viz. Navinon, Novatic, Algol, Arlanon, Arlanthrene, Arlanone, Chemithrene, Cibanone, Helanthrene, Intravat, Mikethrene, Nihonthrene, Palanthrene, Paradone, Patcovat, Sandothrene etc. Structure and C I specification of some commonly used vat dyes are shown in Scheme 8.1 (Colour Index, 1987).

8.6   Application on cellulose

Typical recipe

Vatting and Dyeing:

Dye	– 1 %
Turkey red oil	– a little
$Na_2S_2O_4$	– 'A'g/l (Table 8.2)
NaOH	– 'B'g/l (Table 8.2)

Salt   – as per requirement
Vatting temp. – x°C (Table 8.2)
Dyeing temp.   – (x – 5)°C (Table 8.2)

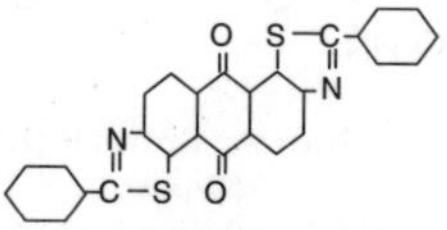

Yellow GCN (IDI), Yellow 5G/GN (Atic)
(Greenish yellow)
C.I. Vat Yellow 2, CI 67300

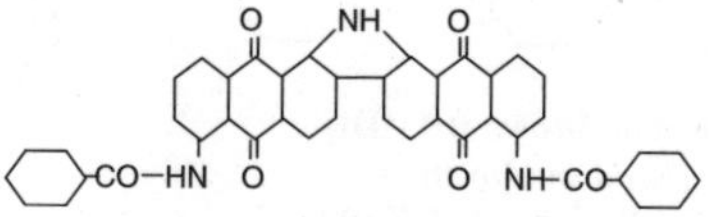

Navinon / Novatic Gold Orange 3G (IDI /Atic)
(Golden yellow)
C.I. Vat Orange 15, CI 69025

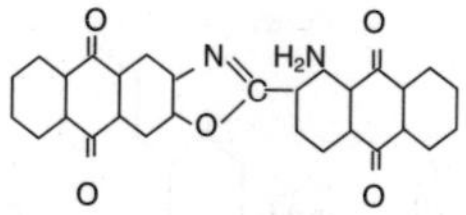

Novatic Red 3B / Navinon Red FBB
(Bluish red)
C.I. Vat Red 10, CI 67000

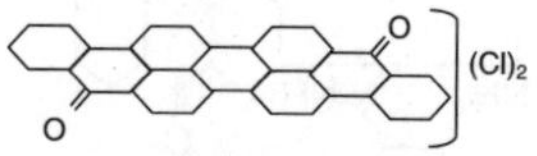

Navinon Violet 2R (IDI) / Novatic Bril. Purple 2R (Atic)
(Bright bluish violet)
C.I. Vat Violet 1, CI 60010

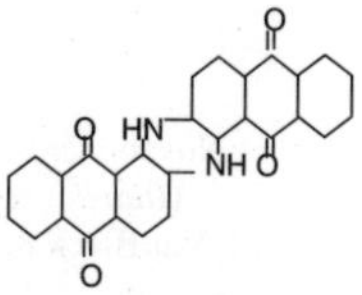

Navinon Blue RSN (IDI)
(Bright reddish blue)
C.I. Vat Blue 4, CI 69800

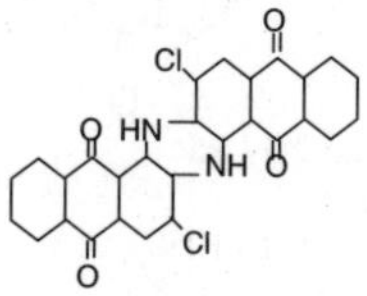

Navinon Blue BC (IDI)
(Bright blue)
C.I. Vat Blue 6, C I 69825

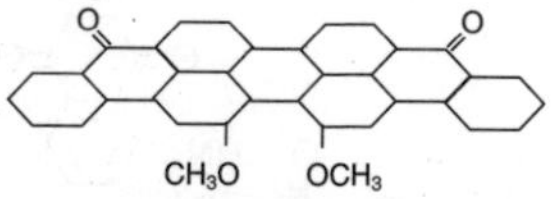

Jade Green XBN (Atic), Green B/Green FFB(IDI)
(Bright green)
C.I. Vat Green 1, C I 59825

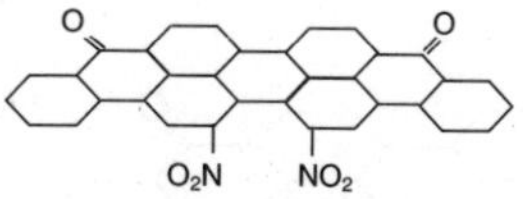

Novatic Black 2B (Atic)
(Dull green→Black)
C.I. Vat Green 9, C I 59850

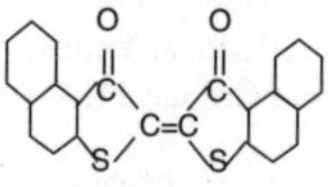

Novatic Brown G (Atic)
(Reddish brown)
C.I. Vat Brown 5, C I 73410

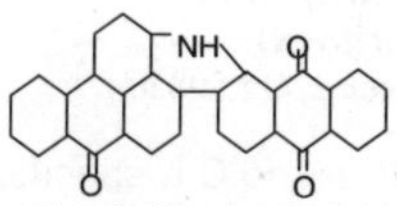

Novatic Olive Green B (Atic)
(Dull green)
C.I. Vat Green 3, CI 69500

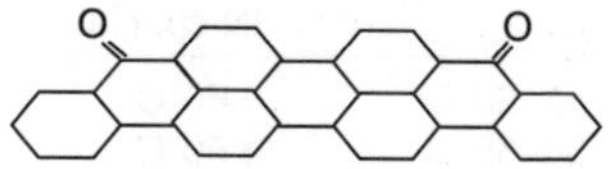

Navinon Dark Blue BO (IDI)
(Reddish navy)
C.I. Vat Blue 20, CI 59800
C.I. Pigment Blue 65 (Violanthrone)

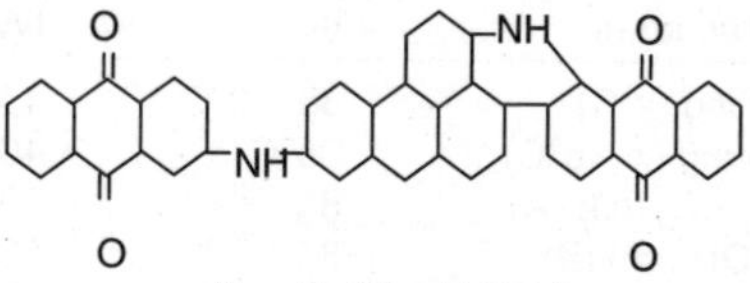

Novatic Olive D (Atic)
(Brownish grey)
C.I. Vat Black 25, C I 69525

Scheme 8.1 Contd. ...

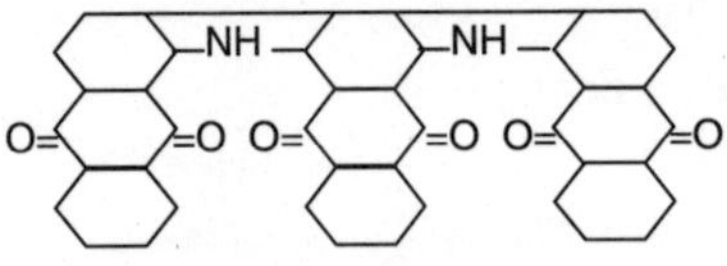

Navinon Brown BR (IDI)
(Reddish brown)
C.I. Vat Brown 1, C I 70800

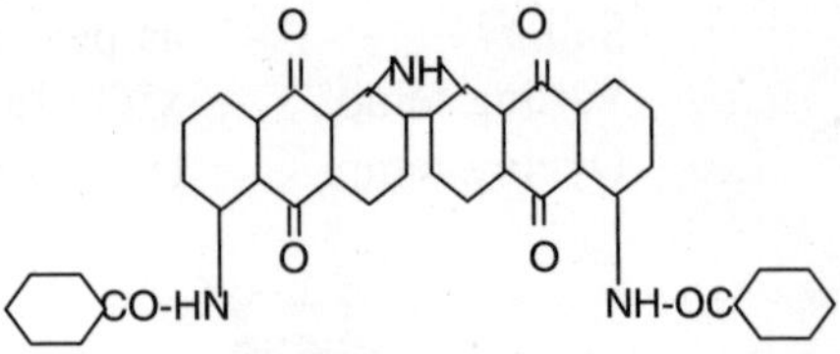

Navinon Brown R (IDI)
(Brown)
C.I. Vat Brown 3, C I 69015
C.I. Pigment Brown 28

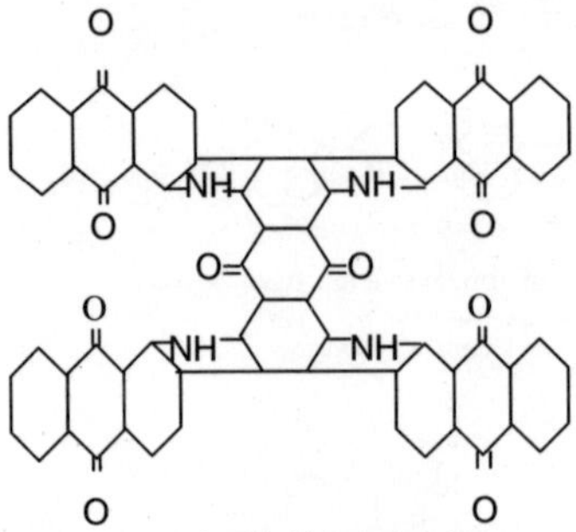

Novatic Khaki 2G (Atic)
(Olive)
C.I. Vat Green 8, C I 71050

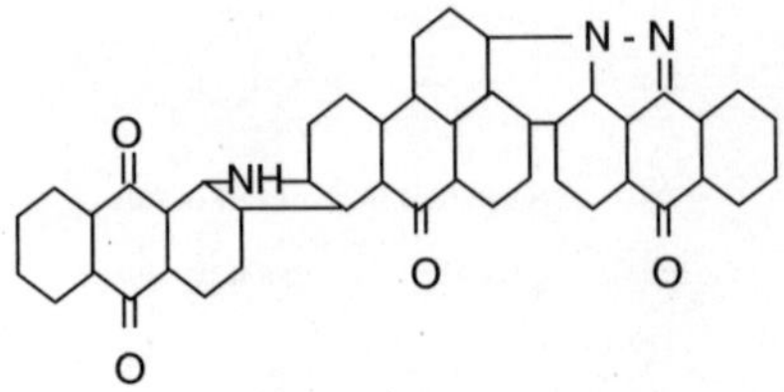

Novatic Grey M (Atic)
(Bluish grey)
C.I. Vat Black 8, C I 71000

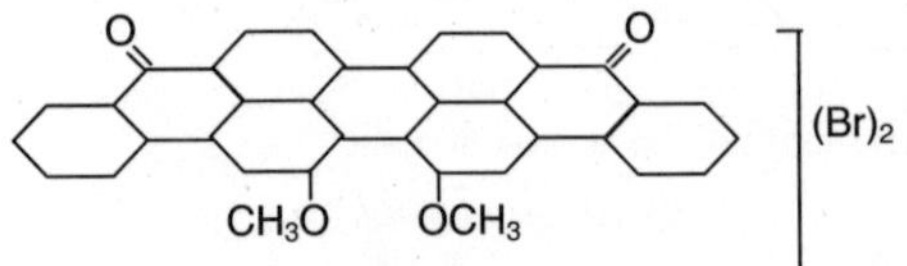

Green 2G (IDI) / (Atic)
(Bright green)
C.I. Vat Green 2, C I 59830

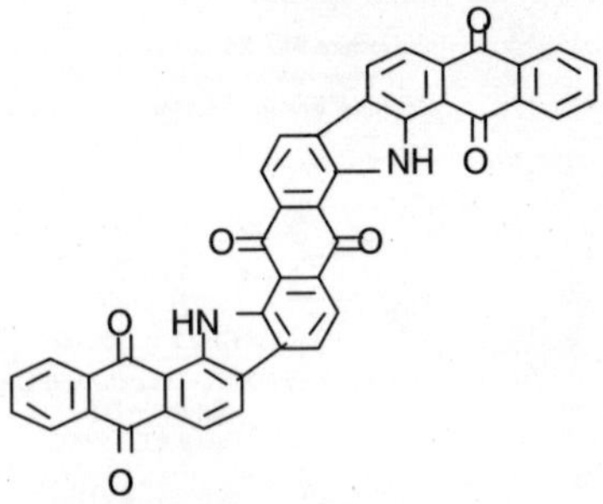

Navinon Yellow 3RT
(Reddish Yellow)
C.I. Vat Orange 11, C I 70805

Scheme 8.1 Structure and C.I. specification of some important dyes

Table 8.2 Dyeing parameters and concentration of chemicals in vat dyeing

Parameters	IK	IW	IN	IN special
Vatting (x°C)	35–50°C	45–50°C	55–60°C	≥ 60°C
Dyeing (x – 5°C)	35°C	40-45°C	50–55°C	≥ 60°C
$Na_2S_2O_4$ (g/l) (A)	8	10	12	15
NaOH (g/l) (B)	8	10	12	15
NaCl (g/l)	20	15	10	–

Oxidation: H_2O_2 (35%) – 1–2 ml/l at 50–60°C for 15–20 min

Soaping: Na_2CO_3 and soap (0.5 g/l each) at boil for 15–30 min or steaming only.

Dye is pasted with turkey red oil, water is added to it and the mixture is heated up to required vatting temperature depending on class of dye, one-third of total NaOH is added followed by addition of one-third of $Na_2S_2O_4$; the solution is stirred till colour of solution is changed (vat dyes on reduction show complete change in colour) with coloured foam at the top layer of solution ; 10 min is allowed for complete reduction.

Dyebath is prepared with reduced and solubilised dye, one-third of total NaOH and $Na_2S_2O_4$ each. Pretreated wet cotton is dyed in this bath for 30 min after which salt is added (if required, as IN special dyes require no salt) and dyeing is continued for further 1–2 hrs depending on depth of shade. Throughout dyeing, remaining one-third of $Na_2S_2O_4$ and NaOH are added in small amounts at regular intervals to retain dye in reduced and solubilised form and to prevent precipitation of oxidized dye on goods. A too low vatting temperature causes incomplete reduction; if high, over-reduction of dye may occur. Dosing of NaOH and $Na_2S_2O_4$ are to be maintained as shown in Table 8.2 (Chakraborty and Chavan, 2004).

IN special dyes are not generally over-reduced if dose of $Na_2S_2O_4$ and NaOH are in excess. After dyeing, a cold wash is imparted followed by oxidation of dyed material at 50–60°C with 1–2 ml/l H_2O_2 (35%) for 15–20 min, soaped or steamed at boil for 15–30 min, finally a thorough wash completes the process.

8.7   Reduction of vat dyes

All vat dyes are to be essentially reduced and solubilised before dyeing. Indigo requires around (-700 to -750) mV and anthraquinoid dyes require (-800 to -1000) mV for complete reduction depending on class of vat dye, i.e. IK, IN etc; IK requires the lowest range of reduction potential while IN special the highest (Bechtold et al, 1991).

8.7.1 $Na_2S_2O_4$ as reducing agent

$Na_2S_2O_4$ (sodium hydrosulphite or sodium dithionite) is the universally accepted reducing agent in vat dyeing. It reduces insoluble vat dye to partially soluble leuco dye, counter-acts effect of dissolved oxygen in water, which otherwise may precipitate a part of reduced dye through oxidation. Little excess of it retains stability of reduced liquor required for levelled dyeing, and it produces a sediment free clear reduction bath. $Na_2S_2O_4$, if in too excess, retards rate of dyeing and large excess results over reduction as well as wastage of dye. It releases nascent hydrogen required for reduction when added in water or alkali.

$$Na_2S_2O_4 + 4H_2O = 2NaHSO_4 + 6H$$
$$Na_2S_2O_4 + 2NaOH = 2Na_2SO_3 + 2H$$

$Na_2S_2O_4$ reduces indigo at 30–50°C, resulting in formation of mostly biphenols (leuco-indigo). The reduced form is quite stable in presence of NaOH to carry out dyeing at room temperature. Other vat dyes are reduced at a temperature depending on class of dye. However, $Na_2S_2O_4$ is very unstable; at the time of reduction of dye, it gets decomposed in water, thermally, oxidatively and in various other ways to produces a list of toxic sulphur products requiring 2–3 times higher amount over the stiochiometric requirement (Nair and Trivedi, 1970 ; Shore, 1995; Shadov et al, 1978; Peters, 1975).

(i) Hydrolytic decomposition

$$2Na_2S_2O_4 + H_2O \rightarrow Na_2S_2O_3 + 2NaHSO_3$$

(ii) Thermal decomposition

$$2Na_2S_2O_4 + 2NaOH \rightarrow 2Na_2S_2O_3 + 2Na_2SO_3 + H_2O$$
$$3Na_2S_2O_4 + 6NaOH \rightarrow 5Na_2SO_3 + Na_2S + 3H_2O$$

(iii) Oxidative decomposition

$$Na_2S_2O_4 + O_2 + 2NaOH \rightarrow Na_2SO_3 + Na_2SO_4 + H_2O$$
$$2Na_2S_2O_4 + O_2 + 4NaOH \rightarrow 2Na_2SO_3 + 2H_2O$$

(iv) Other possible reactions

$$Na_2S + H_2O \rightarrow NaHS + NaOH$$
$$4NaHS + O_2 \rightarrow 2Na_2S_2 + 2H_2O$$
$$2NaHS + 2O_2 \rightarrow Na_2S_2O_3 + H_2O$$
$$2Na_2SO_3 + O_2 \rightarrow 2Na_2SO_4$$
$$Na_2S_2O_3 + H_2 + 2NaOH \rightarrow Na_2S_2O_3 + Na_2S + 2H_2O$$

A few by-products e.g. Na_2S , NaHS etc pollute air forming H_2S ; salts of sulphur, e.g. sulphates and sulphites (Na_2SO_3, $NaHSO_4$, Na_2SO_4, $Na_2S_2O_3$) contaminate sewage, lowers its pH and show corrosive action on concrete pipes (Semet et al, 1995). Other problems related to use of $Na_2S_2O_4$ are its cost and storage stability.

A very attractive feature of $Na_2S_2O_4$ as reducing agent is that it can generate (-700 to -1000 mV) and required reduction potential can be varied by changing its concentration, alkali or temperature.

Consumption of $Na_2S_2O_4$ can be reduced by 15% with 0.25–0.5 g/l dextrin or sodium borohydride, by 30–35% with Na_2S (65–70 parts $Na_2S_2O_4$

and 35–30 parts Na$_2$S) or by 50% with a mixture of dextrin and NaHSO$_3$ (1.5–2.0 g/l each). These additives can not reduce vat dyes rather show a synergistic effect on stability of reduced bath and so are to be added in the bath to start dyeing but vatting must be done with Na$_2$S$_2$O$_4$ only (Gokhale and Shah, 1974).

8.7.2   Over-reduction of vat dyes

When Na$_2$S$_2$O$_4$ is applied in too excess or temperature of vatting is not maintained properly, few vat dyes get over-reduced and pose problem in developing true shade. Over-reduction is of several types and depends on structure of dye. Most of these over-reduction processes are irreversible and cause permanent change in dye structure (Gokhale and Shah, 1974; Srivastava, 1984).

Simple over-reduction

Simple over reduction of some vat dyes occur either at high concentrations of Na$_2$S$_2$O$_4$ or at high temperature. Dyes containing heterocyclic rings with nitrogen atoms and where not all the keto (>C=O) groups present in the molecule are to be reduced are liable to be over-reduced (Fig. 8.2). Indanthrene blue and yellow dyes are susceptible to this type of defect, e.g. only two keto groups of indanthrene blue dyes should be reduced under ideal conditions of reaction, but when over-reduction occurs, all four keto groups are reduced.

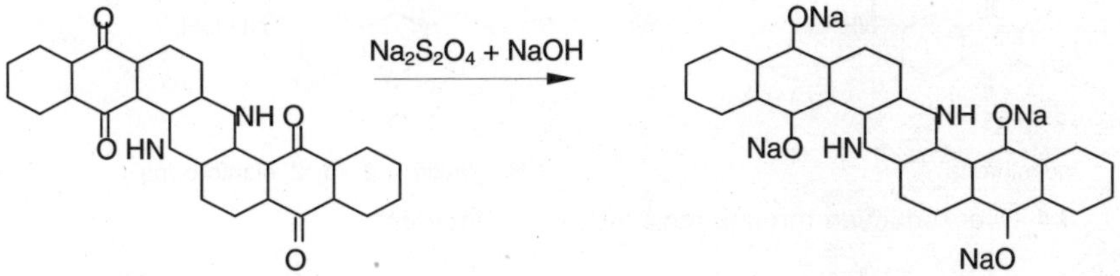

8.2 Simple over-reduction of vat dyes.

Hydrolysis

Vat dyes containing benzyl amino or acyl amino group (yellow GK and gold orange 3G) are hydrolysed at high temperatures and prolonged vatting time resulting duller and less fast shades. During hydrolysis, –NHCOC$_6$H$_5$ group is changed into –NH$_2$ group (Fig. 8.3).

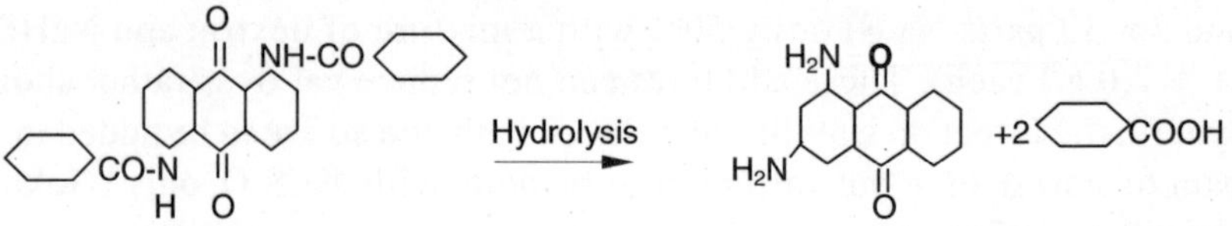

8.3 Over reduction of vat dyes through hydrolysis.

Dehalogenation

Halogenated dyes such as violet 2R, violet 3B, violet 2RN, blue RC, blue BC and green 2G looses halogen in its structure at high temperature vatting/ dyeing. Dehalogenation results poor bleach fastness of dyeings.

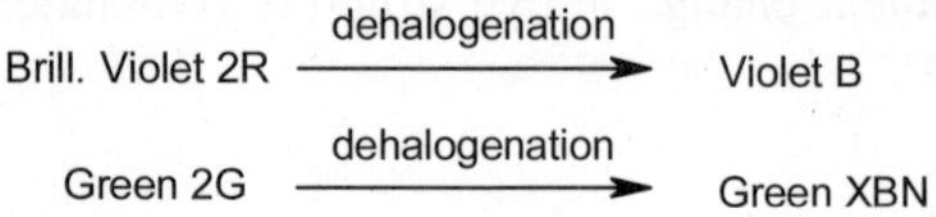

$$\text{Brill. Violet 2R} \xrightarrow{\text{dehalogenation}} \text{Violet B}$$

$$\text{Green 2G} \xrightarrow{\text{dehalogenation}} \text{Green XBN}$$

Molecular rearrangement

Certain vat dyes like khaki 2G, olive T, grey M, when dyed at or above 66°C give permanent change in hue ; this is caused by the irreversible change of some keto groups to >CH or >CH$_2$ groups (Fig. 8.4).

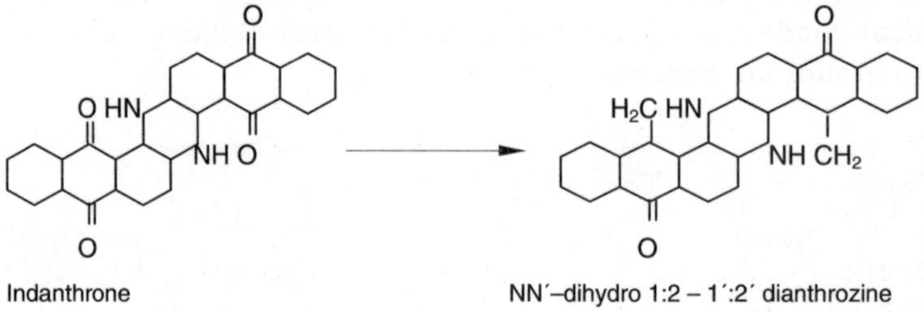

Indanthrone   NN′–dihydro 1:2 – 1′:2′ dianthrozine

8.4 Over reduction through molecular rearrangement.

Chances of over-reduction may be reduced partially by adding HCHO or glucose in bath. Manufacturer's literature is to be consulted to prescribe right dose of reducing agent and dyeing parameters. Different vat dyes possess several number of C=O groups and not all but only a few of these are to be reduced. Also, as a simple rule, if x kgs of Na$_2$S$_2$O$_4$ is used in dyeing, $0.5x$ kg is to be used for reduction, $0.25x$ kg to be added in bath at the start of dyeing and rest at different times of dyeing to retain dye in reduced form; same technique may be applicable for NaOH too.

Presence of little excess Na$_2$S$_2$O$_4$, which is a must, may be tested by dipping vat yellow paper in dyebath – change in yellow colour to deep blue shows presence of excess Na$_2$S$_2$O$_4$. The yellow paper is basically a filter paper dyed earlier with vat yellow 3RT dye [C I Vat Orange 11, C I 70805] and the blue hue is due to reduction of this dye. Based on this fact, other vat dyed papers may also be used to check presence of $Na_2S_2O_4$ accordingly.

8.7.3   Other reducing systems

Wooden vat or copperas method

Iron (II) salts in presence of milk of lime produces iron (II) hydroxide, the latter is the reducing agent. The bath has a reduction potential around (-700 mv to -750 mv), but due to very poor water solubility of $Fe(OH)_2$, partial reduction of indigo is only possible (Chavan and Chakraborty, 2001).

$$FeSO_4 + Ca(OH)_2 = Fe(OH)_2 + CaSO_4$$
$$Fe(OH)_2 + 2H_2O = Fe(OH)_3 + 2[H]$$

Insoluble $Fe(OH)_2$ is precipitated and forms huge sediment, five times bulkier than that of zinc–lime vat (Fox, 1948; Chackravertty and Trivedi, 1979).

Zinc–lime method

$$Zn + Ca(OH)_2 = CaZnO_2 + 2H$$

Nascent hydrogen reduces indigo and the excess slake dissolves reduction product; precipitation of $CaZnO_2$ interferes seriously with dyeing (Fox, 1948; Chackravertty and Trivedi, 1979).

Sodium borohydride method (NaBH₄)

$$NaBH_4 + 2H_2O = NaBO_2 + 8H$$

It is a bi-component reducing system and is suitable for pad – steam process. It can be used only along with accelerator (NaCl). However, it fails to stabilize leuco vat dye (Baumgarte, 1970).

Eco-friendly reducing systems

In recent times, use of eco-friendly reducing systems for indigo and anthraquinone vat dyes have been developed, viz. hydroxyacetone, electrochemical method, glucose–NaOH and iron (II) salt–ligand complex. Hydroxyacetone, (r p ~ -810 mV) results in 20% higher indigo uptake along with less consumption of auxiliary chemicals (Marte, 1989). Other advantages are better quality dyeing, ring dyeing effect, better elasticity of yarns, increased productivity, higher dye uptake and less dyestuff in effluent. Required reduction potential for vatting is obtained at 100°C with higher concentration of NaOH and the produced shade does not correspond with that obtained with $Na_2S_2O_4$. Hydroxyacetone is found to be more suitable for pad-steam method and is not suitable for reduction of anthraquinoid vat dyes (Baumgarte, 1987a).

In electrochemical reduction, vat dye is directly reduced by making contact between dye and electrode. A reducing agent at lower concentration must be added to the reduced dyestuff to ensure corresponding stability of reduced dye–liquor. However, dyestuff requirement for a specific shade is too high (Bechtold et al, 1991; Bechtold et al, 1997a; Bechtold et al, 1997b) 'Glucose in combination with NaOH (r p ~ -600 to -750 mV) can reduce and solubilise sulphur dyes (r p~ -550 to -600 mV) as well as indigo (-700 to -750 mV) at boil, producing reduced indigo baths free from sediments and highly stable for several hours. Padded textile requires more time for oxidation in between two successive dips due to higher stability of reduced dye and indigo uptake is less. This system fails to reduce anthraquinoid vat dyes effectively (Nowack et al, 1982).

Iron (II) salts when react with NaOH produce $Fe(OH)_2$, which is poorly soluble in alkaline conditions and gets precipitated. It can be complexed with weaker ligands, viz. tartaric acid, citric acid, triethanolamine etc to hold $Fe(OH)_2$ in solution and to use it as strong reducing agent for indigo at room temperature (Semet,1995; Chavan and Chakraborty, 2001). Anthraquinoid dyes require higher reduction potential and so a reducing system based on iron (II) salt, NaOH with two suitable ligands were developed, in which instant and complete reduction as well as solubilisation of all vat dyes occurred even at room temperature; dye uptake and the produced hue was comparable to those obtained with hydrosulphite except black (Chakraborty and Chavan, 2004).

8.8   Solubilisation

Reduced leuco dye is converted to its sodium salt with NaOH to make it water soluble and develop affinity for cellulosics. NaOH is the popular and accepted solubilising agent for leuco vat dyes and is multi-functional in nature, viz. (i) it dissolves leuco dye to its Na-salt, (ii) develops affinity of dye for fibre, (iii) neutralizes acidic by-products of $Na_2S_2O_4$ in bath to maintain pH and (iv) suppresses hardness of water, while use in excess increases negative potential of bath and causes lesser colour yield due to more repulsion between dye and cellulose. To keep dye in its water soluble salt form, sufficient excess of both $Na_2S_2O_4$ and NaOH must be present in bath. Oxidation of $Na_2S_2O_4$ by atmospheric oxygen as well as oxygen dissolved in water leads to production of acidic products (Na_2SO_3, $NaHSO_3$) neutralizing a part of alkali, necessitating compensation of it in bath.

$$2NaHSO_3 + 2NaOH = 2Na_2SO_3 + 2H_2O$$

Also, excess alkali is required to combine CO_2 from air getting into liquor to retain reduced form of dye thus making use of NaOH around 2–3 times higher over its stiochiometric requirement throughout dyeing.

In practice, NaOH is added in bath prior to addition of $Na_2S_2O_4$, as reduced dye is very unstable and if NaOH is not available for solubilisation instantly, leuco dye gets oxidized back to the original state.

Presence of excess NaOH in bath can be checked by dipping piece of filter paper soaked in phenolphthalein, turns pink if excess NaOH is present in bath.

8.9   Dyeing

Cellulose has higher affinity for most of sodium salts of leuco vat dyes necessitating control over temperature and salt; dye solution is to be added in two installments. Non-ionic dispersing agents promote diffusion and decrease rate of exhaustion at equilibrium; these are not to be essentially added in bath but can be used to promote level dyeing; no other chemicals are to be added during reduction and solubilisation except $Na_2S_2O_4$ and NaOH.

Dyes in combination for composite shades must belong to identical application class, even after that compatibility is to be ensured before application.

Control over liquor ratio is a must to control strike rate to achieve levelled shades. Too higher a liquor ratio will definitely produce brilliant levelled shades by reducing effective dye concentration in bath as well as strike rate, but in expense of high power, water, $Na_2S_2O_4$ and NaOH consumption; whereas a lower ratio gives a dull shade and chances of unevenness is high due to higher strike rate. The working condition is, therefore, a compromise among all dyeing parameters.

8.10   Oxidation

Dyed textile is oxidized to restore original dye structure, to trap dye molecules inside fibre and to develop actual shade. It is essential to impart a thorough wash to dyeings to remove last trace of alkali for proper oxidation. Oxidation may be carried out in open air, hydrogen peroxide (50–60°C), sodium perborate (50–60°C) or hypochlorite at room temperature for 15–20 min. Air oxidation is suitable for loose structures but time consuming and practicable for smaller production schedule, whereas compact structures are chemically oxidised. Improper oxidation may lead to faulty shades called patchy dyeing, which is mainly due to inadequate washing of dyed textiles – alkali carried forward by textile does not allow oxidizing agents to release free oxygen; shades produced in this way lack desired fastness. It is preferred to oxidize dyeings at lowest possible temperature for minimum time; otherwise cotton may get oxidized, especially when hypochlorite is used as oxidizing agent.

$$2H_2O_2 \xrightarrow{\text{pH} \sim 7} 2H_2O + O_2 \qquad H_2O_2 \xrightarrow{\text{pH} \sim 7} HO_2^- + H^+$$

If the dyed textile is not imparted intermediate wash before oxidation or the wash is not adequate for complete removal of $Na_2S_2O_4$ and NaOH, dose of hydrogen peroxide for oxidation is to be increased accordingly, because a part of it gets wasted to consume $Na_2S_2O_4$ and NaOH (Shore, 1995).

$$Na_2S_2O_4 + 2NaOH + 3H_2O_2 \rightarrow 2Na_2SO_4 + 4H_2O$$

Some vat black dyes contain $-NH_2$ groups and during reduction, these are changed to $-NO_2$ groups. Only single oxidation is not sufficient for restoration of $-NH_2$ groups. Dyeings are oxidized twice with two different oxidizing agents, viz. oxidation with hypochlorite (NaOCl) followed by peroxide treatment to recover original hue and tone. Oxidation with either of $Na_2Cr_2O_7$ or $K_2Cr_2O_7$ along with CH_3COOH or H_2SO_4 causes deposition of chromium on dyeings and is not preferred.

Over-oxidation results in change of tone and shade. Complete washing out of alkali from dyeings prior to oxidation is essential for quality shades. Indanthron blues, viz. C I Vat Blue 4, 6, 16 develop changed hue and tone when air oxidized in presence of high alkali concentration forming yellow-green azine derivative; the right hue and tone can be restored by revatting only (Shore, 1995). Best oxidation is achieved at higher temperature and at around neutral pH for all vat dyes (Baumgarte, 1987a).

Treatment of dyeings in dilute acid produces a greenish tone in final shade. Dyes belonging to IK and IW classes possess moderate affinity for fibre resulting moderate diffusion and so must not be rinsed before oxidation as the processes are slow and dye may remain on the fiber surface. In these cases, if possible, dyeings should be hydro extracted and oxidized in air.

8.11   Soaping

A final treatment with soap and soda ash (0.5 g/l each) at boil or only steaming for 10–15 min is a must for all vat dyed textiles to develop desired bright shade. Only steaming is also efficient but fails to remove superficial dyes. Dye–fibre linkage is established through physical forces. Size of a vat dye molecule is too smaller compared to pore size of cotton implying poor wash fastness but soaping promotes aggregation to remarkable extent to develop excellent wash fastness. This aggregation is accelerated in presence of salt and / or at lower temperature. Tone of shades is attributed to location of dye in respect of fibre axis and when dye molecules are positioned parallel to fibre axis, the true tone of dyeings appear (Gulrajani and Padhye, 1971). This positioning is enhanced by rise in temperature at least for a short period. The function of soaping is thus three folds:

(i)   It removes loosely attached superficial dyes.
(ii) It develops right tone of shade placing dye crystals perpendicular to fibre axis.
(iii) It helps formation of large dye crystals to offer superior fastness.

Individual vat dyes differ in rate of crystallization but the change is more rapid and consistency in shade is achieved at higher temperature that is why soaping is always imparted at boil.

8.12   Photochemical degradation

Few yellow, orange and brown vat dyes show deterioration in strength of dyeings and fading of shade on prolonged exposure to light. Absorbed sunlight, atmospheric oxygen and moisture cause slower decomposition of dye structure; presence of metal in cellulose or dye may catalyze this fading (Fox, 1948). The effect is also pronounced when specific dyes are in mixture – one shows catalytic action on another. Dyed rayon shows intense catalytic oxidation than cotton. However, the overall light fastness remains above satisfactory level.

8.13   Methods for practical application

8.13.1 Exhaust dyeing

Dye solution is prepared outside with dye, $Na_2S_2O_4$ (200–300 g) and NaOH (200–300 g) at required vatting temperature in 8–10 l water and 10–15 min is allowed for complete reduction and solubilisation. A fresh bath is prepared with excess $Na_2S_2O_4$ and NaOH in jigger or winch at dyeing temperature. Cellulosics in wet state is loaded and one-half of the reduced and solubilised dye solution is added to dye bath. After one complete passage of textile or desired time, left dye solution is added and the material is run in this solution for one more passage. Due to higher affinity of dye, jigger is preferred as dyeing is done in open width and maximum surface is offered against dye molecules. Dyeing is done for two turns (~30 min) with half dye at the start of each turn, after which salt (10–20 g/l) is added and dyeing is continued for further 1 hr followed by oxidation and soaping. This method is suitable for light to medium weight fabrics up to around 120 g/m².

8.13.2 Semi-continuous dyeing

In pad-exhaust method, a dye solution is prepared with only ultra-disperse or ultra-fine grade dye (remain in well dispersed state) and cellulose is padded with this solution, dried at lower temperature to avoid migration followed by loading in jigger. A reduced and solubilised dyebath is prepared in jigger with excess pad liquor (in case pad liquor is not available, one-tenth of total dye may be added), $Na_2S_2O_4$ and NaOH. Padded textile is dyed in this bath for desired time after which salt is added. Compact and heavy weight fabrics are best dyed in this method as pressure applied during padding ensures dye penetration and thorough dyeing which otherwise becomes difficult in exhaust method.

8.13.3 Continuous dyeing

Continuous dyeing of cellulose with vat dye could not achieve commercial success on technical grounds. However, cellulose may be padded with ultra-disperse vat dye followed by drying, repad in $Na_2S_2O_4$ and NaOH solution, re-drying and finally passed through a steamer at 100–105°C for 5–10 min for fixation of dye (Baumgarte, 1974; Baumgarte, 1987b). Better dyeing results are obtained if rongolite is included in padding liquor and padding with $Na_2S_2O_4$ and NaOH is omitted.

8.14 Defects in dyeings

Various dyeing faults, reason and probable solutions are listed as follows:

8.14.1 Dark selvedge

This is due to (i) thicker selvedge which absorbs more dye, and (ii) in jigger dyeing, selvedge remains exposed to air causing rapid oxidation, partially dries up and absorbs more dye during subsequent passing through bath. Spraying of $Na_2S_2O_4$ and NaOH solution on selvedge during dyeing reduces this problem by stopping oxidation.

8.14.2 Light spot

This arises due to (i) faulty preparation of fabric, (ii) local oil stain adhered on textile during spinning and was not effectively removed during pretreatment, (iii) partial degradation of cellulose to oxy or hydrocellulose during scouring. A mild scouring in post-dyeing stage followed by redyeing reduces this problem.

8.14.3 Oxidation marks / dark patches

This is the most common fault realized on vat dyed goods. Washing after dyeing should be thorough and eliminate last trace of alkali. Presence of alkali during oxidation with H_2O_2 develops these marks as dissociation of H_2O_2 to release oxygen occurs around a neutral pH. Presence of alkali at places on dyeings forces H_2O_2 to release perhydroxyl ions, not oxygen. Dyeings with such problems are to be treated with a solution of $Na_2S_2O_4$ and NaOH for 15–30 min at reducing temperature of dye followed by thorough washing and re-oxidation. Striping out of a part of dye in bath may be compensated accordingly.

8.14.4 Streaky dyeing

This may be due to (i) reed marks on fabric, (ii) variable tension in weft and warp causing prominence of either of these two and (iii) faulty mercerisation. Dyes possessing good coverage power are to be used to avoid this problem.

8.14.5 Poor rubbing fastness

Main reasons are (i) dyeing in too short liquor causing several dye layer formation on surface of cellulose, (ii) storing solubilised dye for prolonged time causing partial oxidation of dye and subsequent surface deposition on dyeings, (iii) incomplete reduction of dye towards end of dyeing, (iv) inefficient soaping after dyeing and (v) dyeing in hard water when metal salts may react with anionic dye and this complex is deposited on dyeings.

Most of the described faults can be rectified by treating dyeings with NaOH and $Na_2S_2O_4$ at the temperature of vatting in presence of levelling agents, but with a chance of striping of a part of dye in bath. Addition of 10–15% of dye of original recipe is required in bath. A higher liquor ratio at all stages of dyeing and proper control over vatting and dyeing temperatures is a solution to most of the faults occurs in vat dyeings.

8.15   Stripping and correction of faulty dyeings

When uneven dyeing occurs or the desirable shade is not produced, it is required to correct these through partial or complete stripping. Vat dyes produce very fast shades and complete extraction of dye is impracticable due to strong affinity of dye for cotton. Dyeings are treated with $Na_2S_2O_4$ and NaOH at specific temperature of reduction for 30 min for levelling of shade as well as partial stripping. Application of non-ionic surfactant in bath promotes rate of levelling.

For complete stripping, method used for partial stripping may be carried out with addition of a cationic surfactant. The later form complexes with reduced and solubilised anionic dye, looses affinity for cotton and comes out from dyeings. Most of the vat dyes are soluble in hot DMSO and so dyes can be extracted with it from dyeings. However, the does not have commercial importance on cost ground and is only of research interest (Chakraborty et al, 2005)

8.16   References

BAUMGARTE U (1970) 'Questions of reducing agents in the most important processes of vat dyeing', *Melliand Textilber.*, **51**, 11, 1332–1341.

BAUMGARTE U (1974) 'Vat dyes and their application', *Rev. Prog. Color Related Topics*, **5**, 17–32.

BAUMGARTE U (1987a) 'Reduction and oxidation processes in dyeing with vat dyes', *Melliand Textilber.*, **68**, 3, 189–195.

BAUMGARTE U (1987b) 'Developments in vat dyes and in their application 1974-1986', *Rev. Prog. Color Related Topics*, 17, 29–38.

BECHTOLD T, BURTSCHER E, KUHNEL G and BOBLETER O (1997a) 'Electrochemical reduction processes in indigo dyeing', *J. Soc. Dyers Color.*, **113**, 4, 135–144.

BECHTOLD T, BURTSCHER E, TURCANU A, and BOBLETER O (1997b) 'Dyeing behavior of indigo reduced by indirect electrolysis', *Textile Res. J.*, **67**, 9, 635–642.

BECHTOLD T, BURTSCHER E, GMEINER D and BOBLETER O (1991) 'Investigations into the electrochemical reduction of dyestuffs', *Melliand Textilber.*, **72**, 1, January, 50–54.

CHACKRAVERTTY R R and TRIVEDI S S (1979) *A glimpse on the experimental and industrial process laboratory (textiles), Volume 1, Technology of bleaching and dyeing of Textile Fibres*, Ahmedabad, India, Mahajan Brothers.

CHAKRABORTY J N and CHAVAN R B (2004) 'Dyeing of Cotton with vat dyes using iron(II) salt complexes', *AATCC Rev.*, **4**, 7, 17–20.

CHAKRABORTY J N, DAS M and CHAVAN R B (2005) 'Kinetics of Dyeing and Properties of Iron(II) salt Complexes for Reduction of Vat Dyes', *Melliand Int.*,**12**,4, 319–323.

CHAVAN R B and CHAKRABORTY J N (2001) 'Dyeing of Cotton with Indigo using Iron(II) salt Complexes', *Coloration Tech.*, **117**, 2, 88–94.

Colour Index International (1987), 3rd Edition (3rd Revision), Bradford, UK, Society of Dyers & Colourists and AATCC.

FOX M R (1948) *Vat dyestuffs and vat dyeing*, 1st Edition, London, Chapman & Hall.

GOKHALE S V and SHAH R C (1974) *Cotton piece Dyeing*, 1st Edition, India, Ahmedabad Textile Industries Research Association.

GULRAJANI M L and PADHYE M R (1971) Orientation of vat dyes in cellophane on soaping treatment, *Current Sci.*, **40**, 14, 374–375.

PETERS R H (1975) *Textile Chemistry*, Vol-III, New York, Elsevier Scientific Publishing Company.

MARTE E (1989) 'Dyeing with sulphur, indigo and vat dyes using the new RD process: Hydroxyacetone makes it possible', *Textil Praxis Int.*, **44**, 7, 737–738.

NAIR G P and TRIVEDI S S (1970) 'Sodium Sulphide in vat dyeing', *Colourage*, **17**, 27, December, 19–24.

NOWACK N, BROCHER H, GERING U and STOCKHORST T (1982) 'Use of electrochemical sensors for supervising dyeing processes with sulphur dyes', *Melliand Textilber.*, **63**, 2, 134–136.

SEMET B, SACKINGEN B and GURNINGER G E (1995) 'Iron(II) salt complexes as alternatives to hydrosulphite in vat dyeing', *Melliand Textilber.*,**76**, 3, 161–164.

SHADOV F, KORCHAGIN M and MATETSKY A (1978) *Chemical Technology of Fibrous Materials*, Revised English Translation, Moscow, Mir Publishers.

SHORE J (1995) *Cellulosics Dyeing*, Bradford, UK, Society of Dyers & Colourists.

SRIVASTVA S B (1984) *Recent Processes of Textile Bleaching Dyeing and Finishing*, Delhi, SBP Consultants & Engineers Pvt Ltd.

9
Dyeing with indigo

Abstract: Denim is exclusively dyed with indigo to produce brilliant blue shades with desired wash down effect. In spite of being a member vat dye class, indigo possesses no affinity for cellulosics in reduced and solubilised form unlike other vat dyes. Exhaust dyeing does not show any result forcing to introduce a multi-dip-nip padding technique with intermediate airing for gradual build up of shade. Among various dyeing parameters, pH is most crucial to achieve surface intense ring dyeing effect, which helps to develop bio-polishing effect. Extent of indigo uptake may also be synergized through different pre-treatment of denim. Dimethyl sulphoxide is the right solvent to extract indigo with ease. Eco-friendly alternate reducing systems have also been developed.

Keywords: Indigo, pH, affinity, ring dyeing, eco-friendly reducing systems

9.1   Introduction

In spite of high demand of differential coloured jeans in present day market which are produced by dyeing with sulphur and reactive dyes, indigo is the brand dye to produce attractive blue shades with desired fading property and belongs to vat dye class under 'indigoid' category. Chemical structure as well as C I specification of indigo is shown below (Colour Index, 1987).

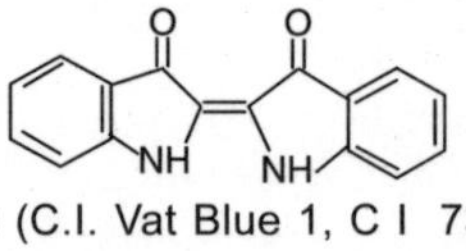

Indigo (C.I. Vat Blue 1, C I 73000)

In reduced and solubilised form, indigo possesses little affinity for cotton – the extent of affinity is too less to exercise exhaust dyeing; rather multi dip/nip padding technique with intermediate air oxidation is followed for gradual build-up of shade (Fox and Pierce, 1990; Chakraborty and Chavan, 2004; Hughey, 1983).

Industries have separate units for dyeing denim with indigo and this exercise is not done in dye-houses, rather in a combined sizing–precise pre-treatment-cum-warp dyeing section. Except denim, indigo is not generally used for dyeing of other cotton articles and so discussions in this chapter will remain focused on dyeing of denim with indigo, though the same technology is to be followed for dyeing cotton textile other than denim.

9.2   Dyeing methods

A suitable method of dyeing, viz. slasher dyeing (sheet dyeing) or rope dyeing (ball warp dyeing or chain dyeing) or loop dyeing is followed for dyeing of cotton warp after which it is being interlaced with white cotton weft to produce a warp faced 3 × 1 or 2 × 1 twill fabric with blue face and white back. During dyeing, preferably a ring dyeing technique is followed in which the dye remains mostly on the outer layer of fibre. Lack of affinity of dye for cotton enhances this surface adsorption. Dyeings are then washed through specific techniques to remove a part of dye to develop unique fading property.

9.2.1   Slasher dyeing

In slasher dyeing, warp sheet is pre-treated precisely in earlier compartments succeeded by multi dip / nip – airing indigo dyeing; the process is completed with after-wash followed by drying, sizing and final drying. Handling of yarn is least as the warp sheet is directly processed and then sent to weaving section for its subsequent conversion to fabric (Parmer et al, 1996; Voswinkel, 1993; Voswinkel, 1994; Haas, 1990).

A slasher dyeing range may consist of either the indigo dyeing range alone or a continuous indigo dyeing range with an integrated sizing range. In the first type, yarn sheet moves forward through one or two pre-treatment boxes with the stages of dipping, squeezing and skiing; washing and drying are the only after-treatments. On requirement, two or more layers of warp sheet may be processed simultaneously. Schematic diagram of a popular slasher dyeing range is shown in Fig. 9.1 (Parmer et al, 1996). The second type includes the steps followed in 1st type; extra attachments are the sizing and washing compartments as described in Fig. 9.2 (Parmer et al, 1996).

Slasher dyeing is suited for better production schedule with a possible problem of centre to side variation. Superior quality of yarn is needed to minimize breakage; broken end roll wrap ups can cause catastrophic shutdown problem as well as undesired waste. In contrast, dipping, squeezing and oxidation require less time as each yarn is independently subjected to treatments.

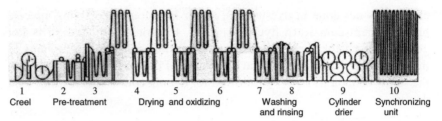

1	2	3	4	5	6	7	8	9	10
Creel	Pre-treatment		Drying	and oxidizing		Washing		Cylinder	Synchronizing
						and rinsing		drier	unit

9.1 Omez SpA continuous slasher dyeing range.

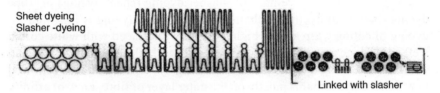

Sheet dyeing
Slasher -dyeing

Linked with slasher

9.2 Slasher dyeing with integrated sizing range.

9.2.2   Rope dyeing

Yarns from creels are pulled up by a ball warping machine; pass through lease stand consisting of a special comb and lease rods to ensure proper sequencing of ends. Drawn ends then pass through a condenser tube assembly to merge into a bundle, called a rope comprising of 350–400 ends. Ropes are then wound on drums; once the beam has been fully wound, it is unloaded and mounted on dye range creel (Stromberg, 1994; Fulmer, 1993). Dyeing is done by passing ropes, up to twelve at a time, through pre-treatment bath followed by multiple dipping in separate indigo baths with intermediate squeezing and skiing, as done in slasher dyeing. Schematic diagram of a rope dyeing unit is shown in Fig. 9.3 (Parmer et al, 1996; Stromberg, 1994). At the exit, each dyed rope is coiled into container and is taken for long chain beaming during which rope of dyed warp is opened to warp sheet and wound under proper tension on warper's beam. A yarn storage accumulator, located around 1.5m away from long chain beamer unwinds the warp from section beam in order to properly repair end breakage with no cross end. The beam is then placed in sizing followed by weaving in air-jet loom.

Ball Warping⟶Rope Creel⟶ Continuous warp dyeing⟶Oxidation⟶After-wash

Finishing◄—Weaving◄—Warp sizing ◄—Long chain beaming◄— Coiling◄— Drying

9.3 Rope dyeing sequence.

Rope dyeing is popular for higher production rate with quality dyeing and good fastness to rubbing and washing. It implies very fewer breakages of ends with better consistency in shade, but in contrast, more handling of yarn to open ropes before sizing. Efficiency of slasher and rope dyeing depends on fortifying spent dyebaths to maintain optimum concentration of indigo and other chemicals in all the six indigo baths simultaneously (Fulmer, 1993); consumption of chemicals and dye is more in first bath with slow fall in subsequent progressive baths.

9.2.3   Loop dyeing

In loopdye 1 for 6" (Fig. 9.4), rather than engaging six different colour troughs, there is only one precise indigo bath with one squeezing unit and is of modular construction (Godau, 1994). It comprises 8–16 warp beams of threading in feature, pre-wetting or pre-mercerising unit, twin-pad colour applicator, an integrated skying passage, two rinsing sections and an accumulator scray. Figure 9.4 shows the sequence of operation used in this technique (Godau, 1994; Parmar et al, 1996; Dierkes and Godau, 1984).

In loop dyeing method, there is substantial fall in concentration of dye and chemical after each dip, necessitating instant fortifying the bath to develop required shade and an indigo bath at higher concentration at the starting is preferred.

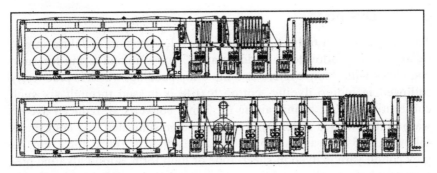

9.4 Principle of Loopdye 1 for 6.

9.3   Application of indigo

As stated, indigo is not applied through exhaust technique due to too poor affinity, rather a reduced pad liquor of indigo (2–3 g/l) is prepared and cellulose is padded with this liquor followed by air oxidation. This sequence of padding and oxidation is repeated for several times for gradual build-up of shade.

9.3.1   Preparation of pad liquor

Dye solution of required concentration is prepared using a stock vat and dilution liquor. For stock vat, required amount of NaOH and indigo are added to 100 ml water, stirred well, heated up to 50°C followed by addition of $Na_2S_2O_4$, kept for 15–20 min to complete reduction of dye. The dilution liquor is prepared separately with required amount of NaOH and $Na_2S_2O_4$ in 1 l water at room temperature. About 567 ml of dilution liquor is added to 100 ml reduced stock vat to get 667 ml final padding liquor. Concentration of dye, NaOH and $Na_2S_2O_4$ in stock vat and dilution liquor according to DyStar guidelines (developed by BASF) are explained below (DyStar, 2009).

(i)   For a final padding liquor with dye concentration: 3g/l

	Stock vat	Dilution liquor	Padding liquor
Dye	20 g/l	nil	3 g/l
NaOH	20 g/l	1.35 g/l	4.416 g/l
$Na_2S_2O_4$	20 g/l	2.0 g/l	4.698 g/l

(ii) For a final padding liquor with dye concentration: 2g/l

	Stock vat	Dilution liquor	Padding liquor
Dye	13.34 g/l	nil	2 g/l
NaOH	20 g/l	1.35 g/l	4.416 g/l
$Na_2S_2O_4$	20 g/l	1.50 g/l	4.273 g/l

9.3.2   Dyeing

This is based on a 6-dip 6-nip padding cycle. Dipping of fabric in dye liquor for 30 s followed by padding at 80% pick-up and airing for 1min completes 1-dip 1-nip cycle. For 6-dip 6-nip padding, which is practically followed in indigo dyeing, cotton textile is passed through six such consecutive cycles; a final airing for 3 min converts all reduced dye on warp to its oxidised state. Dyed cotton is then subjected to 5 cold rinses followed by 5 hot washes at 50–60°C and is dried at 100°C for 2 min. As grey cotton yarn is dyed in sheet, rope or loop form, a mild alkaline pre-treatment before indigo padding enhances colour yield. Selection of right machine is also a pre-requisite factor in which cotton is padded and oxidised several times continuously followed by direct sizing or opening of ropes and sizing.

9.4   Factors affecting indigo uptake

Reduced and solubilised indigo is anionic with negligible affinity for cotton. Depth of shade is primarily governed by number of dips / nips, but dye uptake during each dip is governed by number of other parameters, viz. pH of dyebath, immersion time, time of airing, concentration of indigo and temperature. These also affect build-up of indigo on cotton, degree of penetration and fastness properties.

9.4.1   Ionic nature of cotton and pH of bath

Most crucial factor is pH of indigo bath (Annis et al, 1991; International technical paper competition, 1989; Etters, 1993a; Paul and Naik, 1997; Etters, 1991; Etters and Hou, 1991). Depending on dyebath pH, indigo pre-dominantly exists in four different forms marked in Fig. 9.5, viz. (i) as indigotin at very lower alkaline pH, (ii) reduced but non-ionic form at moderate alkaline pH (<10.5), (iii) mono-phenolate form at relatively higher pH (10.5–11.5) and (iv) bi-phenolate form at very high pH (>11.5) (Parmer et al, 1996; Etters, 1993). Structures (i) and (ii) exist below pH 9–9.5; their relative fraction depends on exact pH of bath. With slow increase in pH, structure (i) collapses and gets converted to either (ii) or (iii) or mixture of both. Further increase in pH above pH 10 slowly converts all the leuco structures (ii) to mono-phenolate structures (iii). At pH around 11.5, almost all indigo molecules are in their mono-phenolate form. Any further addition of alkali beyond pH 11.5 slowly attacks the second C=O group in presence of excess $Na_2S_2O_4$, resulting in transformation of mono-phenolate form (iii) to bi-phenolate form (iv). Extent of conversion strictly depends on pH and excess hydrosulphite in bath. Excess of $Na_2S_2O_4$ is must for successful indigo dyeing, if alkali is added in excess beyond pH 12.5, the structure (iv) pre-dominates. Indigo with structure (iv) shows negligible affinity for cotton causing poor dye uptake. Mono-phenolate form (iii) is the desired status which possesses highest affinity as well as strike rate to give higher dye uptake and restricts dye molecules at the surface hindering diffusion. When cotton is dipped in alkaline bath, it acquires negative charge; higher the pH, higher the anionic charge on fibre with more repulsion between dye and fibre resulting in reduced dye uptake. Out of two soluble forms of indigo, mono-phenolate shows relatively lesser water solubility, higher strike rate and poor diffusion, vice versa of bi-phenolate species explained in Fig. 9.6 (International technical paper competition, 1989).

Ease in washing of dyeings is influenced by pH too. Lower solubility of mono-phenolats hinder penetration inside cotton and give more intensive surface dyeing – the so called 'ring dyeing' where only a few surface layers are dyed leaving interior of yarn undyed due to the absence of adequate time for penetration and thorough distribution. Ring dyed denim possesses characteristics of required wash-down effect (Chong et al, 1995; Etters, 1990, 1992a, 1992b, 1992c; Richer, 1975; Golomb and Shalimova,

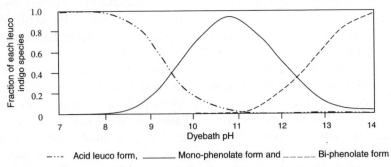

(i)   (ii)

(iii)   (iv)

(i) indigotin, (ii) reduced non-ionic form, (iii) mono-phenolate form, and
(iv) bi-phenolate form

9.5 Different ionic and non-ionic forms of indigo.

-·-- Acid leuco form, ——— Mono-phenolate form and ----- Bi-phenolate form

9.6 Fractional form of each reduced species of indigo as function of pH.

1965; Mishchenko and Artym, 1971; Hauser and Bairathi, 1996; Etters, 1995). Raising pH beyond 11.5 lowers strike rate – this consequently increases time for absorption, resulting in dyeing of few more layers than that obtained with mono-phenolate form. With further increase in pH, mono-phenolates go on diminishing with a substantial rise in bi-phenolates and thorough dyeing of cotton occurs marked by lighter dye yield.

Efficient wash-down property of indigo dyed yarn depends on amount of dye present on yarn surface. It is the 10.5–11.5 pH range which gives ring dyeings with desired washing performance, poor diffusion of dye inside and a deeper shade with lesser amount of indigo. Dyeing at higher pH ranges offers relatively lighter shades due to thorough diffusion as well as distribution of indigo illustrated in Fig. 9.7 (Etters, 1995) with consequence of lower wash-down property (Etters, 1993a, 1993b,1993c, 1994)

Ionic nature of cellulose affects indigo uptake too. With increase in alkalinity of bath, ionisation of cellulose, expressed in g-ions/l, increases with development of more anions, e.g. concentration of Cell-O⁻ is 3 ×10⁻³ g-ions/l at pH 10, which increases to 1.1 g-ions/l at pH 13. This suppresses substantivity of indigo with a consequence of lowering dye uptake. Calculation of different forms of indigo in bath, though cumbersome, can be made through study of equilibrium ionisation constant values (Etters, 1993b).

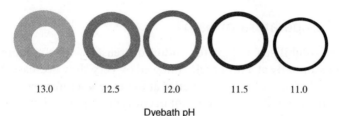

| 13.0 | 12.5 | 12.0 | 11.5 | 11.0 |

Dyebath pH

9.7 Distribution of Indigo in the cross-section of cotton denim yarn at different pH.

9.4.2   Immersion time

With increase in immersion time, dye uptake increases initially, remains nearly same for little further increase in time and beyond certain limit falls considerably. An immersion time of 30 s seems to be adequate – prolonged immersion reduces colour depth mainly due to the re-reduction of oxidised indigo retained by cotton through previous dips and go back to dyebath (Chong et al, 1995).

9.4.3   Time of airing

An air oxidation time of 60 s oxidises all the reduced indigo molecules on cotton, converts these to non-ionics and makes cotton ready for succeeding dip. A lesser time of airing releases a part of reduced dye during next dip. Optimum time of air oxidation is fixed at ≥ 60 s (Chong et al, 1995).

9.4.4   Concentration of indigo

Due to very low affinity of solubilised indigo, increase in concentration of indigo in bath though initially increases colour yield, but beyond certain concentration does not show any remarkable change in it. Increase in indigo concentration up to 3 g/l in bath increases its uptake on cotton but beyond this concentration it shows little impact on dye uptake (Chong et al, 1995).

Optimum depth of shade is obtained in as high as ten dips and beyond that no such increase occurs. But keeping in view the limitation in machinery set up and a lengthy process time, generally a 6-dip 6-nip technique seems to be adequate to get desired shade. An alternate approach may be to use lowest possible concentration of indigo with maximum number of dips. By doing so, colour uniformity, fastness and fading properties improve ; more the number of dips used with low concentration of indigo, the better the colour uniformity, fastness and wash-down properties.

9.4.5   Temperature of bath

Affinity of solubilised indigo for cotton decreases with the increase in temperature, initially at a faster rate followed by very slow decrease (Chong et al, 1995). Affinity of indigo is better at lower temperature and even re-reduction of oxidized indigo does not occur typically. It is better to carry out dyeing without any application of heat – maximum dye uptake occurs if dyeing is carried out at room temperature; with increase in temperature from 40–60°C, dye uptake falls relatively at a slower rate and beyond that remains almost parallel.

9.5   Alternate techniques to enhance indigo uptake

The recipe of padding liquor mentioned possesses a pH around 12.5 and does not fall below 11.0 during multi dip-nip padding. However, only a very small fraction of indigo remains in mono-phenolate state while the most in biphenolate state. NaOH used to control pH in the range of 10.5–11.5 causes efficient dyeing impracticable due to wide fluctuation in pH during dyeing. At the initial alkaline phase, pH is highly sensitive to NaOH concentration – a small change in dose changes pH remarkably, shown in Fig. 9.8 (Etters, 1996). Normal practice is to add NaOH at higher concentration to produce biphenolate indigo with a fear that final dyebath pH does not fall below 10.5 at any time and the consequence of this practice is lesser dye uptake in initial dips. This is because when less NaOH is used in an attempt to achieve a dyebath pH of 11.0, small changes in NaOH concentration lead to very large changes in dyebath pH.

In fact, dyebath pH may drop suddenly into a range in which mainly non-ionised indigo is formed and superficially deposited onto denim yarn. Such deposition inhibits sorption of non-ionised dye, leading to streaking and extremely poor crock fastness. Use of excess alkali makes lighter dyeings of higher alkalinity, lowering fading property (Hauser and Montagnino, 1998).

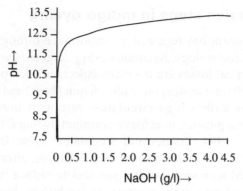

9.8 pH of unbuffered NaOH solution as a function of concentration.

9.5.1   Use of buffered alkali

Attempts to substitute NaOH with other alkalis or bases gave birth of buffered alkali system of undisclosed composition by Virkler (USA), which when applied in indigo bath at recommended concentration produces a stable pH in the range of 11.0 and little fluctuation occurs in dye uptake throughout the period of dyeing (Etters, 1998).

9.5.2   Pre-treatment of cotton with cationic chemicals

Cotton pre-treated with 200 g/l of 12% aqueous Hercosett 57 (cationic water soluble polyamine–epichlorohydrin resin) and 10 g/l of Sandozin NI for 5 min followed by squeezing in padding mangle to achieve 80% pick-up and drying at 100°C for 3 min, when dyed with indigo through 6-dip 6-nip technique showed higher indigo uptake over untreated one (Chong et al, 1995).

9.5.3   Pre-treatment of cotton with metals

Pre-treatment of cotton with metal salts prior to padding with indigo results in substantial increase in indigo uptake. It is also observed that iron and cobalt (II) salts are more effective even at very low concentration. Formation of insoluble metal hydroxide *in situ* cotton in presence of NaOH and air is responsible to attract more anionic indigo molecules from bath, causing increase in dye uptake (Chavan and Chakraborty, 2001).

9.5.4   Mercerization

Introduction of mercerizing step prior to padding with indigo enhances indigo uptake considerably due to enhancement in reactivity of cotton (Lennox, 1989).

9.6   Cost effective steps in indigo dyeing

Multi-vat machine concept has regained its importance through flexible adaptation of machine technology. Accurate dosing, targeted dyestuff use, low dyestuff and chemical losses are the main aspects of such production system. Adequate dwell time in steeping bath (~6 min / bath) and oxidation zones (≥60s) together with a high circulation rate with low dyestuff concentration have been pursued to achieve optimum dyeing (Hauser and Bairathi, 1996; Anon, 1994; Trauter and Stegmaier, 1996; Daruwalla, 1976). Due to lower dyestuff concentration in the effluent, ultra-filtration device can be equipped with smaller filter surfaces to reduce investment and operating cost. Washing water consumption can be kept low at 3 l/kg of warp. Indigo vat, $Na_2S_2O_4$ and NaOH are continuously dosed in proportion to speed of circulation avoiding shade variation. Cost effective dyeing may also be carried out through loop dyeing process using hydroxyacetone as reducing agent to produce waste-free improved quality dyeing at reduced cost.

9.7   Quantitative estimation of indigo

9.7.1   Colorimetry of reduced indigo

Indigo in dyed textile, dyebath and wash liquor is determined spectrophotometrically. Colorimetry of leuco indigo reduced with hydrosulphite in alkaline medium is instantly observed at 412 nm, as reduced indigo has time-bound stability; its spectrum changes considerably at very short intervals and lacks reproducibility – the technique may be dispensed with other methods, if stability of reduced indigo can be increased (Gutierrez et al, 1990). Iron (II) complex with triethanolamine can reduce vat dyestuffs with high degree of reduced dyebath stability against temperature, time and atmospheric oxygen (Bechtold et al, 1992).

9.7.2   Turbidity of indigo

Turbidity of oxidized indigo in water in presence of glycerol shows no linear relation between turbidity and concentration – in some cases higher values of turbidity are also achieved on reducing concentration which may be due to excessive dimensions of indigo particles as well as distribution of uneven sizes; glycerol increases bath density and retards particle fall. Maximum absorbance was found to be at 666 nm but the slope of the calibration curve varies according to status of initial dyeing bath (Bechtold et al, 1992).

9.7.3   Sulphonation of indigo

Incorporation of solubilising groups of sulphuric acid in indigo structure makes indigo water soluble, stable and deep blue coloured; absorbance is maximum at 605 nm. Beer's law is fulfilled upto 15 ppm of concentration and reproducibility of the method is 0.3–2%. Before sulphonation, indigo should be in oxidized and dried state and free from hydrosulphite otherwise a greenish-brown solution is obtained. This analytical technique is hardly suited for instantaneous, continuous control of indigo liquors. Sulphonation of indigo with concentrated sulphuric acid at 75°C possesses trouble (Bechtold et al, 1992).

9.7.4   Extraction with chloroform

Colorimetry of chloroformic extract gives maximum absorbance at 605 nm and good linearity is achieved in the calibration curve up to indigo concentration of 10 ppm. At this concentration, solubility limit of dye in chloroform is reached and a blue inter-phase between the two layers is formed during extraction (Gutierrez et al, 1990).

9.7.5   Extraction with pyridine-water

Indigo resembles structure of pyridine and solubilisation occurs through H-bonds between these two. Oxidized indigo can be dissolved up to a concentration of 40 ppm by boiling in pyridine-water (50:50) solution, which is then photometrically measured at 625 nm. Measurement must be done when the solution is hot to avoid precipitation of indigo. Dye should be completely free from hydrosulphite; otherwise, the later decomposes indigo during heating with reagent solution. Presence of NaOH with dye also changes absorbance as aggregation of dye alters with pH. This method has a reproducibility of 2–4% (Gutierrez et al, 1990).

9.7.6   Extraction with dimethyl sulphoxide

The latest technique is based of solubilising oxidized indigo in dimethyl sulphoxide (DMSO) and the method does not have any problem as with earlier methods. It is time saving, safe, without interference and largely reproducible. Solubility of indigo is excellent even at room temperature; maximum absorbance is achieved at 619 nm; the solution is highly stable for several days and complete dissolution takes place for indigo concentration of upto 250 μg/ml. The only problem lies in handling of DMSO is that it is slightly detrimental to health, causes skin irritation, itching, etc. (Kaikeung and Tsim, 1994).

9.7.7   Assessment of indigo-sulphur dye mixture

Denim is also dyed with a combination of indigo and sulphur or hydron blue dyes. To estimate contribution of these two in the shade developed, indigo is 1st extracted completely from sample at boil in pyridine–water solution; dyed cotton is rinsed with methyl alcohol and dried at 40°C to remove residual pyridine and methyl alcohol. Colour strength is calculated from reflectance value at each wavelength is integrated over the wavelength range; the result is an integrated value (Etters and Hurwitz, 1985).

Stretch denim jeans, the modern days fashionable choice, has acquired immense popularity due to its incomparable fit and comfort, freedom of movement, crease recovery, stretchability, enduring shape retention, consumer brand affinity and quality assurance, etc., and is characterized by elastane lycra in weft with cotton (~5%). However, the warp constitutes pure cotton like that with native denim and the dyeing technology remains unaltered.

9.8   References

Annis P A, Etters J N and Baughman G L (1991) 'Dyeing with Indigo: New data correlating colour yield with ionic form of applied dye, *Canadian Textile J.*,**108**, 5, 20–23.

Anon (1994) 'Cost effective production on indigo dyeing machines', *Melliand Textilber.*, **75**, 3, 214–217.

Bechtold T, Burtcher E and Bobleter O (1992) 'How to determine the indigo concentration of initial and processing baths', *Textile Praxis Int.*, **47**, 1, 44–49.

Chakraborty J N and Chavan R B (2004) 'Dyeing of Denim with indigo', *Indian J. Fibre Textile Res.*, **39**, 1, March, 100–109.

Chavan R B and Chakraborty J N (2001) 'Dyeing of cotton with indigo using iron(II) salt complexes', *Coloration Tech.*, **117**, 2, 88–94.

Chong C L, Poon L W and Poon W C (1995) 'Improving indigo dyeing colour yield', *Textile Asia*, **26**, 1, 55–57.

Colour Index International (1987) 3rd Edition (3rd Revision), Bradford, UK, Society of Dyers & Colourists and AATCC.

Daruwalla E H (1976) 'Saving in the use of chemicals in dyeing', *Colourage*, **23**, 1, 21–27.

Dierkes G, Godau E T (1984) 'Loop-Dye 1 for 6: dyeing of warp yarns for blue jeans', *Chemiefasern / Textil Ind.*, **34 / 86**, 5, 350–352.

DyStar (2009) 'Deutschland KG (1995): Technical Information, T1/T 017 (AJM), Continuous dyeing with indigo (BASF)', September, Textilfarben GmbH & Co.

Etters J N and Hurwitz M D (1985) 'Determining indigo and sulphur dye contributions to denim shade depths', *Am. Dyestuff Rep.*, **74**, 10, 20–21.

Etters J N (1990) 'New opportunities in indigo dyeing', *Am. Dyestuff Rep.*,**79**, 9,19 & 101.

Etters J N (1991) 'New developments in indigo dyeing of cotton denim yarn', *Colourage Ann.*, **37**, 39–42.

ETTERS J N AND HOU M (1991) 'Equilibrium sorption isotherms of indigo on cotton denim yarn: effect of pH', *Textile Res. J.*, **61**, 12, 773–776.

ETTERS J N (1992a) 'Quality control of cotton denim in indigo dyeing', *Am. Dyestuff Rep.*, **81**, 3, 17–24.

ETTERS J N (1992b) 'Sorting dyed denim fabric to possess similar wash-down characteristics', *Am. Dyestuff Rep.*, **81**, 8, 26–29 and 37.

ETTERS J N (1992c) 'Predicting shade depth and wash-down characteristics of indigo-dyed denim: a mathematical analysis', *Am. Dyestuff Rep.*, **81**, 2, 42–46.

ETTERS J N (1993a) 'Reducing environmental contamination by indigo in continuous dyeing of cotton denim yarn', *Am. Dyestuff Rep.*, **82**, 2, 30–33 and 50.

ETTERS J N (1993b) 'Indigo-dyed denim as enhancement', *Textile Asia*, **24**, 11, 34–35.

ETTERS J N (1993c) 'Indigo dyeing of cotton denim yarn: correlating theory with practice', *J. Soc. Dyers Color.*, **109**, 7/8, 251–255.

ETTERS J N (1994) 'Quick-Wash denim: new opportunity for denim garment manufacturers', *Am. Dyestuff Rep.*, **83**, 5, 15–16.

ETTERS J N (1995) 'Advances in indigo dyeing: implications for the dyer apparel manufacturer and environment', *Textile Chem. Color.*, 27, 2, 17–22.

ETTERS J N (1996) 'Recent progress in indigo dyeing of cotton denim yarn', *Indian J. Fibre Textile Res.*, **21**, 1, 30–34.

ETTERS J N (1998) 'pH-controlled indigo dyeing: buffered dyebaths vs caustic dosing', *Am. Dyestuff Rep.*, **87**, 9, 15–17.

FOX M R AND PIERCE J H (1990) 'Indigo: Past and Present', *Textile Chem. Color.*, 22, 4, 13–15.

FULMER T D (1993) 'Rope dyeing a preferred option for denim', *Am. Textiles Int.*, **22**, 8, 58–61.

GODAU E T (1994) 'Dyeing denim warp threads by the Loopdye 1 for 6 process', *Int. Textile Bulletin*, **40**, 1, 23–24.

GOLOMB L M AND SHALIMOVA G V (1965) 'Investigation of the redox and dyeing properties of vat dyes for protein fibres', *Tech. Textile Ind.*, **48**, 5, 105.

GUTIERREZ M C, CRESPI M AND GIBELLO C (1990) 'Determination of indigo in effluents from dyeing and washing baths', *Melliand Textilber.*, **71**, 1, 54–56.

HAAS L (1990) 'Indigo dyeing: process and machine concepts', *Int. Textile Bulletin*, **36**, 2, 45–50.

HAUSER P J AND BAIRATHI A (1996) 'New system for improved denim dyeing', *Am. Dyestuff Rep.*, **85**, 8, 55–57, 69.

HAUSER A AND MONTAGNINO J C (1998) 'Alkalinity studies of commercial denim fabrics', *Am. Dyestuff Rep.*, **87**, 5, 24–29.

HUGHEY C S (1983) 'Indigo dyeing: an ancient art', *Textile Chem. Color.*, **15**, 6, 103 & 108.

INTERNATIONAL TECHNICAL PAPER COMPETITION (1989) 'Effect of dyebath pH on color yield in indigo dyeing of cotton denim yarn', *Textile Chem.Color.*, 21, 12, 25–31.

KAIKEUNG M AND TSIM S T(1994) 'A new method to determine the indigo concentration', *Int. Textile Bulletin*, **40**, 3, 58–59.

LENNOX KERR P (1989) 'New mercerizing unit adds indigo dyeing flexibility', *Textile World*, **139**, 2, 61.

MISHCHENKO A V AND ARTYM M I (1971) 'Properties of aqueous liquors of leuco compounds of vat dyes', *Tech. Textile Ind.*, **4**, 95–98.

PARMER M S, SATSANGI S S AND PRAKASH J (1996) *Denim - A fabric for all*, 1st Edition, India, Northern India Textile Research Association.

PARMAR M S, SATSANGI S S AND PRAKASH J (1996) 'Indigo dyed warp yarn – an essential requirement for manufacturing denim', *Colourage*, **43**, 9, 45–50.

PAUL R AND NAIK S R (1997) 'The Technology of ring dyeing – an economically viable, environment friendly development', *Textile Dyer Print.*, **30**, 7, April 19, 16.

RICHER P (1975) 'Continuous dyeing of cotton warp yarn with indigo', *Textilveredlung*, **10**, 8, 313–317.

STROMBERG W (1994) 'Indigo-rope dyeing for high quality and performance', *Int. Textile Bulletin*, **40**, 1, 19–22.

TRAUTER J AND STEGMAIER T (1996) 'Warp preparation at ITMA 1995', *Melliand Textilber.*, **77**, 1/2, 27–29.

VOSWINKEL G (1993) 'Denim warp dyeing', *Textile Asia*, **24**, 8, 54–55.

VOSWINKEL G (1994) 'Indigo and multicolour dyeing machine', *Int. Textile Bulletin*, **40**, 1, 16.

10

Dyeing with solubilised vat dye

Abstract: Solubilised vat dyes are derivatives of vat dyes. Complications associated with reduction of vat dyes at the shop-floor, few of which are even prone to be over-reduced, has led to manufacture of these dyes from respective vat dye in either of two different methods. Dyes are water-soluble, which is enhances in presence of little soda ash; possess negligible affinity for cellulosics. After application, original vat dye is recovered through oxidation of dyed textile. Solubilised vat dyes produce only light shades with fastness properties parallel to parent vat dye. This dye is very important for dyeing of polyester–cotton blend in single bath.

Keywords: solubilised vat dye, affinity, temperature, oxidation

10.1   Introduction

Reduction of vat dye and subsequent stabilization of reduced bath is a cumbersome work; lack of control over it produces unlevel dyeing. Reduced and stabilized vat dye is called solubilised vat dye, chemically 'sodium salt of sulphuric ester of leuco vat dye' which requires neither reduction nor solubilisation during application. Affinity of this dye class is too less than that of parent vat dye for cotton. The vat dyestuff from which solubilised vat dye is prepared, is reduced by its manufacturer, stabilized and supplied in powder or paste form to the dyer. Solubilised vat dyes are popularly known as indigosol dyes as BASF (Germany) introduced these first in the market with the brand name – indigosol. There are two methods of manufacturing this dye: (i) chloro-sulphonic acid method in which vat dye is reduced with sodium hydrosulphite at required temperature in presence of pyridine. The reduced dye is treated with chloro-sulphonic acid followed by sodium hydroxide (Fig. 10.1)

In second method, parent vat dye is treated with copper-pyridine complex and finally converted into sodium salt of leuco sulphate ester (Fig. 10.2). After dyeing, original vat dye structure is restored using an acidic oxidizing agent, e.g. nitrous acid. The Na-salt of sulphuric ester of vat dye is mixed with a stabilizer, like urea.

(water-insoluble vat dye)   (leuco vat dye)   (sulphonic ester of   (solubilised vat dye or
                                                     leuco vat dye)   Na salt of sulphonic
                                                                            ester of leuco vat dye

10.1 Manufacture of solubilised vat dye in chlorosulphonic acid method.

10.2 Manufacture of solubilised vat dye in copper-pyridine complex method.

After dyeing, H_2SO_4 along with $NaNO_2$ or $K_2Cr_2O_7$ is applied to oxidise solubilised vat dye at 50–70°C when the dye is hydrolysed and oxidized to parent vat dye.

10.2   Properties

Dyes are mostly soluble in hot water and a few are soluble in presence of little Na_2CO_3; solubility can be improved by adding urea at 50–60°C. Dyes are liable to be hydrolysed at higher temperature and are important for dyeing cotton in pale shades only. Exposure of dyebath to sunlight should be avoided as dyes are sensitive to light and ester groups are affected to develop dye prematurely. Shades are brilliant and fast – fastness is equivalent to those with corresponding vat dyes. Due to negligible affinity for cellulose, Na_2CO_3 is added to enhance dye uptake. Though few dyes show levelling at lower temperature, dyeing at higher temperature promotes levelling of shades. Dyes are costly but levelling property is excellent. Range of shade is limited to blue, orange, purple, pink, yellow, green and golden – all in light shades. Tinctorial power of these dyes is too less – increase in concentration by many folds increases shade depth a little.

Dyes are available in two physical forms – powder and paste; powder form is less stable, stabilization with alkaline salts increases stability whereas paste form is stable in dark places, stabilization can be improved by adding Na_2CO_3 and NaOH.

A main use of this dye is in dyeing pale shades on cotton and polyester / cotton blend with excellent wash and light fastness. During dyeing of blend, dye occupies cotton phase only but gets partly sublimised-off and diffuses into polyester in finishing (Guha et al, 1990).

Anthraquinoid derivatives are superior to indigoid due to their superior fastness properties, while indigoids are superior for variety of hues.

Once applied and oxidized, these dyes recover back to original vat dye structure and for any further levelling of shade, the technique used for vat dyeing is to be followed.

In open air, the sulphonic group splits up and colour fading takes place – alkali treatment after dyeing reduces this problem. In some cases, light not only splits-up $-SO_3H$ groups but also brings about a permanent change in dye structure itself – producing finally a bluish shade. Storing of dye in dark room is a must to avoid this.

Dyes possess excellent affinity for protein fibres – the affinity is further enhanced on addition of mild acid, e.g. HCOOH, CH_3COOH in bath. But as the dyes form electro-static bond with protein fibres, original vat dye structure can not be restored and so these are not applied on wool or silk.

$$
\begin{array}{ccccc}
NH_2 & & NH_3^+ & & NH_3^+\cdots\cdots O_3SOR \\
| & \text{Acid} & | & \text{Solubilised vat dye} & | \\
W & \longrightarrow & W & \longrightarrow & W \\
| & & | & ROSO_3Na & | \\
COOH & & COOH & & COOH
\end{array}
$$

10.3   Specification of some solubilised vat dyes

Solubilised vat dyes are manufactured through reduction, solubilisation and stabilization of respective vat dye and obviously these do not have different chemical structure (Colour Index, 1987)

Arlindone Yellow GC (Greenish Yellow), C I Solubilised Vat Yellow 2, C I 67301 is leuco sulphuric ester of Yellow GCN (IDI) or Yellow 5G(Atic), C I Vat Yellow 2, C I 67300.

Arlindone Yellow IGK (Greenish Yellow), C I Solubilised Vat Yellow 4, C I 59101 is leuco sulphuric ester of Arlanthrene Golden Yellow GK (Arlabs), C I Vat Yellow 4, C I 59100.

Arlindone Yellow IRK (Reddish Yellow), C I Solubilised Vat Orange 1, C I 59106 is leuco sulphuric ester of Arlanthrene Golden Yellow RK (Arlabs), C I Vat Orange 1, C I 59105.

Soledon Red 2B (Bluish red), ICI, C I Solubilised Vat Red 10, C I 67001 is leuco sulphuric ester of Novatic Red 3B / Navinon Red FBB, C I Vat Red 10, C I 67000.

Arlindone violet 14R (Bright bluish violet), C I Solubilised Vat Violet 1, C I 60011 is leuco sulphuric ester of Navinon Violet 2R (IDI) / Novatic Bril. Purple 2R (Atic), C I Vat Violet 1, C I 60010.

Arlindone Blue IBC (Bright blue), C I Solubilised Vat Blue 6, C I 69826 is leuco sulphuric ester of Navinon Blue BC (IDI), C I Vat Blue 6, C I 69825.

Arlindone Green IB (Bright green), C I Solubilised Vat Green 1, C I

59826 is leuco sulphuric ester of Jade Green XBN (Atic) or Navinon Green B/Green FFB(IDI), C I Vat Green 1, C I 59825.

Arlindone Brown IRRD (Reddish brown), C I Solubilised Vat Brown 5, C I 73411 is leuco sulphuric ester of Novatic Brown G (Atic), C I Vat Brown 5, C I 73410.

Arlindone Olive Green IBL (Dull green), C I Solubilised Vat Green 3, C I 69501 is leuco sulphuric ester of Novatic Olibe Green B(Atic), C I Vat Green 3, C I 69500.

Arlindone Brown IBR (Reddish brown), C I Solubilised Vat Brown 1, C I 70801 is leuco sulphuric ester of Navinon Brown BR(IDI), C I Vat Brown 1, C I 70800.

Arlindone Green IGG (Bright green), C I Solubilised Vat Green 2, C I 59831 is leuco sulphuric ester of Navinon Green 2G (IDI), C I Vat Green 2, C I 59830.

Arlindone Orange HR (Yellowish orange), C I Solubilised Vat Orange 5, C I 73336 is leuco sulphuric ester of Arlanon Orange RF, C I Vat Orange 5, C I 73335

Arlindone Blue O4B (Blue), C I Solubilised Vat Blue 5, C I 73066 is leuco sulphuric ester of Arlanon Blue 2B, C I Vat Blue 5, C I 73065.

10.4   Application

Dyeing of cotton with these dyes follows a two-stage method, viz.

(i)   Dyeing is affected form an alkaline or neutral bath (for cellulose) or acidic dye bath (for protein fibres); exhaustion is controlled by manipulating temperature and salt.

(ii) Regeneration of dye is carried out through oxidation of dyeings with HNO_2 – the later is produced in bath applying $NaNO_2$ and H_2SO_4. Each dye has its own condition of time, temperature, concentration of H_2SO_4 and $NaNO_2$ (Alat, 1991).

Typical recipe

Dye	– $x\%$
Na_2CO_3	– 2 g/l
$NaNO_2$	– 3–4 g/l
NaCl	– 50 g/l
H_2SO_4 (70%)	– 5–10 g/l

10.4.1 Dyeing in jigger

Dye is dissolved in hot water along with Na_2CO_3 → cellulose is treated with half dye solution and half of $NaNO_2$ (1 turn or 15 min) at liquor ratio below 1:2 → rest dye solution and rest $NaNO_2$ are added (1 turn or 15 min) → NaCl is added (4 turns) → draining of bath → no washing → fresh bath is prepared with H_2SO_4 → oxidation in this acid bath (4 turns) at required temperature (depends on class of dye) → soaping at boil, hot and cold washes completes the process.

10.4.2 Function of chemicals

(i) *Alkali*

Dyes either in powder state or solution ensure better stability in alkaline pH. If alkali is not added, dye gets hydrolysed and premature oxidation occurs in bath itself. Dyes are best solubilised in presence of alkali only. Na_2CO_3 is preferred as a mild alkaline pH is required and it does not cause any remarkable fluctuation in pH. Na_2CO_3 promotes affinity of dye for fibre too.

(ii) *NaNO$_2$*

It works in two ways:

(a) it produces HNO_2, the oxidizing agent for dye through reaction with H_2SO_4
(b) being electrolyte, it promotes exhaustion of dye

   $NaNO_2$ is required to be added after dyeing in oxidation bath along with H_2SO_4 to produce HNO_2, but such a sequence of addition produces a major part of HNO_2 in bath itself, not inside cellulose resulting inadequate oxidation. Under practical conditions, $NaNO_2$ is added during dyeing which is absorbed by cellulose and H_2SO_4 reacts with $NaNO_2$ inside cellulose causing proper oxidation.

$$H_2SO_4 + 2NaNO_2 \rightarrow 2HNO_2 + Na_2SO_4$$

(iii) *NaCl*

Exhausting agent, concentration depends on affinity of dye – higher the affinity lesser the dose (10–50 g/l).

10.5   Affinity of dyes

Depending on substantivity towards cellulose, dyes are classified in five groups (Srivastva, 1984).

Group I < Group II < Group III < Group IV < Group V

The overall affinity of these dyes for cellulose is low except only a very few.

Group I: Indigosol Yellow V, Orange IRK, Pink IR, Violet IRH, Blue IBC, Printing Blue I2G

Group II: Indigosol Orange HR, Scarlet IB, Red IFBB, Green I3G, Brown IRRD

Group III: Indigosol Golden Yellow IGK, Printing Black IB

Group IV: Indigosol Grey IBL, Blue O4B, Brown IBR, Brown I3B, Olive Green IB

Group V: Indigosol Green IB and I2G

To avoid tailing effect during continuous dyeing through padding, selection of a small padding trough, high speed padding and application of little thickening agent in padding liquor are very crucial for efficient dyeing.

Dyes exhaust well at 20–40°C, except Green IB, Brown IBR and olive Green IB – which possess optimum affinity at 50–60°C; few dyes are applied at higher temperature and higher dosing of salt due to low affinity. Application of electrolyte improves overall dye-uptake.

10.6   Continuous dyeing

Cotton is padded with a liquor containing dye (x%), Na_2CO_3 (0.5–1g/l) and $NaNO_2$ (0.5–1 g/l), gum tragacanth thickener (12 g/l) and wetting agent (1 g/l) and is preferably dried. Drying before oxidation ensures better adhesion of dye. Intermediate washing is to be strictly avoided as dye is free to come out from fibre. Drying of padded goods may be carried out in a hot-flue dryer or float drier, especially for dyes possessing lesser affinity for cotton, otherwise precipitation of oxidized dye may occur on fibre causing poor rubbing fastness. Dyes possessing good affinity are wet developed after dyeing. Dyeings may be hydro-extracted before development if dyed with poor affinity dyes succeeded by oxidation. If oxidation is to be carried out at a later stage and drying is not possible by any means, dyed goods are rolled up, covered with polyethylene sheet and not exposed either to light or acid fumes.

10.7   Oxidation

Conversion of indigosol dyes to parent vat dye is affected with a strong oxidizing system, either in nitrite system ($NaNO_2$ + H_2SO_4) or chlorate system ($Na_2Cr_2O_7$ + $NaClO_3$) – the nitrite system is mostly preferred.

Padded and dried (optional) cellulose is developed in either of oxidizing systems when sodium salts of leuco ester (indigosol) is splitted-up through hydrolysis to restore original vat dye. Developing liquor is prepared by using around 2% solution of H_2SO_4 without any further dilution or as stated earlier. After-treatment of dyeings consists of rinsing in water followed by neutralising with Na_2CO_3 and soap (0.5 g/l) each at boil as that is done for parent vat dyes.

Indigosols are classified in three groups based on temperature of oxidation (Srivastva, 1984):

Group I: Rapidly oxidisable dyes require 20–25°C, e.g. Indigosol Blue IBC, Red IFBB, Golden Yellow IGK, Orange HR, Green IB, I2G, Brown IRRD, IBR–I3B and Gray IBL A stabilizer like urea, thiourea, lignin sulphonate or $Na_2S_2O_3$ is essentially added in bath to counter-act tendency of these dyes to be over-oxidized forming dull green azines (over-oxidised vat dye), especially with Blue IBC and Olive Green IB ; if not added, dull azines are formed which may be rectified through treatment with $Na_2S_2O_4$ and NaOH (0.5 g/l each) at vatting temperature of parent vat dye.

Group II: Moderately oxidisable dyes require 50–60°C with higher concentration of H_2SO_4, e.g. Cibantine Yellow V, Yellow RK and Orange RK.

Group III: Slowly oxidisable dyes require 70–80°C, e.g. Indigosol Pink IR, Red Violet RH and Blue O4B.

Ease of oxidation increases with increase in molecular weight of the dye; indigoid and thioindigoids are difficult to oxidize due to presence of halogen in structure of these dyes. Excess of oxidizing agent may further re-oxidise regenerated vat dye to form dull green azines. Nitrite method is suitable for those dyes which possess amino or imino groups and HNO_2 produces change in shade through formation of diazo or nitroso compounds. H_2SO_4 and $NaNO_2$ mixture evolves obnoxious nitrous fumes and so $NaNO_2$ is added to dyebath first followed by passing the dyed textile through hot dilute H_2SO_4.

10.8 · References

ALAT D V (1991) 'Indigosol dyes', *Man made Textiles India*, **34**, 10, 403–404, 410.
Colour Index International (1987), 3rd Edition (3rd Revision), Bradford, UK, Society of Dyers & Colourists and AATCC.
GUHA S B, JARIWALA R and PATEL V (1990) 'Single-stage dyeing and finishing of polyester / cotton blend with Indigosol dyes', *Colourage*, **37**, 13, 17–20.
SRIVASTVA S B (1984) *Recent processes of Textile Bleaching, Dyeing and Finishing*, Delhi, SBP Consultants & Engineers Pvt Ltd.

11

Dyeing with insoluble azoic colour

Abstract: Insoluble azoic colours are produced cellulosics when a coupling component (preferably naphthol) reacts with diazotized aromatic base under specific conditions. The lake thus formed in situ remains trapped in fibre pores and can not come out to result good wash fastness. However, a part of lake formed at the fibre surface remains embedded at the interior to cause poor rubbing fastness. These colours are water insoluble and brilliant, specifically used to develop all shades of red, orange, yellow and maroon.

Keywords: Coupling component, base, diazotization, development, rubbing fastness

11.1   Basic principle of application

Insoluble azoic colours are not manufactured tailor made colours like other dyes but these are lakes or ingrain dyes and are developed *in situ* cellulosics through chemical reaction between diazotized base and naphthol (coupling component). These are popularly known as naphthol dyes as the coupling component is mostly a naphthol (Fig. 11.1). The final azoic pigment is insoluble in water and so the name 'insoluble azoic colours'. Soluble azo colours essentially retain auxochrome while insoluble azoics do not have the same.

Dyeing of cellulose with these colours is basically a two step process, viz. solubilisation of naphthol followed by its application on textile and treatment of this naphtholated textile with a diazotized base. Neither naphthol nor base possesses any such colour, but the pigment developed gives a specific hue depending on naphthol–base combination. A part of pigment is also formed on fibre–solution interface causing poor rubbing fastness; coloured pigments are formed mainly on surface layers of cellulose due to poor affinity of naphtholates developing bright shades.

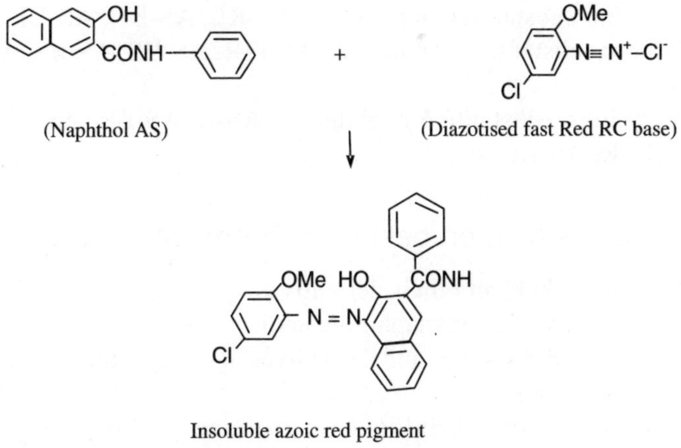

(Naphthol AS)   +   (Diazotised fast Red RC base)

Insoluble azoic red pigment

11.1 Formation of insoluble azoic pigment.

11.2   Colour Index specification of coupling component and base

Out of two essential components to develop these colours, viz. coupling component and base, the later is an aromatic amine whereas the first one is either a naphthol or other products but are applied through same technique, e.g. Brenthols (Naphthol AS-G). In Colour Index, coupling component and bases has been assigned generic names as 'Azoic coupling component' and 'Azoic diazo component' with C I constitution numbers ranging from 37500 to 37625 and 37000 to 37275, respectively (Colour Index, 1987).

11.3   Classification of coupling component

Coupling components or naphthols are marketed worldwide under various brand names, viz. Tulathol, Amarthol, Daito, Dianix, Diathol, Dragonthol, Azonaphthol, Arlanthol, Kako, Kiwa, Naphthanilide, Naphthoide, Naphthol, Naptoelan, Naphtol, Sanatol, Shangdathol, etc. (Colour Index, 1987). Coupling components are insoluble in water and do not have affinity for cellulosics; conversion to sodium salts develop affinity at varying extents. Coupling components are classified in two categories based on application and chemical nature.

11.3.1 Classification based on application (Shenai, 1987)

Based on affinity, these are commercially classified as

(i)   Low substantive naphthols – AS, AS-G, AS-OL, AS-D, etc.

(ii)   Medium substantive naphthols – AS-RL, AS-BG, AS-LT, etc.
(iii) Higher substantive naphthols – AS-ITR, AS-BS, AS-RS, AS-E, etc.,
       and
(iv)   Very high substantive naphthols – AS-S, AS-LB, AS-BT, AS-SG,
       AS-SR, AS-BR, etc.

11.3.2 Classification based on chemical nature

(i)   a-, b-naphthols and their derivatives.
(ii)   Derivatives of hyoxynaphthoic acid.
(iii) Complex of AS-series and derivatives of o-oxyanthracene carboxylic
       acid (AS-GR).
(iv)   Non-naphtholic products to produce yellow azo dyes, e.g.
       Diacetoacetic tolidide and its derivatives (AS-G, AS-LG, AS-L3G,
       etc.) or other typical products, e.g. nigrophor, nigrogene, etc.

11.4   Properties of coupling components

Coupling components (naphthols) possess no auxochrome and does not
show affinity for cellulose; conversion to sodium salts develops affinity at
varying extents. The linear the structure the higher the affinity and vice-
versa; wash fastness also directly varies with molecular size of naphthol,
i.e., larger the size, better the fastness. Ultra-violet rays break down
chromophore causing fading of colour in the long run. Light fastness is
moderate but can be further improved by introducing either of $-NO_2$,
$-CN$, $-COOR$, $-SO_2H$, $-CF_3$ groups in bases. Hue and shade of colour
depends on naphthol–base combination. Keeping the naphthol unchanged,
if only base is changed, enormous number of hues can be developed.
Though sodium salts of different naphthols show affinities at varying
extents, still naphthol combination is practically possible for development
of composite hues with one base. However this must be standardised before
implementation at shop-floor level (Gokhale and Shah, 1981). Shades are
fast to alkali, acid and chlorine treatment; even a few blue, yellow, violet,
black and brown are fast to H_2O_2 bleach.

Naphthol AS-G produces only yellows at varying tones and depths with
any base. The nomenclature of a base is based on the colour it develops
during coupling with naphthol AS, e.g. yellow is obtained through coupling
of naphthol AS with fast yellow GC or RC base, red is obtained with fast
red RC or TR base and orange with orange GC base, etc. Commercially
available 30 naphthols can couple with 50 bases separately to produce
1500 colours, out of which a few combinations develop same colours,
some produce dull shades. Only 4–5 selective naphthols are adequate to
form most of the colours with a very few bases.

11.5   Specification of some important coupling components

Chemical structure and colour index specifications of some important coupling components are shown in Scheme 11.1 (Allen, 1971; Colour Index, 1987).

Naphthol AS
C I Azoic Coupling Component 2
C I 37505

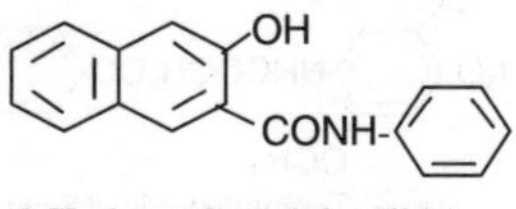

(3-Hydroxy 2-naphthanilide)

Naphthol AS-BS
C I Azoic Coupling Component 17
C I 37515

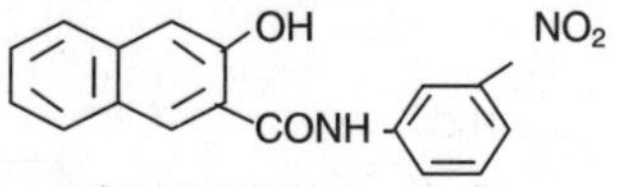

(3-Hydroxy 2-naphthoic *m*-nitroanilide)

Naphthol AS-TR
C I Azoic Coupling Component 8
C I 37525

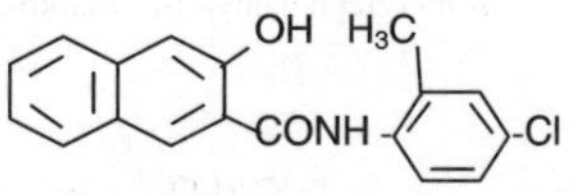

(3-Hydroxy 2-naphthoic 4-chloro 2-methylanilide)

Naphthol AS-SW
C I Azoic Coupling Component 7
C I 37565

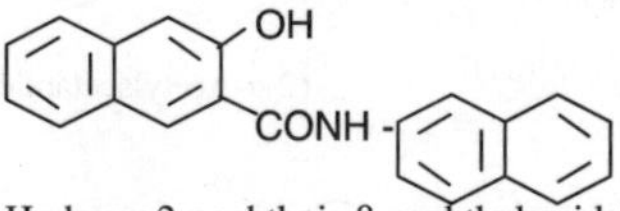

(3-Hydroxy 2-naphthoic β-naphthylamide)

Naphthol AS-OL
C I Azoic Coupling Component 20
C I 37530

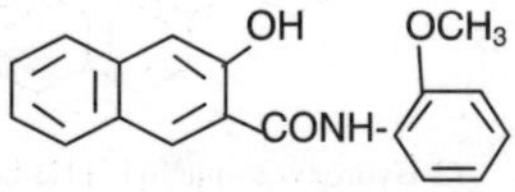

(3-Hydroxy 2-naphthoic *o*-anisidide)

Naphthol AS-BR
C I Azoic Coupling Component 3
C I 37575

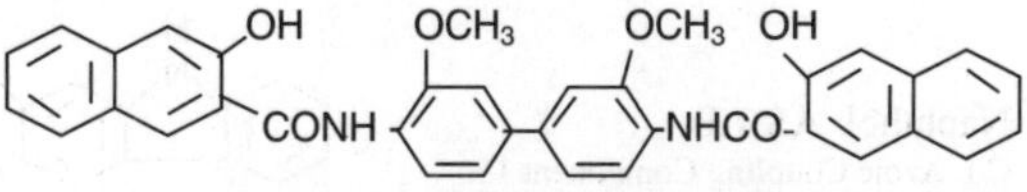

(3-Hydroxy 2-naphthoic dianisidide)

Naphthol AS-BO
C I Azoic Coupling Component 4
C I 37560

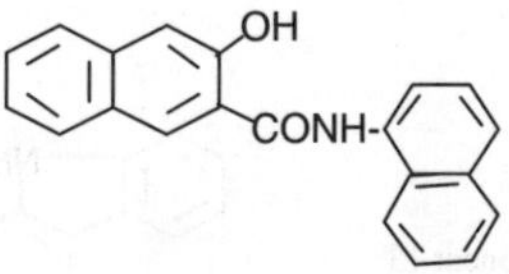

(3-Hydroxy 2-naphthoic α-naphthylamide)

Naphthol AS-LT
C I Azoic Coupling Component 24
C I 37540

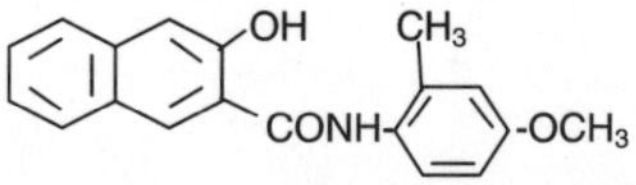

(3-Hydroxy 2-naphthoic 4-methoxy 2-methylanilide)

Scheme 11.1 Contd. ...

Naphthol AS-G
C I Azoic Coupling Component 5
C I 37610

H_3C, CH_3
CH_3COCH_2CONH-⬡-⬡-$NHCOCH_2COCH_3$

(4,4'-Bi-o-acetoacetotoluidide)

Naphthol AS-LG
C I Azoic Coupling Component 35
C I 37615

Cl   Cl
CH_3O-⬡-$NHCOCH_2CO$-⬡-$COCH_2CONH$-⬡-OCH_3
OCH_3 H_3CO

(α- α'-Terephthaloylbis[5-chloro-2,4-dimethoxy-acetanilide])

Naphthol AS-L3G
C I Azoic Coupling Component 33
C I 37620

H_3C, CH_3
Cl-⬡-$NHCOCH_2CO$-⬡-$COCH_2CONH$-⬡-Cl
OCH_3 H_3CO

(α- α'-Terephthaloylbis[4-chloro-5-methyl-o-acetanisidide])

Naphthol AS-L4G
C I Azoic Coupling Component 9
C I 37625

C_2H_5O-⬡-S-C- $NHCOCH_2COCH_3$
                    N

(2-α-Acetylacetamido-6-ethoxybenzothiazole)

Naphthol AS-SR
C I Azoic Coupling Component 25
C I 37590

NH
⬡-⬡-OH H_3C,
CONH -⬡-OCH_3

(2-Hydroxy-2' -methyl -11H-benzo(a)carbazole-3-carbox-p-anisidide)

Naphthol AS-LB
C I Azoic Coupling Component 15
C I 37600

H
N⬡ OH
CONH-⬡-Cl

(4' -Chloro-2-Hydroxy-1-carbazolecarboxanilide)

Naphthol AS-SG
C I Azoic Coupling Component 13
C I 37595

NH
⬡-⬡-OH
CONH -⬡-OCH_3

(2-Hydroxy-11H-benzo(a)carbazole-3-carbox-p-anisidide)

Scheme 11.1 Contd. ...

Naphthol AS-D
C I Azoic Coupling Component 18
C I 37520

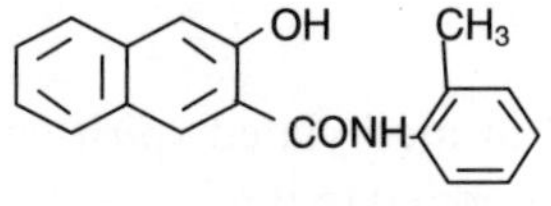

(3-Hydroxy 2-naphthoic *o*-toluidide)

Naphthol AS-KN
C I Azoic Coupling Component 37
C I 37608

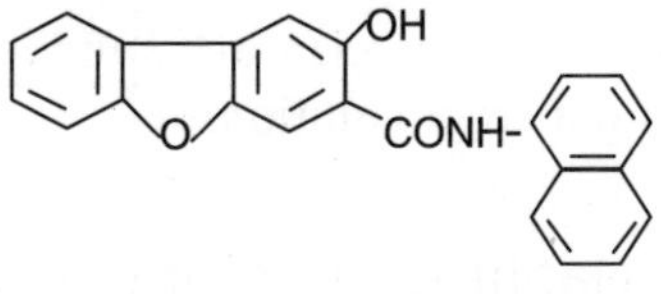

(2-Hydroxy-N-1-Naphthyl-3-dibenzofurancarboxamide)

Scheme 11.1 Chemical structure and C I specification of few coupling components.

11.6   Practical application

The dyeing process consists of six main stages, viz. (i) solubilisation of naphthol, (ii) naphtholation of cotton, (iii) diazotization of base, (iv) coupling or development, (v) soaping or steaming, and (vi) washing.

11.6.1 Solubilisation of naphthol

Naphthols are solubilised in two methods: hot and cold.

Hot method

Coupling component is pasted with little Turkey red oil, hot water is added to it followed by slow addition of NaOH with constant stirring and boiled till a clear solution is obtained. Most of the naphthols on solubilisation develop yellow mustard oil like colour, while brenthols develop a turbid curdy colour.

If the naphthol solution is to be used instantly or within a short time, no additives are added; otherwise the solution is cooled and HCHO is added for stabilization. Addition of HCHO reduces affinity of solubilised naphthol and not to be added if naphtholation is not delayed too much. It is laborious than cold method but cost of production is too less making the hot method commercially popular. Dosing of NaOH is to be controlled properly; if in excess, precipitates naphthols. A typical recipe for solubilisation of naphthol AS consists of naphthol AS (10 g/l), T R oil (a little), NaOH (6.5 g/l) at boil for 15 min. Concentration of NaOH depends on type of naphthol; it is around 60–65% wrt weight of most of naphthols but for AS-G and its derivatives, the dose is around 35%. The dose is not to be added at one time, rather 50% should be added first followed by slow addition with constant stirring till a clear solution is obtained.

Cold method

Naphthol is added in methylated spirit and stirred with NaOH (at lower dose). Cold water is added to it when a clear naphthol solution is obtained. This solution is added in the naphtholating bath containing Turkey red oil and NaOH. It is expensive, and cotton is to be naphtholated just after dissolution – as in open air, soluble naphthols show a tendency to be separated out from solution.

11.6.2 Naphtholation of cotton

Cellulose is treated in naphtholate solution in jigger at boil for 30 min followed by addition of 20 g/l NaCl and naphtholation is continued for further 1.5–2 h. Heating up of bath reduces substantivity of naphthols.

In semi-continuous method, the solution is directly used for padding of cellulose followed by drying in a float drier or hot flue and stored for development in batches in jigger. In continuous method, cellulose is padded with solubilised naphthol at higher dose and is continuously passed through a coupling bath for development of shade followed by open soaping and drying (Atul, 2009; Wilcoxson, 1966; Aspland, 1992a, 1992b; Shore, 1995). If development is done just after naphtholation, no drying is required. If coupling with diazotized base is delayed, padded cotton must be dried and are to be covered with polyethylene sheet to avoid contact with acid fumes which may separate out naphthol from cotton. Any water drop is to be prevented from coming on padded goods which removes a part of naphthol from the spot producing lighter shades.

Most of commonly used naphthols possess lower affinity for cotton and so naphtholation through padding is preferred followed by either drying or passed through coupling bath for development. For lower substantive naphthols, subsequent drying is essential at controlled temperature. Adequate care is to be taken to avoid migration of naphthol.

11.6.3 Diazotisation of base

Fast bases are aromatic primary amines ($ArNH_2$) and are marketed as either parent fast bases or their hydrochlorides under various brand names, viz. Tulabase, Diazo, Kako, Diabase, Sugai, Daito, Shangda, Natasol, Amarthol, Atlantic, Azanil, Fast, Spectrolene, Azoene, Durgasol, Hindasol, Mitsui, Sanyo, Variamine, etc. (Colour Index, 1987).

If the base is in its hydrochloride form, HCl is to be added first followed by $NaNO_2$, because prior addition of $NaNO_2$ will start diazotization of base with the help of HCl already present, which is too little for the purpose. This will cause insufficient diazotization, even diazotized base will react with undiazotised base resulting loss of base and colour value. Addition of HCl after applying NaNO$_2$ will not improve the situation as diazotization and coupling are too fast processes. Bases marketed as hydrochlorides are diazotized in direct method where acid is first added to base followed by NaNO$_2$. In contrast, fast bases are diazotized by adding NaNO$_2$ first followed by HCl, known as indirect method of diazotization.

There are two methods for diazotization of bases, viz. direct and indirect. A specific base can only be successfully diazotized in a specific method only to achieve desired coupling. Inclusion of 'C' in nomenclature of base denotes its hydrochloride form and needs to be diazotized in direct method, e.g. Fast Red RC base, Fast Orange GC base, Fast Yellow GC base, etc., with few exceptions, e.g. Fast Red KB/TR base, Fast Scarlet G/GG base etc.

Direct method

Base is mixed with 5 times hot water (owb) followed by addition of HCl. Ice is added to cool down the mixture to around 5°C. Sodium nitrite, elsewhere dissolved in water (1:2) is slowly added to acidified cold base and stirred well till all foams subside and is waited till completion of reaction.

Indirect method

Base is pasted with cold water, sodium nitrite is added and is stirred till complete dissolution occurs. HCl cooled with sufficient ice is added to this dissolved base bath and stirred till foam subsides completely. To complete diazotization of base, 10 min is allowed.

For a specific amount of naphthol, optimum requirement of base will be as shown in Table 11.1 (Atul, 2009).

Table 11.1 Combining ratio of some coupling components and bases

For 1 kg base	Naphthol (kg)			
	AS	AS-G	ASBS	AS-TR / AS-SW
Fast Yellow GC base	0.95	1.05	0.80	0.80
Fast Orange GC base	0.95	1.05	0.85	0.80
Fast Scarlet G base	0.75	0.80	0.65	0.65
Fast Scarlet RC base	1.20	1.25	1.05	1.05
Fast Red B base	0.90	0.95	0.75	0.75
Fast Red RC base	1.10	1.20	0.95	0.95
Fast Violet B base	1.52	1.63	1.30	1.30

Diazotization reaction takes place between hydrochloride of the base and HNO_2. Presence of both HCl and $NaNO_2$ should be tested by Congo red paper (turns blue if excess of HCl is present) and by starch-iodine paper (turns blue if excess of HNO_2 is present). Deficiency in any of these components will result incomplete diazotization.

If HCl is insufficient, diazotized base reacts with undiazotized base.

$$Ar-N^+ \equiv N - Cl^- + Ar-NH_2 \rightarrow Ar-N = N-NH-Ar$$

The final pigment develops a colour different from that of the desired one. Successful diazotization is indicated by the clarity of final solution and absence of froth.

Some bases are marketed as their hydrochlorides, solubilisation of these in water favours backward reaction by hydrolysis. Excess HCl is added to favour the back reaction.

$$ArNH_3Cl \leftrightarrow ArNH_2 + HCl$$

The decomposition of diazotised base is generally accompanied by turbidity and formation of a scum. CH_3COONa and CH_3COOH are added in coupling bath just before development.

Stability of diazotized bases vary at different extents, e.g. Red R / RC / TR, Scarlet R / RC and Blue BB possess good stability; Yellow GC, Scarlet G, Violet B. Red B, Boedeaux GP and Orange GC have medium stability while Blue B, Garnet GBC and Red KB possess poor stability necessitating use of it without further delay.

A typical recipe of diazotization may be formulated as follows:

Fast Scarlet RC base	–	10g/l
$NaNO_2$	–	3 g/l
HCl	–	5 ml/l
Temperature		~ 5°C (with ice)
Time	–	10–15 min

11.6.4 Coupling or development

Insoluble pigment is formed *in situ* fibre via reaction between coupling component (naphthol) and diazotized base; an instantaneous and faster reaction. During coupling, the diazonium ion replaces a hydrogen ion located on the ring in a position ortho or para to the activating amine or hydroxyl group based on whether the coupling component possesses a – NH_2 or –OH group, i.e. the chemical structure of lake can be ascertained by attaching the N=N of diazotized base at the -*o* or -*p* position of naphthol.

Coupling power of diazonium chloride gets reduced with fall in temperature and around 15–20°C is ideal for development.

In order to prevent migration of coupling component in the liquor from impregnated fabric when dipped and before the reaction has taken place, NaCl (25 g/l) is added. Concentration of diazotised base should not go down beyond 0.5 g/l, otherwise rate of coupling will be slower. Addition of sulphated fatty alcohol or ethylene oxide condensate speeds up coupling and keeps in suspension any pigment formed in the liquor. Concentration of chemicals required for diazotization and coupling have been shown in Table 11.2 (Atul, 2009).

Table 11.2 Concentration of chemicals required for diazotization and coupling

Base (1 kg)	Method of diazotisation	HCl (32°Tw) (l)	NaNO$_2$ (kg)	CH$_3$COONa (kg)	CH$_3$COOH (50%) (l)
Fast Blue B base	Indirect	2.50	0.65	1.25	–
Fast Blue BB base (25%)	Indirect	1.00	0.25	0.50	–
Fast Bordeaux GP base (50%)	Indirect	1.75	0.50	0.85	0.75
Fast Garnet GBC base (32%)	Direct	1.20	0.32	1.25	–
Fast Orange GC base (50%)	Direct	1.20	0.50	1.00	0.75
Fast Red B base (50%)	Indirect	1.75	0.50	0.85	0.75
Fast Red KB base	Direct	1.00	0.40	0.75	0.55
Fast Red RC base (50%)	Direct	1.00	0.50	0.75	0.25
Fast Red TR base	Direct	1.00	0.40	0.75	0.15
Fast Scarlet G base (50%)	Direct	2.00	0.50	1.00	0.75
Fast Scarlet GG base	Direct	3.60	0.55	3.75	–
Fast Scarlet RC base (40%)	Direct	1.00	0.40	0.75	0.50
Fast Violet B base (30%)	Indirect	1.15	0.30	0.55	–
Fast Yellow GC base (50%)	Direct	1.12	0.50	1.00	0.75
Fast Orange GR base	Indirect	2.25	0.55	1.25	0.75

[Percent shown in parenthesis denotes purity of commercial products those are supplied to industries. For Blue B, BB and violet B bases, CH$_3$COONa is partly added in diazotized base, rest is added in coupling bath. After development, dyed cellulose is to be boiled with Na$_2$CO$_3$ (10 g/l) and soaping is omitted. Shades of AS-G with any base will be brighter if 4 g/l CH$_3$COOH in excess is added to coupling bath]

Fast bases are classified as per coupling energy required for reaction at specific pH. Group-I bases possess high coupling energy; pH is maintained at 4–5 with CH$_3$COOH, no buffer is needed but an alkali-binding agent (CH$_3$COOH) is essential. Group-II bases possess medium coupling energy, optimum pH for coupling is 5.5–6.5, CH$_3$COOH and CH$_3$COONa buffer is added. Group-III bases have low coupling energy, optimum coupling occurs at pH 6–7, bath should be neutralized with NaHCO$_3$, then buffered with mono and disodium phosphate buffer mixture (4 and 2 g/l respectively at 1:20 liquor ratio). Group-IV bases have very low coupling energy, a pH around 7–8.2 is best for fast coupling. Coupling is also retarded due to separation of NaOH in bath used in naphtholation. Excess of alkali can be neutralized with acetic acid or Aluminium sulphate [Al$_2$(SO$_4$)$_3$] – the so called alkali-binding agents.

11.6.5 Steaming or soaping

One inherent problem associated with insoluble azoic colours is that pigments are formed *in situ* fibre, in bath and at the fibre–liquor interface; the last factor influences poor rubbing fastness. This happens, when naphthols, preferably those remain nearer to fibre surface and react with diazotized base at the fibre–bath interface. Soaping causes removal of superficial pigments and their dispersion in the bath, some sort of removal of pigments formed at the interface and promotes aggregation of pigments inside. High pressure steaming serves all the purposes as those with soaping, except the problem of superficial deposition. Thorough and repeated soaping can improve rubbing fastness.

11.7   Specification of some important bases

Chemical structure and C I specification of a few important aromatic bases, which are frequently used to produce naphthol colours are shown in Scheme 11.2 (Allen, 1971; Colour Index, 1987).

11.8   Function of chemicals

11.8.1 Naphtholation

Coupling component

A naphthol or brenthol which develops the hue on reaction with diazotized base.

NaOH

Naphthols are insoluble in water and have no affinity for cellulosics. NaOH converts these into sodium salt of naphthols, called naphtholates, which are soluble and show affinity for cellulose. When used in excess, causes precipitation of naphthols (Shenai, 1987).

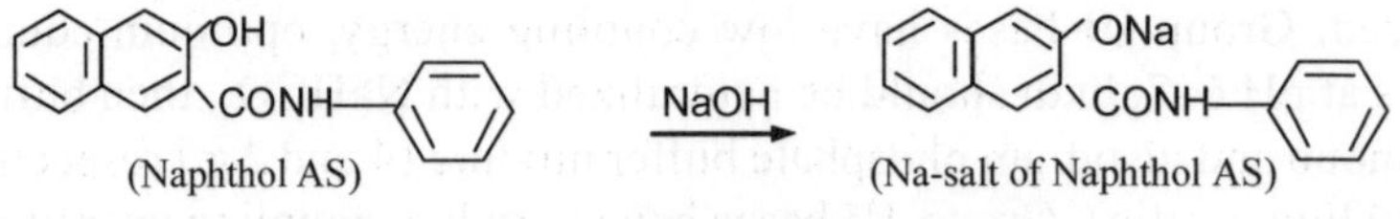

(Naphthol AS) → NaOH → (Na-salt of Naphthol AS)

Addition of excess NaOH produces disodium salts and the naphthol starts precipitating.

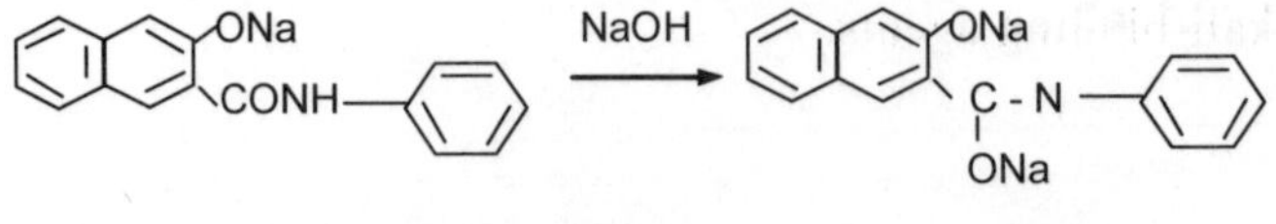

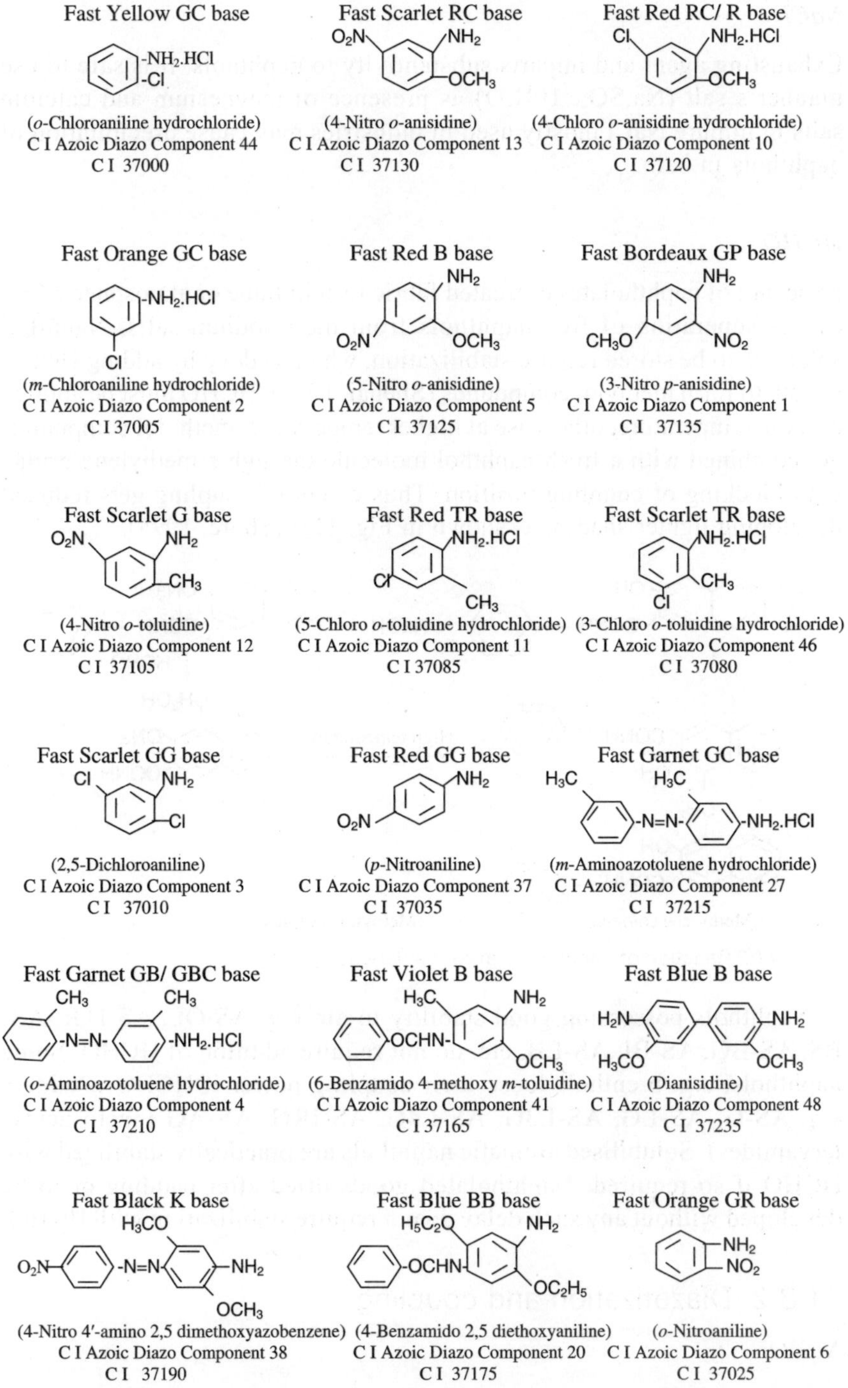
Fast Yellow GC base
-NH2.HCl
-Cl
(o-Chloroaniline hydrochloride)
C I Azoic Diazo Component 44
C I 37000
Fast Scarlet RC base
O2N
NH2
OCH3
(4-Nitro o-anisidine)
C I Azoic Diazo Component 13
C I 37130
Fast Red RC/ R base
Cl
NH2.HCl
OCH3
(4-Chloro o-anisidine hydrochloride)
C I Azoic Diazo Component 10
C I 37120
Fast Orange GC base
-NH2.HCl
Cl
(m-Chloroaniline hydrochloride)
C I Azoic Diazo Component 2
C I 37005
Fast Red B base
NH2
O2N
OCH3
(5-Nitro o-anisidine)
C I Azoic Diazo Component 5
C I 37125
Fast Bordeaux GP base
NH2
CH3O
NO2
(3-Nitro p-anisidine)
C I Azoic Diazo Component 1
C I 37135
Fast Scarlet G base
O2N
NH2
CH3
(4-Nitro o-toluidine)
C I Azoic Diazo Component 12
C I 37105
Fast Red TR base
NH2.HCl
Cl
CH3
(5-Chloro o-toluidine hydrochloride)
C I Azoic Diazo Component 11
C I 37085
Fast Scarlet TR base
NH2.HCl
CH3
Cl
(3-Chloro o-toluidine hydrochloride)
C I Azoic Diazo Component 46
C I 37080
Fast Scarlet GG base
Cl
NH2
Cl
(2,5-Dichloroaniline)
C I Azoic Diazo Component 3
C I 37010
Fast Red GG base
NH2
O2N
(p-Nitroaniline)
C I Azoic Diazo Component 37
C I 37035
Fast Garnet GC base
H3C
H3C
-N=N-
-NH2.HCl
(m-Aminoazotoluene hydrochloride)
C I Azoic Diazo Component 27
C I 37215
Fast Garnet GB/ GBC base
CH3
CH3
-N=N-
-NH2.HCl
(o-Aminoazotoluene hydrochloride)
C I Azoic Diazo Component 4
C I 37210
Fast Violet B base
H3C
NH2
OCHN-
OCH3
(6-Benzamido 4-methoxy m-toluidine)
C I Azoic Diazo Component 41
C I 37165
Fast Blue B base
H2N
NH2
H3CO
OCH3
(Dianisidine)
C I Azoic Diazo Component 48
C I 37235
Fast Black K base
H3CO
O2N
-N=N
NH2
OCH3
(4-Nitro 4′-amino 2,5 dimethoxyazobenzene)
C I Azoic Diazo Component 38
C I 37190
Fast Blue BB base
H5C2O
NH2
-OCHN
OC2H5
(4-Benzamido 2,5 diethoxyaniline)
C I Azoic Diazo Component 20
C I 37175
Fast Orange GR base
NH2
NO2
(o-Nitroaniline)
C I Azoic Diazo Component 6
C I 37025

Scheme 11.2 Chemical structure and C I specification of a few important aromatic bases.

NaCl

Exhausting agent and imparts substantivity to naphthols. It is safe to use glauber's salt (Na_2SO_4, $10H_2O$) as presence of magnesium and calcium salts in impure NaCl mostly used in industries may cause precipitation of naphthols in bath.

HCHO

Exposure of naptholates or treated fabric to acid fume or atmospheric CO_2 causes separation of free naphthols from their sodium salts. Naphthol solutions to be stored require stabilization, which is done by adding HCHO in bath to form methylol compounds (Shenai, 1987). HCHO must be applied at room temperature, otherwise at higher temperature, methylol compounds get combined with a fresh naphthol molecule through a methylene bridge with blocking of coupling position. Thus extent of coupling gets reduced developing lighter shades, as shown in Fig. 11.2 (Shore, 1995).

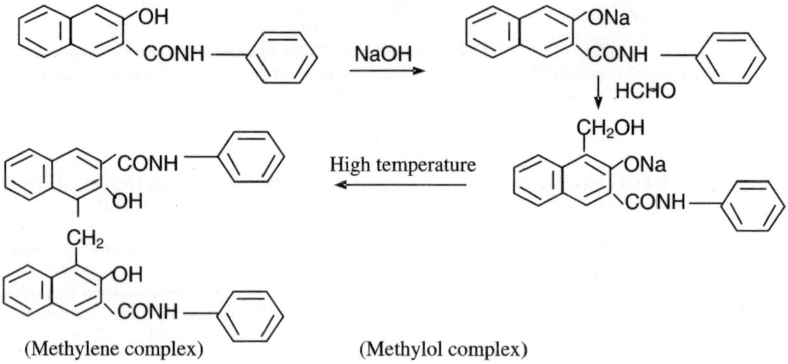

(Methylene complex)   (Methylol complex)

11.2 Reaction of naphthol with formaldehyde.

Naphthols possessing good stability to air, e.g. AS-OL, AS-ITR, AS-BS, AS-BG, AS-BI, AS-LB, etc. do not require addition of HCHO. Some naphthols, e.g. brenthols loose their coupling power if HCHO is added, e.g. AS-G, AS-LG, AS-L3G, AS-L4G, AS-IRG, AS-AG (aceto acetyl laryamides). Solubilised aromatic naphthols are practically stabilized with HCHO if so required. Naptholated goods dried after padding or to be developed without any such delay do not require stabilization with HCHO.

11.8.2 Diazotization and coupling

NaNO₂

It produces HNO_2 after reaction with HCl, which in turn diazotizes the base.

$$NaNO_2 + HCl \rightarrow HNO_2 + NaCl \text{ (at 5°C)}$$

HCl

It reacts with $NaNO_2$ to produce HNO_2 as well as converts free bases into water soluble salts

$$Ar-NH_2 + HCl \longrightarrow ArNH_3^+Cl^-$$

This hydrochloride of base reacts with HNO_2 to produce diazonium salt.

$$ArNH_3^+Cl^- + HNO_2 \longrightarrow Ar-N^+ \equiv N-Cl^- + 2H_2O$$

H_2SO_4 is not preferred for diazotization because sulphates are less soluble than chlorides.

Generally, a molar ratio of base, $NaNO_2$ and HCl is maintained at 1: 1: 3. One molecule of base requires one molecule of HNO_2 for diazotization which comes from one molecule of $NaNO_2$. Out of 2.5 molecules of HCl, one forms base hydrochloride, one reacts with $NaNO_2$ to produce HNO_2 while the half maintains pH during diazotization.

Ice

Diazotization reaction is an exothermic process; ice is added to maintain temperature (0–5°C). If ice is not added, especially above 30°C, diazotized salt decomposes, react with undiazotized salt and water resulting loss of a part of base.

$$Ar-N^{+o} N - Cl^- + Ar-NH_2 \rightarrow Ar-N= N-NH-Ar$$
$$Ar-N^{+o} N - Cl^- + H_2O \rightarrow Ar-N= N-OHCH_3COONa$$

After diazotization, the solution has a strong acidic pH (2–3) due to presence of excess HCl; for efficient coupling at pH~5–6, the pH is raised by adding CH_3COONa (3–4%), which converts inorganic HCl to organic CH_3COOH, thus reducing acidity of bath.

$$CH_3COONa + HCl \rightarrow CH_3COOH + NaCl$$

CH_3COONa and CH_3COOH combination act as buffer to maintain desired pH; however, some extra CH_3COOH needs to be added to bath to maintain pH.

Alkali binding agent

Naphthol treated fabric before development has strong alkaline pH and the coupling bath has a pH~5–6. When naptholated fabric is dipped in coupling bath for reaction, alkali attached with naphthol becomes free and migrates to the bath, neutralizes acid and coupling is either retarded or stopped. Application of $Al_2(SO_4)_3$ or more commonly used CH_3COOH as alkali binding agent neutralizes this NaOH and maintains acidity of bath required for coupling.

$$Al_2(SO_4)_3 + NaOH \rightarrow Al(OH)_3 + Na_2SO_4$$
$$CH_3COOH + NaOH \rightarrow CH_3COONa + H_2O$$

Soap

It promotes aggregation of dye molecules within fibre and removes superficial pigments.

11.9   Precautions to be taken

11.9.1 Naphtholation

- Falling of water drops on impregnated fabric before coupling leads to formation of lighter spots and it should be kept away from source of water.
- Sodium salts of coupling components (naphthols) are unstable in open air (CO_2) and storing of impregnated wet goods or standing naphthol bath result in break down of salt linkages, reducing the coupling efficiency and resulting a loss in colour value. The problem may be minimized by drying naphtholated fibre or immediate coupling with diazotized base or HCHO may be added in naphthol solution for stabilization.
- Naphtholated cellulose should be dried at 80–90°C, otherwise a part of naphthol will be sublimised-off.
- Substantivity of coupling components is low and so a lower liquor ratio is preferred.
- Rise in temperature hinders absorption of naphthols; hence room temperature impregnation is preferred.
- Salt must be added in bath to enhance diffusion of naphthol for good rubbing fastness.
- Application of HCHO must be restricted.

11.9.2 Diazotisation

- Diazonium salts are very unstable beyond 20°C necessitating addition of ice in bath, otherwise these react with water and gets decomposed liberating nitrogen.
- Use of metal vessels enhances decomposition of diazotised salts.

- Diazotization must be carried out in diffused light, as direct sunlight promotes decomposition.
- Due to evolution of HNO_2 during diazotization, standing just before the bath should be avoided.

11.9.3 Soaping

- Over-soaping produces dull shades.
- Soaping with 0.5–1 g/l $Na_2S_2O_4$ on yellow combinations of AS-G brings out true shades.
- Some combinations should not be soaped beyond 65–70°C, e.g. combinations of Yellow GC base with naphthol AS, AS-BO, AS-SW, AS-OL and combinations of Orange GC with naphthol AS, AS-BO, AS-SW.

11.10 Rubbing fastness

All insoluble azoic colours show poor to moderate rubbing fastness due to formation of pigments at the outer layer of cellulose, a part of which remains projected in bath / air and leave a part of colour on objects against which dyed cotton is rubbed. The problem can be handled in various ways:

(a) Pretreatment of cellulose with metal salts followed by naphtholation results better diffusion causing better rubbing fastness.
(b) Minimizing soluble naphthol deposition at the outer layer of cotton during naphtholation by allowing better diffusion of naphthols with application of NaCl at higher dose during naphtholation or running naphtholated fabric for shorter time in water containing NaCl at higher dose.
(c) Thorough washing with severe agitation to detach pigments at the outermost layer.
(d) Selecting high substantive naphthols in exhaust dyeing.
(e) Low substantive naphthols must be applied only through padding under higher pressure for better penetration.
(f) Migration of naphthols in coupling bath should be restricted using NaCl.
(g) Soaping must be carried out instead of steaming.

11.11 Limitations

(a) Build-up of shade, i.e. application of naphthol or base at any stage for shade match is not possible
(b) Naphthol or base combination difficult to exercise:

(i)   different coupling components have affinity towards cellulose at varying extents, still combination of naphthols is somehow possible with only one base.

(ii)   combination of base is difficult as coupling speed typically varies from one diazotized base to other.

(c) Production of light shades is difficult as well as risky.

(d) Shades are invariably bright.

(e) Mainly restricted to red, maroon, orange and yellow colours.

(f) Applied to products where rubbing fastness does not pose a problem.

(g) Percent shade formation impossible as colouring materials such as dyes or pigments are not applied directly rather produced in bath.

11.12   Stripping

Stripping of insoluble azoic colours is not possible but these can be discharged at boil for 1 min with $Na_2S_2O_4$ in combination with NaOH.

11.13   References

ALLEN R L M (1971) *Colour Chemistry*, 1st Edition, London, Thomas Nelson & Sons Ltd.

ASPLAND J R (1992a) 'A series on dyeing: Chapter 6: Azoic combinations: chemical principles', *Textile Chem.Color.*, **24**, 8, 26–28.

ASPLAND J R (1992b) 'A series on dyeing: Chapter 6 / Part 2: azoic combinations: practical applications', ***Textile Chem.Color.***, **24**, 9, 74–77.

ATUL LTD (2009) 'Naphthol dyeing user's manual', Gujarat, India.

Colour Index International (1987), 3rd Edition (3rd Revision), Bradford, UK, Society of Dyers & Colourists and AATCC.

GOKHALE S V and SHAH R C (1981) *Cotton Piece dyeing*, India, Ahmedabad Textile Industries Research Association.

SHENAI V A (1987) *Chemistry of dyes and principles of dyeing*, 3rd Edition, Mumbai, Sevak Publications.

SHORE J (1995) *Cellulosics Dyeing*, Bradford, UK, Society of Dyers and Colourists.

WILCOXSON W C (1966) 'Dyeing of vats and naphthols on cellulosic fibres', *Am. Dyestuff Rep.*, **55**, 6, 38–44.

12

Dyeing with mineral colour

Abstract: Mineral colours are formed in situ cellulosics when water soluble metal salts are applied on fibre followed by conversion to insoluble metal oxides. No dye or pigments are applied rather colour of specific metal oxide alone or in combination shows the final hue. The metal oxides occupy free volume in fibre making it stiffer. Mineral colours are acid sensitive and less brilliant; invariably topped with vat or sulphur colours. Important member is mineral khaki for military uniforms.

Keywords: Metal salt, oxidation, metal hydroxide, metal oxide

12.1   Introduction

Metal salts can be applied on cotton either alone or in combination followed by treatment with alkali to form metal hydroxides and finally to metal oxides through oxidation when coloured pigments are formed *in situ*. This can not be termed as dyeing, rather formation of coloured metal oxide complexes. Shades lack brightness, cheaper, dyed cotton develops harsh feel, gains excessive weight, fast to wash and light but sensitive to acid. Viscose is not dyed with mineral colours due to problem of loosing lustre.

Metal oxides e.g. TiO_2, ZnO, ZnS, etc. are used as white pigments, while Fe_2O_3 is red brown, etc. Water insoluble metal oxides, alone or in combination, can be formed at specific proportion for a specific hue. Metal salts used possess no affinity for cotton and a padding method is evident for thorough application.

Mineral khaki, the most important colour in this category, is a mixture of chromium oxide and ferric oxide at right proportion for colouration of military uniforms (Kramrisch, 1986). It is developed through padding cotton with a solution of chromium sulphate and ferrous sulphate, dried and passed through alkali solution at boil to form respective hydroxides; on drying, chromium hydroxide gets oxidized to green chromium oxide (Cr_2O_3) and ferrous hydroxide to red ferric oxide (Fe_2O_3) finally to produce khaki. Insolubility of both these oxides in water develop excellent wash fastness; post-treatment with Na_2SiO_3 enhances perspiration fastness too.

12.2   Fundamentals of application

Chromium oxide (Cr_2O_3) and ferric oxide (Fe_2O_3) are the final components required to form khaki shade. These can be produced from water soluble higher salts (-ic) of respective metal sulphates but higher cost of these has led to use of $FeSO_4$ and $Na_2Cr_2O_7$ as starting materials.

$FeSO_4$ with alkali forms $Fe(OH)_2$; the latter releases water to form FeO and then to Fe_2O_3 through oxidation. $Na_2Cr_2O_7$ is reduced in presence of H_2SO_4 and jaggery to $Cr_2(SO_4)_3$, excess of acid is neutralized with stiochiometric Na_2CO_3 as little excess of it forms $CrCO_3$; $Cr_2(SO_4)_3$ in turn forms $Cr(OH)_3$ with alkali and finally to Cr_2O_3. In practice, $Na_2Cr_2O_7$ is reduced to $Cr_2(SO_4)_3$, the latter is mixed with $FeSO_4$ and both are treated with alkali to form respective oxides.

Cotton fabric, padded with mixture of chromium sulphate and ferrous sulphate solution is dried in a float drier or hot flue followed by treatment in a boiling solution of alkali, preferably combination of NaOH (10%) and Na_2CO_3 (5%) when corresponding metal hydroxides are precipitated in situ fabric, which is then aired for conversion of chromium hydroxide and ferrous hydroxide into chromium oxide and ferric oxide, respectively. Use of NaOH alone forms colloidal metal hydroxides which are difficult to oxidise; if soda ash is also present, metal hydroxide and carbonate are formed which makes conversion into oxides much easier. Following are the related chemical reactions (Shenai, 1993).

(i)   Conversion of $Na_2Cr_2O_7$ to $Cr_2(SO_4)_3$

$$8Na_2Cr_2O_7 + 3H_2SO_4 + C_{12}H_{22}O_{11} \rightarrow 8Na_2SO_4 + 8Cr_2(SO_4)_3 + 43H_2O + 12CO_2$$
$$H_2SO_4 + Na_2CO_3 \rightarrow Na_2SO_4 + H_2O + CO_2$$

(ii) Padding with $FeSO_4$ and $Cr_2(SO_4)_3$ at 75–80% expression at a speed of 15–20 m/min.

(iii) Passing of padded cotton through alkaline boiling bath for 30 min, when respective hydroxides are formed

$$FeSO_4 + 2NaOH \rightarrow Fe(OH)_2 + Na_2SO_4$$
$$Cr_2(SO_4)_3 + 6NaOH \rightarrow 2Cr(OH)_3 + 3Na_2SO_4$$

(iv) Oxidation of hydroxides to form oxides *in situ*

$$Fe(OH)_2 \rightarrow FeO + H_2O$$
$$2FeO + O_2 \rightarrow Fe_2O_3$$
$$2Cr(OH)_3 \rightarrow Cr_2O_3 + 3H_2O$$
$$Fe_2O_3 + Cr_2O_3 \rightarrow Khaki\ shade$$

12.3   Practical application

12.3.1 Preparation of pad liquor

Dichromate method

$Na_2Cr_2O_7$ (80 kg) is solubilised in water (160 l) in lead lined wooden tank followed by gradual addition of H_2SO_4 (166°T_w, 77 kg) with constant stirring; the solution is cooled, jaggery (20 kg) dissolved in 45 l of water is added with constant stirring. Complete oxidation of jaggery and conversion of $Na_2Cr_2O_7$ to $Cr_2(SO_4)_3$ are brought over a time of 24 h. Excess H_2SO_4 is neutralized with Na_2CO_3 (~4 kg) dissolved in water (15 l) when CO_2 is evolved with vigorous effervescence. $FeSO_4$ (28 kg) dissolved in of water (45 l) is added to the above solution. The solution is diluted depending upon depth of shade required (Doshi and Dixit, 1975).

Chromium sulphate method

A solution is prepared with 25% chromium sulphate $[Cr_2(SO_4)_3]$ (35 kg), Na_2CO_3 (3 kg), 95% $FeSO_4$ (5.5 kg), CH_3COONa (1 kg), dispersing agent (0.25 kg) and wetting agent (0.06 kg) dissolved in rest of water to make finally 100 l.

$Cr_2(SO_4)_3$ powder is stirred well to dissolve in cold water in a plastic drum. The solution is filtered through sieve or jute fabric and taken in a stainless steel container. Pre-dissolved Na_2CO_3 is added slowly to this $Cr_2(SO_4)_3$ solution, which follows an exothermic reaction. Air is injected in this solution and allowed to stand overnight. $FeSO_4$ is dissolved in boiled water, allowed to stand for 24 h followed by addition of it in the stainless steel container. Pre-dissolved CH_3COONa, wetting agent and dispersing agent are added to this with injection of air and allowed to stand for a few hours; pH and specific gravity of the solution are checked (~44°T_w). Dichromate method is cheaper to chromium sulphate method (Doshi and Dixit, 1975).

12.3.2 Padding of cotton

Cotton fabric is padded with this solution for 75–80% expression, dried or covered with polyethylene sheet to avoid oxidation of selvedge and developed in a boiling solution of NaOH (40 g/l) and Na_2CO_3 (20 g/l) for 30 min, aired, washed in cold and hot water, treated with Na_2SiO_3 solution (2.5 g/l), washed thoroughly and dried. A greener or redder khaki is produced by adjusting chromium or iron content of pad liquor. To develop brilliancy of shade, dyeings are often topped with either vat olive green B, khaki GG or sulphur khaki dyes.

12.3.3 Precautions to be taken

- Well pretreated cotton fabric with equal moisture content through prolonged exposure in dye-house conditions is to be taken for padding.
- No water drops should fall on cloth till its development and padded fabric should be protected from acid fumes.
- Rate of drying on both sides of padded fabric should be uniform to avoid migration of metal salt, which otherwise forms uneven shade.
- Process conditions and concentration of chemicals must be controlled accurately to avoid variation from lot to lot.

12.4   Other mineral colours

Chrome yellow, chrome red, chrome orange, chrome green, manganese brown, prussian blue, iron oxide etc are the other examples (Shenai, 1993; Allen, 1971).

Chrome colours are derivatives of lead chromate ($PbCrO_4$) and so are invariably poisonous. These colours are of least textile importance and are mainly used in paints, printing inks etc.

12.4.1 Chrome yellow ($PbCrO_4$, C I Pigment Yellow 34, CI 77603)

$$2(CH_3COO)_2Pb + K_2Cr_2O_7 + H_2O = 2PbCrO_4 + 2CH_3COOK + 2 CH_3COOH$$

The yellow varies from lemon yellow to orange yellow ; fast to wash, light and acids. Alkali solubilises it with change to orange while H_2S turns it to brown due to formation of PbS.

$$PbCrO_4 + 4NaOH = Na_2PbO_2 + Na_2CrO_4 + 2H_2O$$

Cotton is padded with lead nitrate or acetate, treated with Na_2SO_4 to develop lead sulphate, oxidised with dilute solution of $K_2Cr_2O_7$ or $Na_2Cr_2O_7$ to form $PbCrO_4$; depth of shade depends on strength of lead solution and quantity of lead oxide formed on fibre.

12.4.2 Chrome orange (C I Pigment Orange 21, C I 77601)

It is a mixture of lead chromate (chrome yellow, $PbCrO_4$) and basic lead chromate (chrome red, $PbCrO_4.PbO$). It is produced by precipitating basic solution of basic lead salt with $Na_2Cr_2O_7$ or $Na_2Cr_2O_4$.

$$2PbNO_3(OH) + Na_2CrO_4 = Pb_2CrO_5 + 2NaNO_3 + H_2O$$
$$4PbNO_3(OH) + Na_2Cr_2O_7 + Na_2CrO_4 = Pb_2CrO_5 + 2PbCrO_4 + 4NaNO_3 + 2H_2O$$

Practically chrome orange is developed by reacting lime with chrome yellow which is fast to alkali but not to acids.

12.4.3 Chromium oxide (Cr_2O_3, C I Pigment Green 17, C I 77288)

It is produced by heating $Na_2Cr_2O_7$ with sulphur or charcoal.

$$Na_2Cr_2O_7 + S \rightarrow Na_2SO_4 + Cr_2O_3$$
$$Na_2Cr_2O_7 + 2C \rightarrow Cr_2O_3 + Na_2CO_3 + CO$$

Green powder is produced with very high melting point (2000°C); used in oil and water based paints, tinting glass, distempers and painting porcelain. Fast to light, heat, acid and alkali. It is soulblised by fusion with either of Na_2O_2 or $KHSO_4$.

$$Cr_2O_3 + 3Na_2O_2 + H_2O \rightarrow 2Na_2CrO_4 + 2NaOH$$
$$Cr_2O_3 + 6KHSO_4 \rightarrow Cr_2(SO_4)_3 + 3 K_2SO_4 + 3H_2O$$

Hydrated chromium oxide is called Guignet's green (Cr_2O_3, $2H_2O$). It is a bright pigment (C I Pigment Green 18, C I 77289) and is obtained by fusing $K_2Cr_2O_7$ with boric acid.

12.4.4 Manganese brown (MnO.OH)

$$MnCl_2 + 2NaOH = Mn(OH)_2 + 2NaCl$$
$$2Mn(OH)_2 + O = 2MnO(OH) + H_2O$$

The brown is fast to light, soap, dilute acid and alkali. To enhance precipitation and oxidation simultaneously, NaOCl is mixed with NaOH solution. The brown is very readily discharged by reducing agents.

12.4.5 Iron oxides

Hydrated ferric oxides are prduced by precipitating a ferrous or ferric salt with alkali; in case a ferrous salt being used, the resulting oxide is oxidised by airing or oxidising agents.

$$2FeSO_4 + 4NaOH = 2Fe(OH)_2 + 2Na_2SO_4$$
$$2Fe(OH)_2 + O = Fe_2O_2(OH)_2 + H_2O$$
$$Fe_2(SO_4)_3 + 6NaOH = Fe_2O_2(OH)_2 + 3Na_2SO_4 + 2H_2O$$

Yellows (C I Pigment Yellow 42 & 43, C I 77492), reds (C I Pigment Reds 101 & 102, C I 77491), browns (C I Pigment Browns 6 & 7, C I 77491) and black (C I Pigment Black 11, C I 77499) shades can be produced by precipitating iron salts followed by oxidation. These are fast to light, soap, alkali; durable and chemically inert with low cost; brightest shades are obtained using ferric nitrate.

12.4.6 Prussian blue (C I Pigment Blue 27, C I 77510)

It is prepared by reacting $FeSO_4$ with potassium ferrocyanide to give a complex 'white paste' which on further oxidation with acidic $K_2Cr_2O_7$ forms blue pigments.

$$FeSO_4 + K_4Fe(CN)_6 \rightarrow K_2Fe.Fe(CN)_6 + K_2SO_4$$
$$K_2Fe.Fe(CN)_6 + H_2SO_4 + O \rightarrow 2KFe.Fe(CN)_6 + K_2SO_4 + H_2O$$

Bright blue shades with excellent fastness to light, stable upto 120°C but sensitive to alkali.

12.5   Stripping of mineral colours

Most of the minerals colours can be stripped out completely through treatment in 10% HCl at 70–80°C while NaOCl has no effect on these colours (Acharya and Venkataraman, 1994).

12.6   References

ACHARYA B S and VENKATARAMAN K (1994) 'Determination of barium activity number of mineral khaki dyed cotton fabrics', *BTRA Scan*, **25**, 4, 16–17.

ALLEN R L M (1971) *Colour Chemistry*, 1st Edition, London, Thomas Nelson & Sons Ltd.

DOSHI S M and DIXIT M D (1975) 'Dyeing of mineral khaki on cotton', *Colourage*, **22**, 1, January 9, 28–31.

KRAMRISCH B (1986) 'Pigment printing and dyeing of cotton', *Am. Dyestuff Rep.*, **75**, 2, 13 & 43.

SHENAI V A (1993) *Technology of Dyeing*, Mumbai, Sevak Publications.

13

Dyeing with oxidation black

Abstract: Aniline in its hydrochloride form goes on attaching with multiple such hydrochlorides stepwise through progressive oxidation finally to develop jet black shades, called oxidation black. The process is cumbersome but important to produce cheap bright black shades on cellulosics. All the three methods of formation of oxidation black, viz. direct black, aged black and steam black are associated with tendering of cellulose. Diphenyl black is another superior black with less or no tendering, but costlier, is produced from diphenyl amine.

Keywords: Aniline, oxidation, catalyst, oxidizing agent, oxygen carrier

13.1　Introduction

When simple chemicals goes through progressive oxidation to develop colouring material or ingrain dye or lake inside cotton, is called oxidation colours. Superior and cheaper fast black shades are developed on cotton through oxidation. Aniline black is produced *in situ* when cotton is impregnated with a solution containing aniline hydrochloride, oxidizing agent, acid liberating agent and a catalyst (oxygen carrier) followed by drying and ageing; condensation occurs at different stages to develop ingrain jet black which is fast to wash and light – the cheapest method of dyeing cotton with black shade. In all these structures, basic chromophore is dibenzopyrazine or phenazine; cotton must be completely free from alkali to develop ingrain jet black.

Dibenzopyrazine　　　　　　Phenazine

13.2   Mechanism of formation of lake

The stages through which gradual oxidation and condensation of aniline occurs to reach to jet black are (i) Caro's Yellow Imide, (ii) Willstater's Blue Imide, (iii) Willstater's Red Imide, (iv) Colourless Leucoemeraldine, (v) Violet Protoemeraldine (yellowish green salt), (vi) Blue Emeraldine (green salt), (vii) Dark Blue Emeraldine, (viii) Greenable Black Pernigraniline and finally to (ix) Aniline Jet Black. The various reaction steps, through which aniline forms oxidation black through progressive oxidation are explained in Scheme 13.1 (Venkatraman, 1952; Shenai, 1987).

13.3   Application

Aniline, the starting material, is insoluble in water and is converted to its hydrochloride for solubilisation. Aniline molecules are condensed in stages to produce black hue *in situ* fibre releasing HCl in proportion to number of aniline molecules condensed (Kramrisch, 1986).

$$C_6H_5NH_2 + HCl \rightarrow C_6H_5NH_2 \cdot HCl$$
$$C_6H_5NH_2 \cdot HCl + C_6H_5NH_2 \cdot HCl \rightarrow (C_6H_5NH_2)_2 \cdot HCl + HCl$$
$$(C_6H_5NH_2)_2 \cdot HCl + C_6H_5NH_2 \cdot HCl \rightarrow (C_6H_5NH_2)_3 \cdot HCl + HCl \text{ and so on.}$$

The process is carried out at elevated temperature by steaming or curing when liberation of acid degrades cellulose; to reduce release of HCl as well as rate of condensation of aniline under control, the process is started with only a little of aniline hydrochloride followed by addition of sufficient pure aniline (aniline oil, $C_6H_5NH_2$). Excess of aniline does not get solubilised due to insufficient HCl – remains isolated in bath and when HCl is released due to condensation of aniline earlier applied in the form of its hydrochloride, pure aniline molecules slowly enter into solution in salt form, gets condensed with release of HCl which converts another aniline molecule to its soluble form and so on. This in turn also reduces degradation of cellulose through diversion of HCl to solubilise aniline rather than acting on cotton.

The colourant acts as basic dye, is retained by negative ionic centres in cellulose and can form salt linkages. There are three methods to produce aniline black on cotton, viz. (i) direct black, (ii) aged black and (iii) steam black. In direct black method, cellulose is treated with aniline, HCl and dichromate in a single bath for 1 hr at boil followed by washing and soaping. In aged black method, aged aniline black is produced by treating cotton with aniline hydrochloride, oxidizing agent ($NaClO_3$) and oxygen carrier ($CuSO_4$ or VCl_5). NH_4Cl is added to develop acidic pH; impregnated cotton is dried at 60°C and then steamed for 6 hrs to develop a bottle

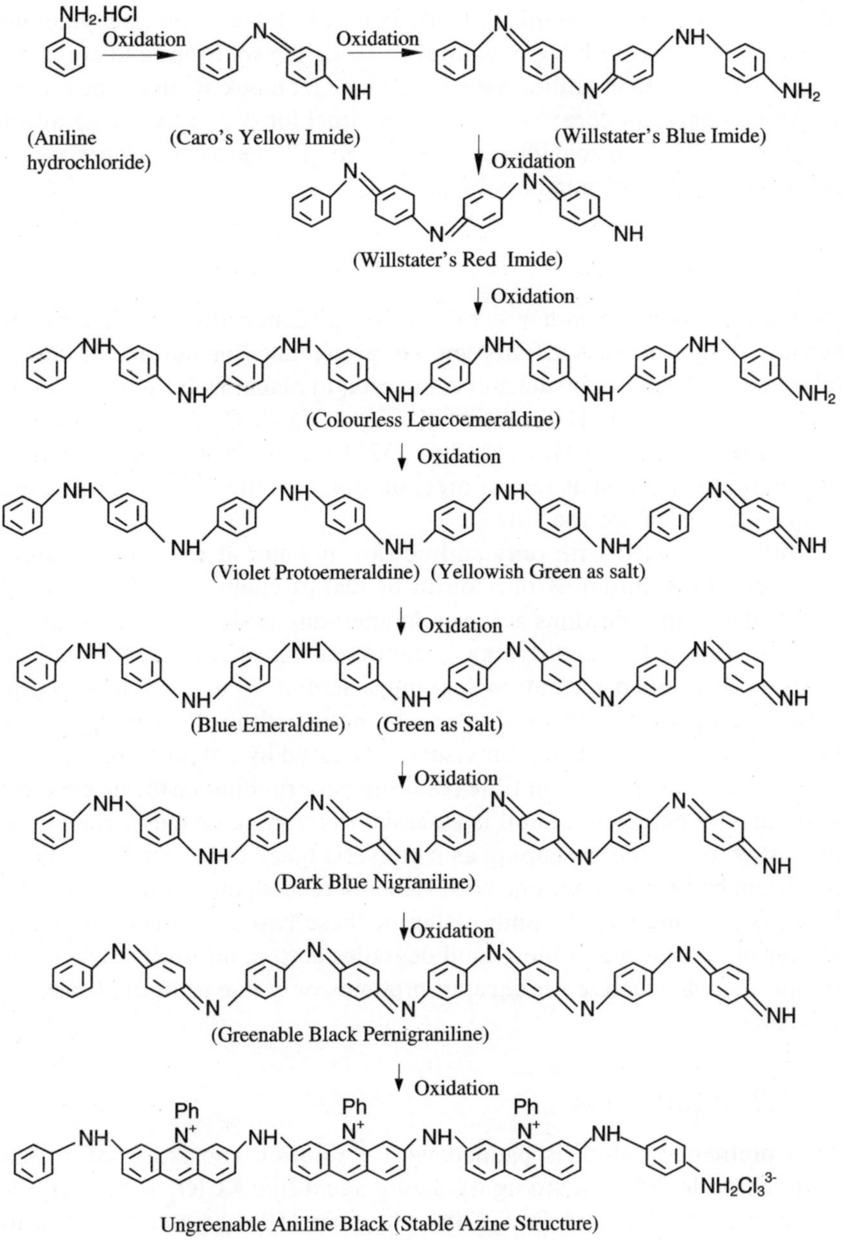

Scheme 13.1 Various steps of formation of aniline black.

green colour, which on further oxidation with acidic dichromate forms jet aniline black. In steam black method, the fabric is padded with aniline and other chemicals, dried and steamed – final oxidation is achieved with acidic dichromate. A green shade is mostly developed with some sort of degradation of cotton. Sulphamine derivatives of 4-aminodiphenylamine or other amines are soluble in neutral and alkaline solutions and are more readily oxidized than aniline salts to produce jet black shades. The recipe of dyeing generally consists of aniline hydrochloride, an oxidizing agent ($K_2Cr_2O_7$, $KClO_3$), an acid liberating agent (NH_4Cl) and a catalyst or oxygen carrier like cupric nitrate or sulphate.

13.3.1 Direct black

Dyeing is done in one bath with relatively high concentration of chemicals as these do not possess affinity and are precipitated on surface of cotton followed by diffusion inside and condensed to black. Dye liquor contains aniline hydrochloride (10–15%), $K_2Cr_2O_7$ or $Na_2Cr_2O_7$ (10–15%), $CuSO_4$ (1–2%) with excess of HCl (15–20%, 32°Tw). No fresh aniline is added in bath in this method as rate of precipitation on cotton surface, diffusion and condensation are too slow.

Cotton is treated with only aniline salt in water at room temperature succeeded by addition of one-fourth of rest of chemicals after each 15 min. Once all the additions are over, temperature is slowly raised to about 80°C or little below and dyeing is continued for at least 2–3 h. A deep black is developed on cotton leaving a lemon yellow coloured bath. However, a part of condensation occurs in the bath itself forming aniline black molecules outside cotton visually assessed by colour of bath which are partially precipitated on fibre resulting poor rubbing fastness; efficient soaping with detergent at boil is essential to improve fastness. Na_2CO_3 at no cost is to be used in soaping as it converts black to brown. The tone of black can be kept greener one with HCl and reddish one with H_2SO_4; a jet black is possible with a combination of these two acids. The method is slower one, presence of more acid degrades cotton, uniform shade is not produced and effective for cheaper products only due to limited fastness of shade.

13.3.2 Aged black

Well pretreated cotton is padded with a solution consisting of aniline hydrochloride (95 g), a strong oxidizing agent like $KClO_3$ or $NaClO_3$ (35 g), aniline oil (5 g) and Cu_2S (30%, 30 g) in 150 ml water followed by squeezing and development in moist and humid chamber for 45 min or at 35°C for 24 h (for yarn or small pieces of cotton) or with oxidizing agents (woven fabrics) to produce a greenish black which is converted to ungreenable black or jet black with $K_2Cr_2O_7$ and finally soaped hot.

Cu_2S being a strong oxygen carrier absorbs evolved oxygen from $KClO_3$ and slowly releases this oxygen for condensation of aniline through oxidation. Only $KClO_3$ or $NaClO_3$ is not sufficient enough to produce a jet black shade as efficiency of either of these is rapid upto only emeraldine stage beyond which it becomes too slow necessitating application of another oxidizing agent in bath like $K_2Cr_2O_7$. The method is useful to develop jet black on pieces of cotton and yarn with high rate of production. Cotton gets degraded with released HCl and also oxidized with oxygen carrier.

13.3.3 Steam black or prussiate black

This is the widely accepted method for production of quality jet black shades on umbrella cloth and to resist prints with superior fastness and less damage of cotton. A padding liquor is prepared with gum tragacanth (2%, 500 g), aniline hydrochloride (95 g), aniline (5 g), $NaClO_3$ or $KClO_3$ (25 g dissolved in 150 ml water), potassium ferrocyanide (50 g dissolved in 150 ml water) with rest of water to make 1 kg padding liquor. Each ingredient is separately dissolved in cold water and all of these are mixed in equal proportion just before application.

Cotton free from trace of alkali is padded in 2-dip 2-nip technique at nearly 100% expression, dried in stenter or hot flue at around 60–70°C followed by passing through an aniline ager (an ageing chamber exclusively used for this dyeing) in open width for 2–3 min. A hood and a fan drive out all acid fumes formed during ageing. Expelling acid from ageing chamber is essential to prevent damage of cotton, though free aniline in recipe also absorbs a part of HCl liberated to enter in reaction. A blackish green produced in ageing is passed through a bath of $K_2Cr_2O_7$ (4%) and H_2SO_4 (1%) at 60°C for 15 min to develop a jet black shade. Dyeings are finally thoroughly washed, soaped at boil for 10–15 min with detergent and rinsed.

The prerequisite to produce jet black or ungreenable black is that condensation should occur in acidic pH up to only emeraldine stage (Scheme 13.1) beyond which the pH must be neutral to form azine structure through 'azine condensation' – susceptible to reduce green shade is produced if an acidic pH is maintained throughout.

A weak oxygen carrier like potassium ferrocyanide is used as the process is longer enough. In contrast, Cu_2S is a strong oxygen carrier and is suitable only for short time processes for quick oxidation of condensed product. Tendering of cotton is least due to presence of weak oxygen carriers (catalyst). During whole of this operation, colour of the material changes from yellow to green, to deep green and to black progressively.

13.4   Problems in dyeing with aniline black

- Tendering of cellulose cannot be fully avoided due to liberation of HCl; development in lime water instead of dichromate resists tendering.

- A separate dyeing line is essential to eliminate repeated cleaning of machines and chances of uneven dyeing because presence of trace of alkali inhibits formation of jet black shade. Alkali, if present, converts jet black to brown or green shades. Soaping is to be done with non-ionic detergents.
- Handle of dyed cotton becomes stiffer due to filling of fibre pores with pigments.
- After treatment with CH_3COONa avoids tendering. Diphenyl black, chemically 4-amino diphenylamine, is costlier but free from any tendering.

13.5   Dyeing with diphenyl black

To produce diphenyl black (C I Oxidation base 2, C I 76085), cotton is padded with a solution of diphenyl amine (35 g), CH_3COOH (9°Tw, 130 g), lactic acid (50%, 45 g), $NaClO_3$ (30 g dissolved in 150ml water), $AlCl_3$ (52°Tw, 18 g), Cu_2S paste (30%, 10 g), ammonium vanadate (10 g/l) and a little of gum tragacanth (2%) to make 1 l solution (Allen, 1971). Padding and after-treatment are exactly same as those are applied for aniline black.

HCl used in dyeing plays dual role – it forms aniline hydrochloride and during ageing gets liberated to activate $NaClO_3$ to release oxygen required for oxidation. Excess of HCl facilitates formation of more hydrocellulose where lesser amount leads to incomplete development of colour. Optimum concentration of chlorates and chromates of sodium and potassium are to be used as oxidising agents; any excess use will increase chances of formation of oxycellulose. Ferrocyanides, vanadates and Cu_2S are some of the oxidation catalysts or oxygen carriers often used to form black shades.

13.6   Continuous dyeing with aniline black

Aniline black may be continuously developed on cotton using large volume of padding liquor. A huge stock solution may be prepared with aniline hydrochloride (80 kg), aniline (5 kg) and gum tragacanth (2%, 150 kg) in 200 l water in the first vessel. In second and third vessels, potassium ferrocyanide (50 kg) and $NaClO_3$ (30 kg) are dissolved separately in 200 l of water each. The three solutions are thoroughly mixed, pH is adjusted to 7 with CH_3COOH followed by its dilution to 1000 l. Cotton is padded with this solution, dried to green followed by ageing (100°C, 20 s) and chroming ($Na_2Cr_2O_7$, 3g/l, 50°C) as done with steam black; rinsing, soaping and washing complete the process.

During steaming, aniline is condensed and oxidized to form the black shade; a part of aniline is lost through its vaporization due to prolonged oxidation at higher temperature. The loss can be effectively reduced through reduction during oxidation or increasing rate of condensation of aniline. A catalyst like p-amino azobenzene in bath absorbs oxygen and gets oxidized to form p-amino azoxy benzene, which releases oxygen for rapid oxidation of aniline to black and p-amino azobenzene is regenerated for re-oxidation. Loss of aniline upto 20% can be saved through use of this catalyst.

Desired p-amino azobenzene can be prepared in the shop-floor level by reacting diazotized aniline with fresh aniline.

Application of p-amino azobenzene ensures a black shade during ageing itself through oxidation of more aniline molecules, though final chroming is essential to produce jet black. In absence of p-amino azobenzene, a deep green shade is formed after ageing.

In spite of taking all possible precautions, HCl dissociated during condensation degrades cotton. The risk can be reduced, if concentration of HCl can be lowered down or the process can be carried out at nearly neutral pH (Allen, 1971). Solanile Black (C I Oxidation base 3), chemically sulphamino derivatives of 4-aminodiphenylamine is applied in neutral steam to produce all round fast jet black shades, but costlier. The comparative features of various methods for production of aniline black are shown in Table 13.1.

Table 13.1 Comparative features of various methods for production of aniline black

Direct black	Aged black	Steam black
1. One bath exhaust method, slower in nature	1. Continuous padding and ageing method	1. Continuous padding and steaming method
2. Due to absence of affinity and substantivity, concentration of chemicals is higher.	2. Lack of affinity and substantivity are no problem. Less concentration of chemicals used	2. Lack of affinity and substantivity are no problem. Less concentration of chemicals used
3. Exhaustion is too poor causing a unlevelled green black and not a jet black	3. Levelled jet black is formed	3. Levelled jet black is formed

Table 13.1 Contd. ...

Direct black	Aged black	Steam black
4. Moderate damage of cotton occurs due to a relatively low temperature process	4. Typical damage of cotton due to use of strong oxidizing agent	4. Least damage of cotton due to use of weak oxygen carrier
5. Cost of installation is too less due to use of least number of machines	5. Heavy padding mangle and separate ageing machine are essential	5. Heavy padding mangle and high pressure steamer are essential
6. This is an exhaust process and aniline has no affinity for cotton, causes surface deposition and too poor rubbing fastness	6. A padding method and penetration of aniline is good causing moderate rubbing fastness	6. A padding method and penetration of aniline is adequate causing excellent rubbing fastness

13.7   References

ALLEN R L M (1971) *Colour Chemistry*, 1st Edition, London, Thomas Nelson & Sons Ltd.

KRAMRISCH B (1986) 'Dyeing cotton with Aniline Black', *Am. Dyestuff Rep.*, **75**, 3, 30–31, 41.

SHENAI V A (1987) *Chemistry of dyes and Principles of dyeing*, 3rd Edition, Mumbai, Sevak Publications.

VENKATRAMAN K (1952) *The chemistry of synthetic dyes*, Volume-II, New York, Academic Press.

14

Dyeing with phthalocyanine dye

Abstract: Phthalocyanine dyes are popular to develop cheaper brilliant blue shades on cellulosics. The shades produced are exceptionally resistant to all acid, alkali, oxidising and reducing agents. Phthalic anhydride, in combination with other chemicals, specifically copper complex form the blue hue where four phthalocyanine molecules remain attached with one copper atom. The complete process includes, preparation of liquor, padding, drying, curing and developing. However, the dyeing technology is complicated and requires huge attention.

Keywords: Phthalocyanine, copper complex, padding, polymerizing

14.1 Formation of phthalogen blue

Phthalogen blue is a lake or ingrain dye and is neither available nor applied as such, rather developed *in situ* cotton through reaction of a few chemicals followed by development under suitable conditions. The constituent chemicals applied for the purpose do not have affinity for cotton and so are applied essentially by padding. A subsequent controlled drying, curing or polymerizing, development followed by soaping and washing completes the process. Due to lack of affinity, concentration of pad liquor remains unchanged and high throughout padding.

The main component, water insoluble amino-imino-isoindole, formed through reaction between phthalic anhydride and urea, is emulsified into a fine dispersion with a solvent followed by addition of a copper complex. Cotton fabric is padded with this emulsion, dried and cured at 150°C for 4–5 min when a chelated cyclic product copper phthalocyanine is formed as shown in Fig. 14.1. Dyed fabric is treated with $NaNO_2$ and HCl to remove various byproducts formed and to develop true tone of hue (Schofield, 1978; Allen, 1971).

14.2 Chemistry of phthalocyanine dyes

Phthalocyanine based colouring materials are highly useful in developing blue and green hues, which are the most stable organic colourants due to

159

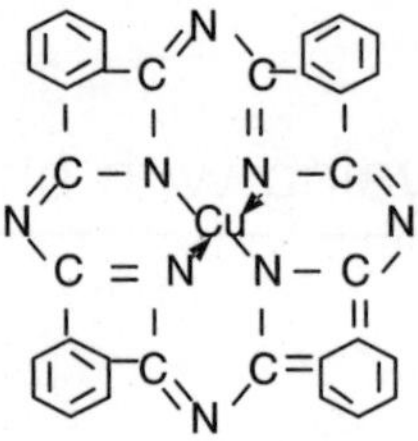

(Phthalogen Brilliant Blue IF3G, C I Ingrain Blue 2, C I 74160)

14.1 Formation of phthalogen blue complex.

extensive structural resonance stabilization. It is resistant to almost all acids, alkalis, oxidizing and reducing agents; planner and tetraaza derivative of tetrabenzoporphin, consisting of four isoindole units connected by four nitrogen atoms that form together an internal 16-membered ring of alternate carbon and nitrogen atoms. The blue hue is somehow dull and a brilliant intense hue with λ_{max} 670–685 nm is developed only when it reacts with a metal atom, preferably that of copper, known as copper phthalocyanine which contains three benzonoid and one p-quinonoid outer rings (Christie, 2001).

Different stages of formation of phthalocyanine blue can be described as a chain of reactions ultimately to coordinate a copper atom by four phthalocyanine molecules. Phthalic anhydride on reaction with urea produces phthalamic acid, which on heating above 140°C gives phthalimide. The complete scheme of formation of complex is shown in Scheme 14.1 (Allen, 1971; Booth, 1977; Lifentsev et al, 1973).

14.3   Properties

In copper phthalocyanine, a copper atom is chelated by four phthalocyanine molecules firmly through balanced co-ordination imparting compactness (Gruen, 1972). It is highly resistant to action of chemicals and is characterized by a highly tinctorial brilliant blue almost fast to alkali, acid, light, heat, etc.; moderately soluble in boiling solvents, such as chloronapthalene and quinoline. It is invariably used in chemical industries to produce paints, plastics, printing inks, etc.; in resist printing through emulsion technique and in continuous colouration of cellulose. Copper phthalocyanine sublimes without decomposition at 550–580°C, dissolves in conc. H_2SO_4 and original colour is restored on dilution. Sulphonation of copper phthalocyanine converts it to water soluble direct dye, e.g. Durazol Blue 8GS (ICI), shown in Fig. 14.2, is fast to light but not to wash (Allen, 1971). It is applied to cellulose, wool, silk, etc., but it is its poor wash fastness which has led to its maximum use in colouration of paper only.

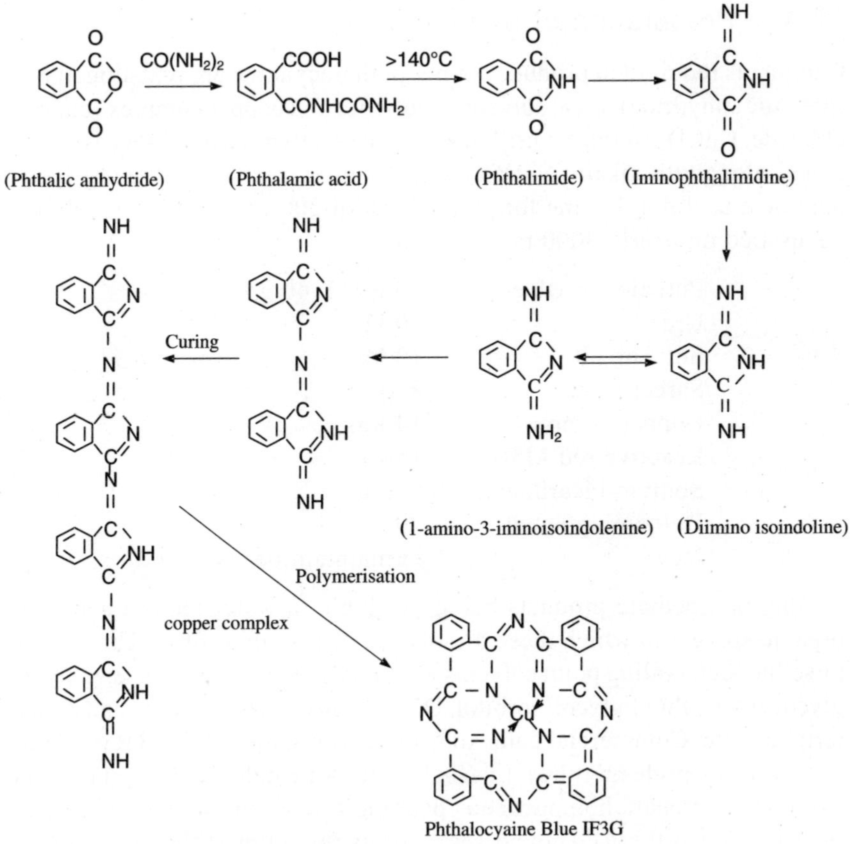

Scheme 14.1 Mechanism of formation of phthalogen blue complex.

14.2 Durazol Blue 8G (C I Direct Blue 86, CI 74180).

14.4   Application

The application technique consists of a few steps, viz. (i) preparation of pad liquor, (ii) padding of cotton, (iii) drying, (iv) polymerizing/curing, (v) developing and (vi) soaping and washing (Hiuke et al, 1994 , 1996; Zysman, 1971).

14.4.1 Preparation of pad liquor

Chemicals required to produce copper phthalocyanine are phthalogen blue (phthalic anhydride), urea, solvent, emulsifier and copper complex (cuprous chloride, CuCl). To impart brilliancy, little reactive red dye is occasionally added alongwith alkali (NaHCO$_3$) in the mixture. The following recipe may be a useful guideline for preparation of 200 l pad liquor which may be applied on nearly 3000 m of poplin.

Phthalogen Blue	– 7 kg (35–40 g/l)
Urea	– 20 kg
Solvent TR	– 10 kg
Sarcet paste	– 8 kg
Copper complex	– 14 kgs
Reactive red M5B	– 150 g
Sodium bicarbonate	– 300 g
Defoaming agent	– 200 g
Ice	– to maintain below 20°C

The intermediate products being insoluble in water require a suitable organic solvent in which these dissolve to diffuse in to fibre. These water miscible high boiling point solvents also act as reducing agents, e.g. glycol, glycol ethers, thiodiglycol, sorbitol, DMF, formyl methyl aniline and diethyl tartarate, etc. Commercial name of such solvents are Solvent TR (ethylene glycol–formamide mixture), Levasol P, etc. An emulsifier (sarcet paste) is essential to prepare homogeneous padding liquor and to retain chemicals in dispersion as these do not possess affinity for cotton (Eibl, 1975; Jordan, 1975).

In practice, he pad liquor is prepared through the following cumbersome sequences, viz.

(i) In 1st container, 200 l of water is cooled with sufficient ice.

(ii) In 2nd container, 20 kg urea is dissolved in water followed by cooling with ice.

(iii) In a 3rd container, Solvent TR (10 kg) is mixed with Sarcet paste (8 kg) and stirred for 5 min with an electrical stirrer.

(iv) Approximately half of cold urea solution is added to Solvent TR–Sarcet paste mixture with constant stirring followed by addition of phthalogen Blue (7 kg) alongwith Reactive Red M5B and is stirred.

(v) Rest urea solution is added succeeded by addition of copper complex and stirred

(vi) Finally NaHCO$_3$ and defoaming agent are added and stirred for 5 min for thorough mixing

(vii) The final volume is adjusted to 200 l with ice cold water and excess ice is added to maintain temperature below 20°C. The padding liquor is stable at or below 20°C for 3–5 h.

The sequence of addition may be changed a little bit e.g. after adding solvent TR and sarcet paste, phthalogen blue is added followed by whole of the urea solution, copper complex, red M5B, NaHCO$_3$ and defoaming agent. However throughout mixing, an electrical stirred is to be preferably taken in to action. The pad liquor has limited storage stability even after cooling and so can not be stored rather consumed.

14.4.2 Padding of cotton

Well prepared dry bleached cotton fabric, free of metallic deposits with good absorbency (~5 s), is padded with phthalogen blue solution at 70–80% thorough pick-up through a 3 bowl 2-dip 2-nip padding mangle, best synchronized with a float drier, pin stenter or hot flue for immediate drying of padded cotton.

14.4.3 Drying

Padded cotton is dried uniformly at around 100°C in crease free state with low air velocity and moderate heat with no contact with metal parts. Any chances of migration are to be minimized by maintaining same temperature on both side of fabric. It is to be ensured that no water drops fall on the dried cloth before curing, as this leads to lighter spots.

14.4.4 Curing / Polymerizing

Dried fabric is cured at 150°C for 4–5 min in a polymeriser. Increase in temperature beyond this produces a charry on the fabric with dulling of shade while curing below 150°C does not produce good colour value and dyeings are greener in tone. The curing unit should have an interlocked exhaust and it should be sufficiently powerful to displace all the fumes from the chamber, even at increased speed.

14.4.5 Development and soaping

After curing, the cloth is developed in a jigger to remove unreacted chemicals as well as by-products to produce a brilliant blue shade. Cotton fabric is treated with NaNO$_2$ (1 g/l) at 70°C for two ends followed by addition of HCl (5 g/l) with four more ends in jigger. Development is succeeded by rinsing and soaping at boil with soap (2 g/l) and soda ash (1 g/l) with four ends followed by washing. Development in open soaper requires five times of HCl (25 g/l) and NaNO$_2$ (5 g/l).

The copper complex releases copper only during curing, so that intermediate product diffuses inside fibre before the big cyclised copper phthalocyanine is formed (Blagrove et al, 1973). Any reduction in quantity of solvent may cause floating of particles as well migration during drying.

14.5   Phthalocyanine green

Though this can be produced through change in metal, the green shade is not so brilliant and fast. Commercially, a brilliant green is produced by dyeing cotton with copper phthalocyanine followed by topping with a yellow colour from other series, e.g. Naphthol ASG in combination with yellow GC/RC base or Vat yellow G or Reactive yellow MGR dyes, etc.

14.6   Precautions to be taken

From preparation of liquor to development, every stage of dyeing with phthalocyanine dyes is complicated, unhygienic and cautious approach is essential for development of levelled true colour.

- Mask should be used during preparation of pad liquor to escape inhalation of pungent smell of solvent and fly-outs of phthalic anhydride.
- The pad liquor is stable for at most 5 h at or below 20°C; it should be used without late to avoid precipitation of phthalic anhydride and consequently to avoid loss in colour value.
- Pre-treatment of cotton fabric should be efficient to show good absorbency (≤ 5 s) for good colour strength of dyeings.
- As the components have no affinity for cotton, only padding method is suitable.
- Padded goods are to be dried at or just below 100°C immediately otherwise separation of components starts; drying at higher temperature starts partial polymerization with loss in shade depth due to simultaneous drying and curing.
- Falling of water drops on white or padded goods develops light spots.
- Care should be taken to avoid charring or burning of fabric with no leakage in polymerizer for radiation of heat during curing at 150°C.
- Thorough development of fabric may be ensured by treating with $NaNO_2$ first followed by addition of HCl. If both are added simultaneously, a part of HNO_2 will escape in to atmosphere and will not be available for development.

14.7   References

ALLEN R L M (1971) *Colour Chemistry*, 1st Edition, London, Thomas Nelson & Sons Ltd.

BLAGROVE R J, GROSSMAN V B and GRUEN L C (1973) 'Copper analysis of phthalocyanine dyes', *J. Soc. Dyers Color.*, **89**, 1, 25.

BOOTH G (1977) 'Manufacture of dyes and intermediates', *Rev. Prog. Color Related Topics*, **8**, 1–10.

CHRISTIE R M (2001) *Colour Chemistry*, Great Britain, Royal Society of Chemistry.

EIBL J (1975) 'New opportunities for the application of Phthalogen dyestuffs without solvents', *Bayer Farben Revue*, **24**, 47–57.

GRUEN L C (1972) 'Aggregation of copper phthalocyanine dyes', *Australian J. Chem.*, **25**, 8, 1661–1667.

HIUKE T, INOUE Y and SOGA H (1994) 'Dyeing methods to produce deep dyeings with phthalocyanine dyes', Sandoz Ltd, GB 2 275 693 A.

HIUKE T, INOUE Y and SOGA H (1996) 'Dyeing methods to produce deep dyeings with phthalocyanine dye', Sandoz Ltd, USP 5 484 456.

JORDAN H D (1975) 'Acramin and Phthalogen dyestuffs in piece-dyeing, *Bayer Farben Revue*, **24**, 11–30.

LIFENTSEV O M, EGOROV NV and MEL'NIKOV B N (1973) 'Mechanism of formation of copper phthalocyanine on the fibre in dyeing of cellulosic fabrics', *Tekh. Tekstil'noi Prom.*, **95**, 4, 83–87.

SCHOFIELD M (1978) 'Fifty years of phthalocyanines', *Int. Dyer*, **159**, January 6, 24.

ZYSMAN J (1971) 'Phthalocyanine dyes', *Ind. Textile*, **997**, January, 49–54.

15

Dyeing with acid dye

Abstract: Acid dyes are applied from acidic bath, are anionic and soluble in water; applied on wool, silk and nylon when ionic bond is established between protonised amine group of fibre and acid group of dye. Overall wash fastness is poor, though light fastness is quite good. Due to opposite electrical nature of dye and fibre, strike rate and uptake of acid dye on these fibres is faster; electrolyte at higher concentration is added to retard dye uptake and to form levelled shades. Acid generates cation on fibre and temperature helps to substitute negative part of acid with anionic dye molecules.

Keywords: Acid, affinity, electrolyte, levelling, fastness

15.1   Basic principle

Acid dyes are mostly sulphuric or carboxylic acid salts and are essentially applied from an acidic bath and hence the name 'acid dye'. The dye anion is the active coloured component in these dyes; invariably synthesized as sodium salts as free dye acids are more difficult to isolate. Dyes possess affinity for protein fibres and are used in dyeing of wool, silk and nylon; ionic bond is formed to retain dye with fibre. Fastness properties vary with chemical constitution as well as application class of dyes.

15.2   Classification

15.2.1 Based on chemical nature

Chemically, acid dyes are classified as (i) Monoazo, e.g. Metanil Yellow and Acid Orange II; Bisazo, e.g. Red G and 2B, (ii) Nitro, e.g. Picric acid, (iii) Nitroso, e.g. Naphthol Green B, (iv) Triphenylmethane, e.g. Acid Magenta, (v) Xanthene, e.g. Rhodamine B, Acid Violet B, (vi) Azine, e.g. Wool fast Blue EB, (vii) Quinoline, e.g. Quinoline Yellow PN, (viii) Ketonimine, e.g. Sulfonine Grey G, (ix) Anthraquinone, e.g. Carbolan Green G, Kiton fast Blue 4GL, Polar Blue 4GL, Solvay Violet BN, Alizarine light Blue SE, Alizarine light Red R and (x) Phthalocyanine, e.g. Coomassie Turquoise Blue 3G (Trotman, 1994; Shenai, 1987).

15.2.2 Based on application

From application point of view, acid dyes are of three types, viz. (i) strong or levelling or equalizing acid dyes, (ii) milling or weak acid dyes and (iii) very weak or super milling or neutral or aggregated acid dyes.

Strong acid dyes

Dyes are adsorbed rapidly at a pH 2–3 even at 40°C and are applied from a dye bath containing H_2SO_4 (3.5%, 168°Tw) and glauber's salt (10–20%). Goods are entered at 60°C in dyebath, raised to boil in 30 min and continued at this temperature for further 45–60 min. These dyes are also known as 'levelling acid dyes' or 'equalizing acid dyes' due to their capability to produce mostly levelled shades, possess poor wash fastness because of their low molecular weight with higher water solubility which can be improved through suitable post-dyeing treatments.

Milling acid dyes

Dyes belonging to this group possess better wash fastness than that with strong acid dyes but do not migrate well and are called 'milling acid dyes' as these are fast to milling process imparted to woollen textiles. Dyebath is set with CH_3COOH (1–3%) and glauber's salt (10%). Dyeing is started at 50°C with pH around 4.5–5.5, raised to boil in 45 min and kept at boil for further 30–60 min at pH 5.2–6.2. Wash fastness of dyeings is good due to high molecular weight of dye rendering low solubility. Levelling is moderate as affinity of dye anion for fibre is quite high.

Super milling acid dyes

Dyes are applied from a neutral bath with acid liberating agent, e.g. ammonium acetate, sulphate or phosphate which liberates respective acid beyond 80°C to develop required pH. Dye anions possess higher affinity for fibre even at neutral pH requiring a minimum of acid. These are commonly known as 'super milling dyes' due to their high fastness to milling. Dyeing is started at 60°C with ammonium acetate (CH_3COONH_4) or ammonium sulphate (2–5%) at pH around 6–7, raised to boil in 45 min and dyeing is continued at boil for further 45 min. Dyes possess poor levelling property due to very high affinity and consequent higher strike rate for fibre. Wash fastness is excellent due to larger molecular size of dye due to which remains insoluble in water. These are also known as 'aggregated acid dyes' due to their availability in the form of aggregates, necessitating dyeing at higher temperature by breaking down these aggregates to individual dye molecules. Characteristics of various types of acid dyes are summarized in Table 15.1 (Vickerstaff, 1950).

Table 15.1 Characteristics of various classes of acid dyes

Criteria	Levelling dyes	Miling dyes	Super-milling dyes
Wash fastness	Poor	Good	Excellent
Acid system	H_2SO_4	CH_3COOH	CH_3COONH_4
pH of dyeing	2-4	4-6	6-7
Levelling property	Good	Moderate-poor	Very poor
Dye characteristics	Low molecular weight with very high solubility, molecular solution	High molecular weight with low solubility, colloidal solution	High molecular weight with very low solubility, colloidal solution
Affinity of anions	Low	High	Very high

15.3   Mechanism of dyeing

Wool retains each of $-NH_2$ and $-COOH$ groups at either ends which are capable of taking part in chemical reaction. Based on this chemical structure of wool may be better represented as $H_2N-W-COOH$, where 'W' denotes rest part of wool structure.

When wool is immersed in water, H-atom attached to the carboxylic group at one end of wool is transferred to $-NH_2$ group at the other end of the macromolecules so that the two ends of wool chain acquire opposite electrical charges, called zwitter ions:

$$H_2N - W - COOH \xrightarrow{H_2O} H_3N^+ - W - COO^-$$

On addition of acid in bath containing wool with zwitter ions, some of the negatively charged carboxylate ions ($-COO^-$) take up hydrogen ions released by acid into solution and are transformed to electrically neutral carboxylic acid groups ($-COOH$). At the same time, the acid anions, released from acid are absorbed by the positively charged amino ends of keratin macromolecules.

$$CH_3COOH \leftrightarrow H^+ + CH_3COO^-$$
$$H_3N^+ - W - COO^- + H^+ CH_3COO^- \leftrightarrow H^+_3N\ CH_3COO^- - W - COOH$$

Amount of acid absorbed (g acid / kg of fibre) by wool increases with time, until it reaches to a specific value called equilibrium sorption and thereafter remains unchanged. Cationic nature of wool increases with increase in time of acid treatment with generation of more and more positive sites ($-NH_3^+$) in wool through absorption of excess acid.

When an acid dye (R-SO$_3$Na) is added to this bath containing cationised wool, the dye anion gets attached with cation of wool through electrostatic force with liberation of salt.

$$H_3N^+ CH_3COO^- - W - COOH + R- SO_3^- + Na^+ \rightarrow R\text{-}SO_3^- {}^+H_3N \text{-}W\text{-}COOH + CH_3COONa$$

Some of the groups present in dye are bound to wool by hydrogen bonds too. Since the dye anion is held on the protein molecules more strongly than the acetate anion, the acetate ions taken up by the positive sites prior to addition of dye are continuously replaced by dye anions. As dyeing proceeds, transfer of different ions from solution to wool and from wool to solution takes place.

Dye uptake also depends on nature and concentration of acid (strong or weak) in bath. With increase in concentration of acid in bath, exhaustion of dye bath also increases. H$_3$PO$_4$ has been recommended for dyeing of wool with acid dyes in place of H$_2$SO$_4$ [1% H$_2$SO$_4$=2% H$_3$PO$_4$(75%)] or CH$_3$COOH [1%CH$_3$COOH (80%) = 0.8% H$_3$PO$_4$ (75%)] as H$_3$PO$_4$ causes less damage to wool and any cellulosic fibres present in blend, if buffered with Na$_2$CO$_3$, H$_3$PO$_4$ also produces better levelled shades (Trotman, 1994).

Acid dyes possess hydrophobic or non-polar head (benzene rings) and hydrophilic or polar tail (–COOH, –SO$_3$H). In contrast, wool contains a hydrocarbon or non-polar backbone (keratine) which feels higher affinity for the non-polar head of dye causing higher affinity of acid dye for wool; dyeing is thus analogous to extraction of non-polar solvent from an organic solvent. It also has been found that introduction of non-polar groups in dye structure, such as –CH$_3$, increase its affinity for wool many-folds while introduction of polar groups (–COOH, –SO$_3$H) reduce affinity. In addition to a electrostatic bond (NH$_3^+$D$^-$), the dye anion is attached to wool with co-ordinate linkage – the strength of which increases from equalizing dyes to aggregated dyes causing improved fastness when the latter is used (Rabinowitz, 1984; Shahed et al, 1983; Rush, 1980). Kinetics of dyeing wool with acid dyes and nylon has been disclosed via existing explanations of wool–dye isotherm data, i.e. Gilbert-Rideal model, Donnan model and the law of mass action (Bruce et al, 2000; Bird and Boston, 1975).

15.4   Selection of dye

This is based on fastness and levelness required. Strong acid dyes easily develop well levelled shades with poor wash fastness, while milling and super-milling dyes produce fast shades. If fastness is to be improved in post-dyeing stage through after-treatment with metal salts, dyes with chelating sites are preferred.

15.5   Influence of dyeing parameters

15.5.1 Effect of electrolyte

Electrolytes act mainly as retarding or levelling agent for strong acid dyes (Yiqi, 1998):

$$CH_3COO^- \, {}^+H_3N\text{-W-COOH} + R\text{-}SO_3Na \leftrightarrow RSO_3^- \, {}^+H_3N\text{-W-COOH} + CH_3COONa$$

Dyeing proceeds with elimination of CH_3COONa. At the initial stages of dyeing, due to higher attraction between cation of fibre and dye anion, more and more salt will be liberated due to faster dye–fibre attachment creating chances of uneven dyeing. A levelled shade can only be produced through reduction in formation of salt in bath by adding excess salt in bath at the start of dyeing to ensure backward process, i.e. stripping; higher the concentration of salt added, higher is the rate of stripping. In other words, if an excess of salt is added to bath, stripping is favoured and it acts as retarding or levelling agent.

For super-milling dyes, which are often applied from a neutral bath, salt plays opposite effect (similar effect as dyeing of cellulose with direct dyes), i.e. salt promotes dye uptake on protein fibres by reducing zeta potential. Wool probably acquires a negative electrical potential in neutral solutions, which normally repels the negatively charged acid dye anions present in dye bath. Presence of an electrolyte reduces this surface negative charge difference between dye and fibre through absorption of positive sodium ions released form salt.

15.5.2 Effect of acid

Levelling and milling acid dyes are not taken up by protein fibre unless acid is added to bath. As stated earlier, immersion of protein fibre in water transfers the terminal H atom from $-COOH$ to another end retaining $-NH_2$ group. During this process, carboxylic ion is formed at one end of fibre (COO^-) while amine ion (NH_3^+) at the other making the fibre oppositely charged (zwitter ion). Addition of acid makes the fibre cationic in nature through attachment of H atom of acid to the COO^- ion making it COOH again. Absence of acid in bath, marked by pH at or above 7.0 will cause repulsion of more dye anions with little dye uptake, as shown in Fig. 15.1 (Shenai, 1987).

Not only the total amount of the dye adsorbed is influenced by amount of acid, but the rate of exhaustion is also dependent upon acidity or pH of bath. When different acids are used in equivalent quantities, these produce similar extent of exhaustion on wool only if dyeing time is short; if dyeing is done below boil then weaker acid is preferred in larger quantities than the stronger ones at lower concentration.

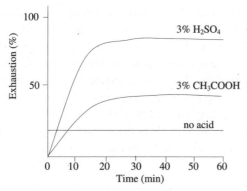

15.1 Rate of dyeing in presence and absence of acid with Solway Blue BS.

15.5.3 Effect of temperature

Protein fibres are ionized when dipped in bath.

$$H_2N - W - COOH \xrightarrow{H_2O} H_3N^+ - W - COO^-$$

Due to this ionization, fibres possess inherent negative tendency to absorb acid when the acid is added to bath ; anionic COO^- group at one end attracts H^+ released by acid and gets neutralized thereby making the fibre highly cationic in nature (explained earlier). Acid dyes added to bath can attach themselves with cationic site of fibre through replacement of anion released by acid (CH_3COO^-) which remains attached with the cationic site of fibre.

$$H_3N^+CH_3COO-W-COOH + R- SO_3^- + Na^+ \rightarrow R-SO_3^- {}^+H_3N -W-COOH + CH_3COOH$$

This replacement is not possible at room temperature but only when the bath is heated up causing acceleration of dye molecules to generate the required momentum. Efficient dyeing results are obtained if dyeing is started at about 40°C, raised slowly to boil and dyeing is further carried out at boil for desired time; slow rise in temperature causes replacement of CH_3COO^- ions at a specific rate avoiding unlevelled dyeing. Acid dyes are not transferred from bath to fibre below 39°C; beyond this temperature, rate of adsorption increases and the trend varies from one type of dye to other. Milling acid dyes have a minimum temperature of exhaustion at 60°C, but at 70°C, transfer of dye is fast. Super-milling dyes cause level dyeing only at boil.

15.5.4 Effect of levelling agents

Levelling agents are occasionally used in dyeing with acid dyes (Riva et al, 1996; Shenai et al, 1981). Pyridine can break and stop aggregation of dye molecules promoting levelling through formation of pyridine-dye complex which retards ionization of dye and releases dye from complex at constant rate. Non ionic surface active agents can break down dye aggregates and retain single dye molecules in well dispersed form while anionics compete with acid dye for fibre ensuring level dyeing. Selective cationic-non ionic surface active agent mixtures can promote levelling – while cationics form complex with anionic dye molecule, the non ionic retain these complex molecules in dispersion; the complex is stable at room temperature but on heating it breaks down to release dye anions slowly for level dyeing.

15.6   Specification of some acid dyes

The C I specification and chemical structure of a few commonly used acid dyes are shown in Scheme 15.1 (Colour Index, 1987).

Acid Black 10A / 10B / 12B / 10BR
(Bluish Black)
C I Acid Black 1, C I 20470

Acid Brilliant Red GF
(Bright Bluish Red)
C I Acid Red 1, C I 18050

Acid Brown SRN
(Reddish Brown)
C I Acid Brown 15, C I 20190

Acid Orange II
(Bright Reddish Orange)
C I Acid Orange 7, C I 15510

Acid Yellow G (Greenish yellow)
C I Acid Yellow 18, C I 19020

Acid Scarlet 3R / 4R / 4RS (Bright Red)
C I Acid Red 18, C I 16255

Scheme 15.1 Contd. ...

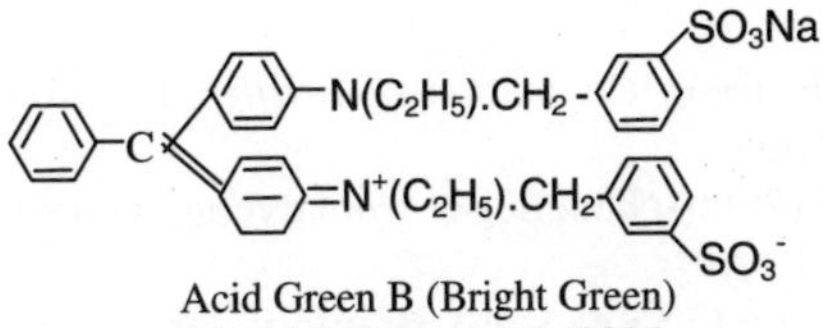

Acid Green B (Bright Green)
C I Acid Green 3, C I 42085

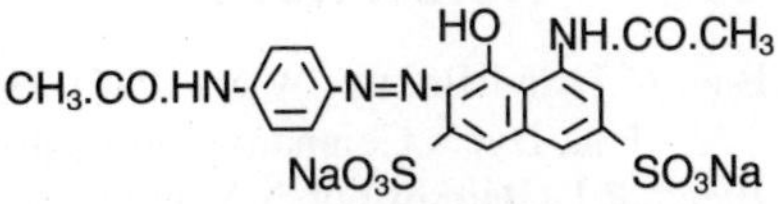

Acid Red 6B (Magenta)
C I Acid Violet 7, C I 18055

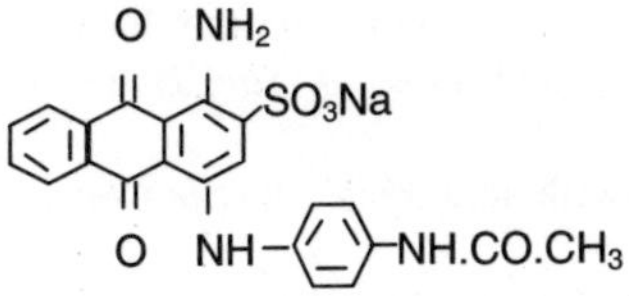

Acid Blue 2G (Greenish Blue)
C I Acid Blue 40, C I 62125

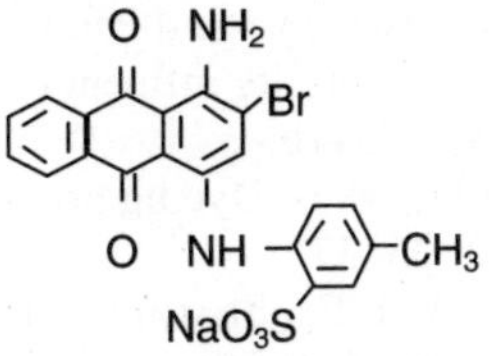

Acid Alizarine Blue B (Blue)
C I Acid Blue 78, C I 62105

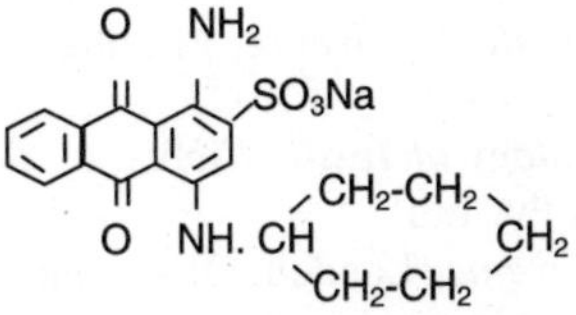

Acid Anizarine Blue R (Bright Reddish Blue)
C I Acid Blue 62, C I 62045

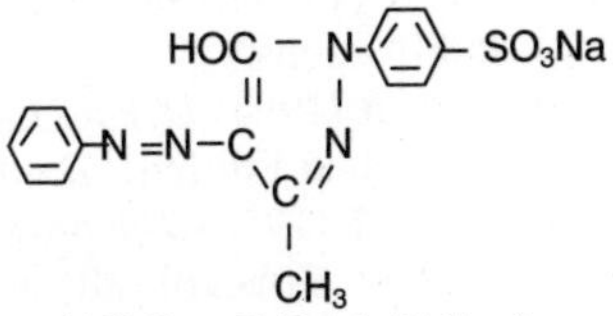

Acid Yellow G (Bright Yellow)
C I Acid Yellow 11, C I 18820

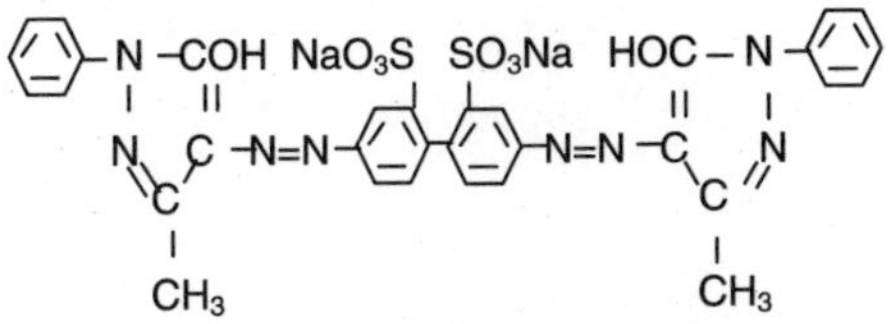

Acid Yellow R (Reddish Yellow)
C I Acid Yellow 42, C I 22910

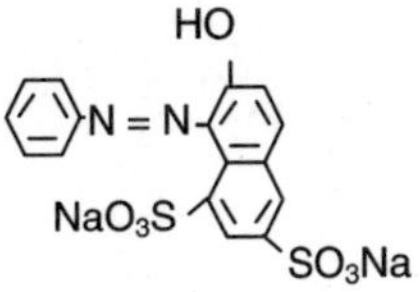

Acid Orange G (Bright Orange)
C I Acid Orange 10, C I 16230

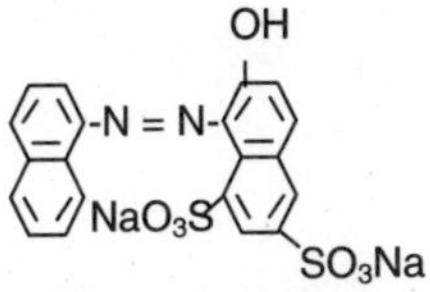

Acid Red 6A (Bright Red)
C I Acid Red 44, C I 16250

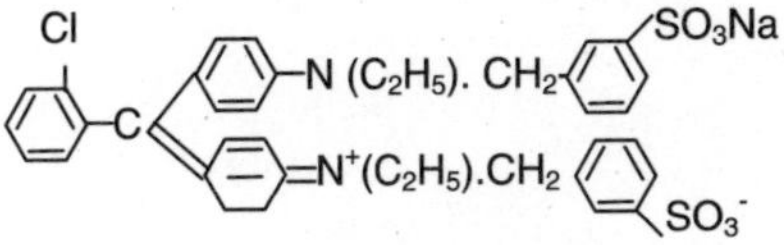

Acid Miling Green B (Bright Bluish Green)
C I Acid Green 9, C I 42100

Scheme 15.1 Chemical structure and C I specification of some important acid dyes.

15.7   References

BIRD C L and BOSTON W S (1975) *The Theory of Coloration of Textiles*, Bradford, UK, Dyers Company Publications Trust.

BRUCE R L, BROADWOOD N V and KING D G (2000) 'Kinetics of wool dyeing with acid dyes', *Textile Res. J.*, **70**, 6, 525–531.

Colour Index International (1987), 3rd Edition (3rd Revision), Bradford, UK, Society of Dyers & Colourists and AATCC.

RABINOWITZ B (1984) 'Dyeing with acid dyestuffs', *Am. Dyestuff Rep.*, **73**, 2, 26–35.

RIVA A, CEGARRA J and PRIETO R (1996) 'Sorption of nonionic auxiliary products by wool and its influence on dyeing with acid dyes', *J. Soc. Dyers Color.*, **112**, 4,114–116.

RUSH J L (1980) 'Dyeing primer-II: Dyeing with acid dyes', *Textile Chem.Color.*, 12, 2, 35–37.

SHAHED M F EL, SHALABY S E and KAMEL M M (1983) 'Dyeing properties of PAN/ nylon 6 copolymers-I: Dyeing with acid dyes', *Am. Dyestuff Rep.*, 72, 11, 42– 45, 49.

SHENAI V A, PANDIAN P S and PARWANA R S (1981) 'Use of sulphated alkyl oleates in dyeing with acid dyes', *Textile Dyer Print.*, **14**, 12, 10th June, 29–31.

SHENAI V A (1987) *Chemistry of dyes and Principles of dyeing*, Mumbai, Sevak Publications.

TROTMAN E R (1994) *Dyeing and Chemical Technology of Textile Fibres*, 6th Edition (1st Indian Edition), Delhi, B I Publications Pvt Ltd.

VICKERSTAFF T (1950) *The physical Chemistry of Dyeing*, London, Oliver and Boyd.

YIQI Y (1998) 'Effect of salts on physical interactions in wool dyeing with acid dyes', *Textile Res. J.*, **68**, 8, 615–620.

16

Dyeing with metal–complex dye

Abstract: Acid dyes possess smaller structure; in spite of formation of ionic bond with fibre, wash fastness remain on poor side causing staining of adjacent garments during washing. Selective acid dyes possessing chelating sides in structure may co-ordinate with various metal atoms to form metal–acid dye complex, named metal complex dye. Due to increase in overall size of dye, wash fastness is greatly improved with a problem of poor affinity and substantivity of dye for fibre.

Keywords: Acid dye, metal, metal complex dye, mordant dye, pre-metallised dye

16.1   Introduction

In dyeing wool with acid dyes, the ionic bond between $-NH_3^+$ and DSO_3^- is weaker and as a result, these bonds are easily broken and reformed under favourable circumstances allowing dye molecules to migrate. Stripping out of colour or staining of adjacent white wool during domestic washing remains a problem with acid dyeings. This nature of migration of acid dye can be reduced or arrested, if the bond strength between dye and fibre can be remarkably enhanced, i.e. the dye structure can be made sufficiently larger. Metal complex dyes are mostly acid dyes possessing chelating sites to enable these to be combined with metal atoms; invariably used for dyeing of wool, silk and nylon to produce fast shades (Shenai, 2002). The dye–metal complex, when produced during dyeing is called a mordant dye and when produced itself at the dye manufacturing plant, is called a pre-metallised dye. Structure as well as other technical details of metal complex dyes have been reviewed elsewhere (Szymczyk and Freeman, 2004).

16.2   Mordant dyes

A mordant is a simple chemical which possesses affinity for both fibre as well as dye. If a dye has less or no affinity for fibre then mordant already applied improves affinity of that dye and makes a dye–mordant–fibre complex. Mordant are basically metal salts and electrically cationic. In contrast, all acid dyes are anionic in nature and their affinity is increased many folds for fibre when mordant is applied on fibre before application of dye. Various natural dyes, e.g. logwood black, madder, etc., are also applied on wool with the help of mordant, e.g. alum, chrome, iron and tin salts – brighter colours and better fixation are obtained only by pre-mordanting. A few substantive or non-mordant dyes, e.g. turmeric, berberis, dolu, annato and henna contain tannin in the colouring matter itself as natural mordant and produce fast colours at boil.

In this complexion, out of six co-ordination sites of chromium, three are used by dye and rest three by water molecules in acidic / neutral pH or hydroxyl ions in alkaline pH (Race et al, 1946).

Poor wash fastness of dyeings with acid dyes (specifically strong acid dyes) is due to smaller size of dye molecule as the ionic attachment is weaker in nature. Acid dyes possessing chelating sites in its structure can be reacted with metal salts in various ways to develop bigger dye–metal complex through co-ordinate and covalent bond formation, is known as mordant dye (Fig.16.1). All mordant dyes are acid dyes but the reverse is not true. In simpler techniques, chromium salts are invariably used for complex formation and the reactions are carried out in shop-floor level to make dye–metal complex, also called chrome dyes. Chromium salts act as efficient mordant, e.g. $Na_2Cr_2O_7$, $K_2Cr_2O_7$, etc., and do not only promote dye–metal complex formation to improve fastness, also enhance acid dye

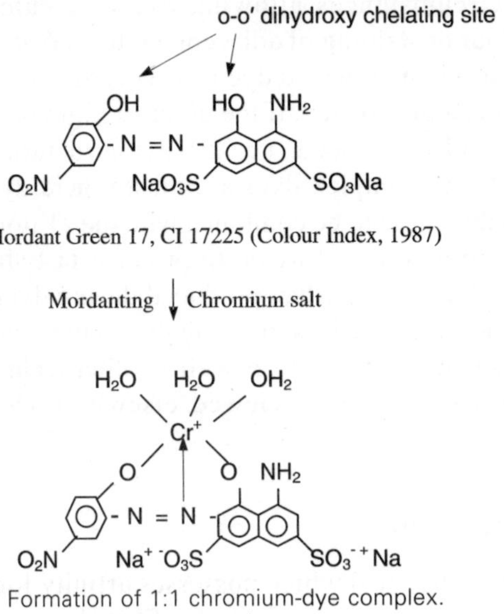

o-o' dihydroxy chelating site

CI Mordant Green 17, CI 17225 (Colour Index, 1987)

Mordanting ↓ Chromium salt

16.1 Formation of 1:1 chromium-dye complex.

uptake due to its cationic nature when applied on wool before dyeing (Gills, 1944). These mordant dyeing methods are useful tools for cottage industries to dye wool with superior wash fastness. A single mordant with different mordant dyes or vice-versa produces a long range of hues.

Mordant dyes are classified chemically as azo, anthraquinone, oxazine, xanthene, triphenyl methane, nitroso and thiazine types, in which only azo dyes cover whole range of spectrum and are of paramount importance (Shenai, 1987).

Alizarin

Alizarin, chemically 1, 2 dihydroxyanthraquinone, C I Mordant Red 11, CI 58000 (Colour Index, 1987) is one of the important anthraquinone mordant dyes, which is used with various mordants to develop different hues, e.g. aluminium (red), ferrous iron (deep violet), ferric iron (brownish black), stannous tin (reddish violet), stannic tin (violet) and chromium (brownish violet). It is a non-acid dye and has chelating site required for complex formation with mordants, frequently derived from madder (Allen, 1971).

Mordanting is taken into action before, after or concurrently with the dyeing process with efficient mordants. Depending on the stage at which mordants are used, chrome dyeing methods are classified as (i) onchrome, (ii) metachrome, and (iii) after-chrome.

16.2.1 Onchrome method

Chroming of fibre is done at the start followed by dyeing with selected mordant dye. Chromium in higher valency state is applied on woollen material, by treating in $K_2Cr_2O_7$ solution either in neutral condition known as sweet chrome or in acidic condition known as sour chrome method. Reduced chrome method is used for dyes which are discharged by oxidising agents.

In 'sweet chrome' method, wool is treated with $K_2Cr_2O_7$ solution (2%) at neutral pH at 60°C for 30 min, when chromium in higher valency state is absorbed by wool forming a wool–chromium complex consuming only half the amount of chromium. In 'sour chrome' process, $K_2Cr_2O_7$ (1–2%) is applied with H_2SO_4 (1%) at 60°C for 30 min, when all the chromium is absorbed by wool. In both these cases, chroming is succeeded by dyeing with selective mordant dyes.

'Reduced chrome' method is especially useful for dyes susceptible to oxidation. The process is carried out using $K_2Cr_2O_7$ (2%), lactic acid (3–4%) at 60°C for 15–20 min.

All 'on chrome' methods are a two bath process, expensive, usually only one lot of material can be chromed, washed-off and dyed in a day; more suitable for light and medium shades. In contrast, matching of colour is relatively easy, regular building of shade occurs and not complicated by sharp changes in shade of material during dyeing.

16.2.2 Metachrome method

Milling and super-milling acid dyes with chelating sites are most suitable for this process, which exhaust well at pH around 6–8.5. In this single bath process, dye and mordant are simultaneously applied from the same bath. In spite of being electrically opposite in nature, dye and mordant do not react initially rather both are exhausted on the substrate after which dye–mordant reaction occurs *in situ* under favourable conditions to develop the hue. If conditions are not maintained properly during exhaustion, dye and mordant react in bath and get precipitated.

Chromium is deposited on or combines with fibre followed by reduction of CrO_3 to Cr_2O_3 at boil. At the same time, dye combines with wool in the same way as that with a milling acid dye. Dye and reduced chromium then combine with fibre to give dye– chromium–wool complex. All three reactions probably occur simultaneously in bath (Shenai, 1987).

$$(NH_4)_2SO_4 \rightarrow H_2SO_4 + 2NH_3$$
$$2Na_2CrO_4 + H_2SO_4 \rightarrow Na_2Cr_2O_7 + Na_2SO_4 + H_2O$$
$$Na_2Cr_2O_7 + H_2SO_4 \rightarrow Na_2SO_4 + 2Cr_2O_3 + H_2O$$
$$Cr_2O_3 + dye \rightarrow Cr\text{-}dye\ lake$$

Chances of formation of other metal–dye–fibre complex due to presence of other metals in water used for dyeing rather than with chromium is prevented by chromates which form protective oxide film on the surface of the metal and does not allow it to take part in reaction.

Dyebath contains dye, $(NH_4)_2SO_4$, Na_2CrO_4 (sodium chromate) or $Na_2Cr_2O_7$ usually. $(NH_4)_2SO_4$ promotes deposition of CrO_3 on wool without damaging it and generates a pH around 6–8.5.

$$Na_2CrO_4 + H_2O \rightarrow 2NaOH + CrO_3$$
$$2NaOH + (NH_4)_2SO_4 \rightarrow Na_2SO_4 + 2H_2O + 2NH_3$$
$$Na_2CrO_4 + (NH_4)_2SO_4 \rightarrow Na_2SO_4 + 2NH_3 + CrO_3 + H_2O$$

Due to generation of more and more NH_4OH, bath is likely to be more alkaline with time in a closed dyeing machine.

Pre-treated wool is treated with chrome mordant (2–5%), glauber's salt (5–10%) and a surfactant (0.5 g/l) at 40–50°C for 20 min. Dissolved dye is added, temperature is raised to boil over a period of 45 min and treatment is continued at this temperature for further 60–90 min. Complete exhaustion may be affected by adding 0.5–1% of CH_3COOH (40%) or half the amount of HCOOH (85%), 30 min before dyeing is completed. $(NH_4)_2SO_4$ may be substituted with CH_3COONH_4 when using dyes known to cause uneven dyeing. CrO_3 is reduced to Cr_2O_3 by wool during this process.

Metachrome process is a single bath process, simple to apply with higher rate of production. Shade matching is easier because the final shade is produced during dyeing. Chromium does not form complex with dye in chromate form and so chromate and dye can be applied from the same bath; pH of dye bath is around 6–8.5 causing minimum damage to wool too. Dulling of shade due to iron or copper is absent. One pre-requisite is that dyes which do not react with Cr^6 can be used in this method.

However, range of shade is limited as only a very few dyes are suitable for this method. Uneven dyeing can not be rectified. In deep shades, the dye is apt to rub due to precipitation of it. Due to relatively higher pH of bath, exhaustion of dye remains incomplete, particularly with deeper shades; blacks, navy blue and other deep shades are seldom dyed in this method (Trotman, 1994).

16.2.3 After-chrome method

Chrome dyes are first applied on wool in presence of acid followed by $K_2Cr_2O_7$ (1–2%) in the same bath at 60–80°C. Dyeings show superior fastness to milling and potting (Hannemann, 2001). Shade matching is not easy and any change in shade at the last moment causes trouble.

16.3   Specification of some mordant dyes

Colour Index specification and chemical structure of a few mordant dyes are illustrated in Scheme 16.1 (Colour Index, 1987).

All dye baths used in dyeing with mordant dyes are mostly two-bath or two-stage processes. Produced shades lack brightness, shade change is evident causing trouble in matching, maintenance of pH, poor rubbing fastness due to deposition of a part of complex on dyed fibre, etc. List of dyes suitable for various crome methods have been detailed elsewhere (Shenai, 1987).

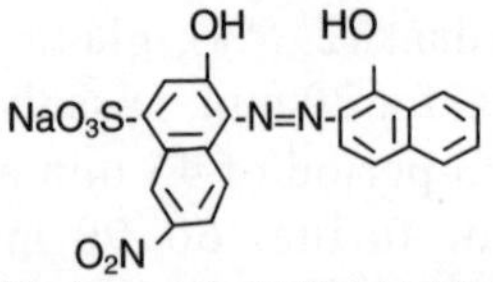

Eriochrome Black T (Bluish black)
C I Mordant Black 11, C I 14645

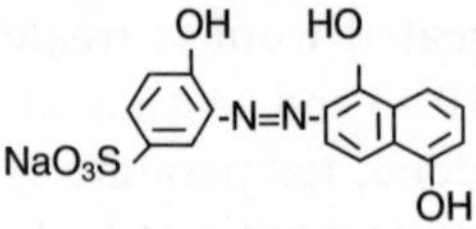

Atlantichrome Black PV (Black)
C I Mordant Black 9, C I 16500

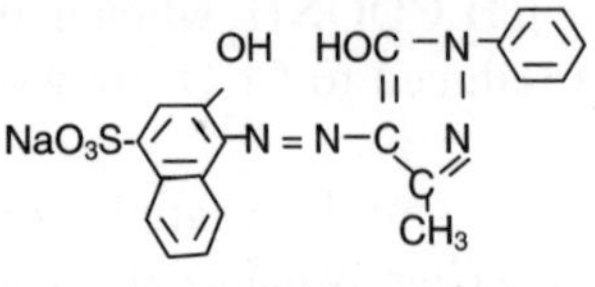

Acid Chrome Bordeaux B (Bluish red)
C I Mordant Red 7, C I 18760

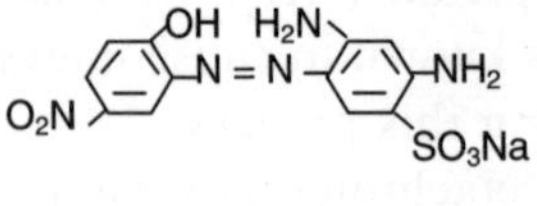

Acid Chrome Brown RH (Yellowish brown)
C I Mordant Brown 33, C I 13250

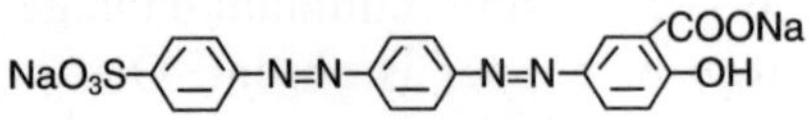

Acid Chrome Orange G (Reddish orange)
C I Mordant Orange 6, C I 26520

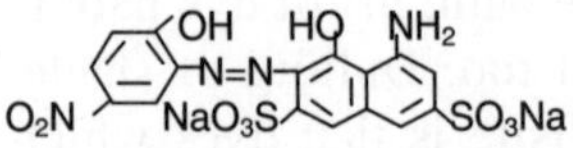

Acid Chrome Green G (Bluish green)
C I Mordant Green 17, C I 17225

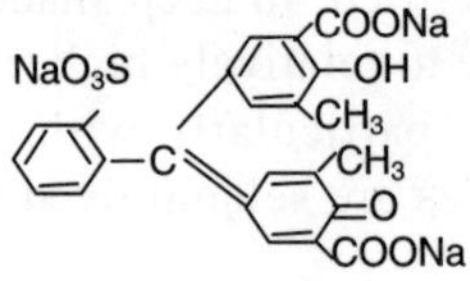

Chrome Cyanine RLA (Bright reddish blue)
C I Mordant Blue 3, C I 43820

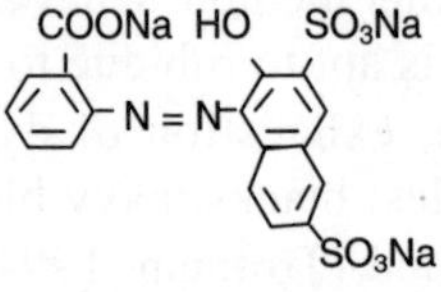

Acid Chrome Red B (Bluish red)
C I Mordant Red 9, C I 16105

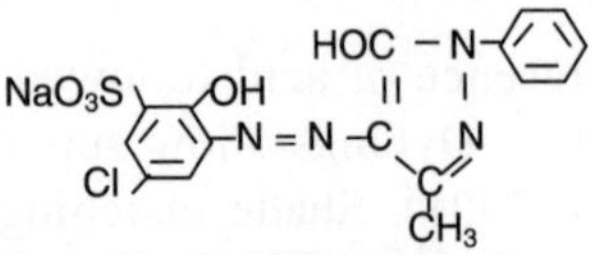

Chrome Fast Red GW (Bright red)
C I Mordant Red 19, C I 18735

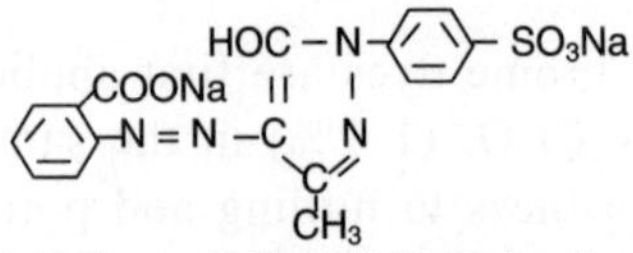

Acid Chrome Yellow FR (Reddish yellow)
C I Mordant Yellow 8, C I 18821

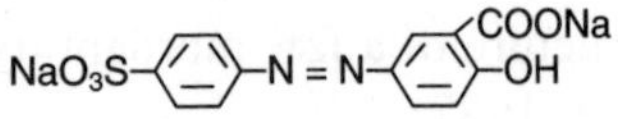

Chromefast Yellow 2G (Greenish yellow)
C I Mordant Yellow 10, C I 14010

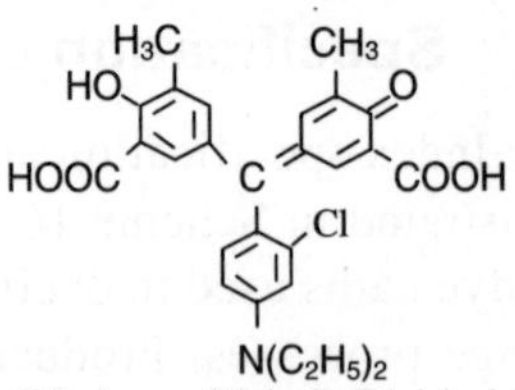

Eriochrome Violet B (Bluish violet)
C I Mordant Violet 28, C I 43570

Scheme 16.1 C I specification and chemical structure of a few mordant dyes.

16.4   Pre-metallised dyes

To avoid obvious complications associated with reaction of metals with dye possessing chelating sites in shop-floor level, in many cases, the reaction is carried out in the dye manufacturing plant itself and the dye–metal complex is supplied to the dyeing sector, known as pre-metallised dye. Selective acid dyes are complexed with copper, chromium, cobalt or nickel at suitable proportions to develop these dyes. Reaction of 1 atom of metal with 1 molecule of dye produces 1:1 metal–complex dye, whereas reaction of 1 atom of metal with 2 molecules of dye produces 1:2 metal–complex dyes. All the valencies of a metal are not co-ordinated with only one dye molecule and so unsaturation exists in the 1:1 complex, leading to formation of 1:2 unsymmetrical complex with two different dye molecules, if such scopes exist.

Pre-metallised dyes are produced from mainly o-o'-dihydroxyazo, o-carboxy o'-hydroxyazo, o-amino o'-hydroxyazo, o-hydroxy o'-methoxyazo types of acid dyes (Allen, 1971). Neolan (Ciba) and Palatine (BASF) dyes are 1:1 metal complex dyes through reaction of chromium salts with o-o' dihydroxy azo dyes. These dyes are applied from a very strong acidic bath (8% H_2SO_4). Due to absence of dulling effect caused by wool–chromium complex, shades are brighter and matching is simple compared to mordant dyes. All light to heavy shades can be easily produced with these dyes. Light fastness is parallel to mordant dyes with slight deterioration in wash fastness. Dyes are water soluble because of $-SO_3H$ groups in structure. Dyeing at boil only gives the true shade and tone. Cycolan (Geigy), Ultralan (ICI), Chromacyl (Du Pont), Inochrome (Francolor) and Vitrolan (Clarient) are some other brand names of these dyes. Higher dosing of acid may be reduced to just half with application of cationic levelling agents, e.g. Palatine fast salt O and Neolan Salt II, which are fatty alcohol-ethylene oxide condensates having general formula $CH_3-[CH_2]_x-O-[CH_2-CH_2O]-H$. Levelling agents form loose complex with dye as these are of opposite electrical nature. The complex is adsorbed by the fibre and at boil; the complex breaks to release levelling agents (Shenai, 1987). Thorough wash and neutralisation at the end are essential to remove acid from fibre. Dyeings can be discharged to white and so are extensively used in fancy prints.

1:1 dyes exhaust too less from a neutral bath with poor rubbing, milling and light fastness. But due to lighter in structure, possess affinity for vegetable and acetate fibres too. Strong acidic pH which damages wool, poor exhaustion due to higher solubility of $-SO_3H$ groups (oxygen in $-SO_3H$ is electron donor to H of water) of 1:1 metal complex dyes and prolonged boiling of dyebath led birth of 1:2 metal complex dyes with the pre-condition that (i) the complex should carry a negative charge, (ii) any residual valency of the metal atoms must be fully or nearly satisfied, (iii) solubilising groups should be less hydrophilic, may be by introducing sulphonamide groups in place of $-SO_3H$, (iv) complex should retain no or minimum hydrophilic groups and (v) all dyes in range should have identical exhaustion property. These dyes have too little affinity for fibre; being salts of strong acids, these ionize in water with lesser solubility and applied from mild acidic or neutral baths. The degree of exhaustion and wash fastness are function of molecular weight of dye and around a weight of 600, both these functions remain fairly good. Inclusion of water solubilising groups, e.g. $-SO_2-CH_3$, $-NHCOCH_3$, $-SO_3-CH_2$, etc., improves their solubility in bath.

1:2 metal complex dyes are simple to apply, shorter dyeing time with perfect levelled shade, possess excellent fastness and ease in exact reproduction, marked in Fig. 16.2 (Allen, 1971). These dyes have practically no migrating power necessitating proper control over pH and slow rate of heating as rate of dyeing is rapid just below boil; the degree of migration depends on temperature, nature of dye and fibre. Remarkable migration occurs below 82°C, beyond which it decreases. Rate of exhaustion depends on pH, temperature and agitation of bath. An optimum pH around 6.5 at boil may be sufficient for level dyeing with maximum dye uptake; to increase rate of dyeing, pH may be lowered down to 4 or below it. Due to poor migrating power, agitation of bath is essential from start of dyeing for even distribution of dye. Matching of shade and required addition may be done at boil after adding little $-NH_3$ to raise pH just to slow down rate of exhaustion. Salt alone is not capable of showing any retarding or levelling action but in presence of levelling agents, e.g. Lyogen SMK (Sandoz), it promotes levelling effect.

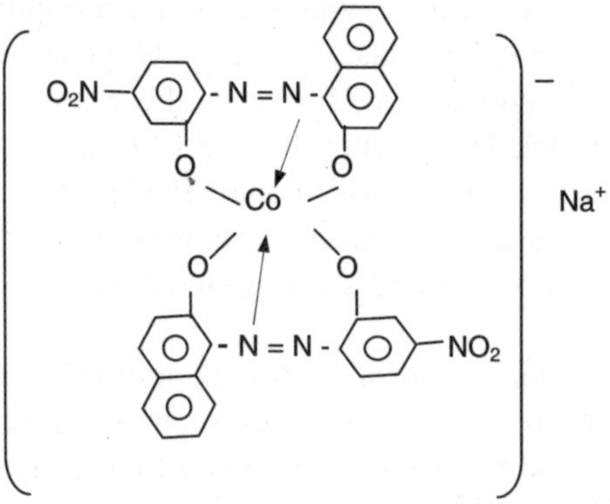

Perlon Fast Violet BT (IG), CI 12196

16.2 Formation of 1:2 pre-metallised dye (cobalt complex).

To improve water solubility of 1:2 dyes, a new series was developed by introducing non-hydrophilic methylsulphonyl, sulphonamide or acetylamino groups without affecting dyeing and fastness properties. These substituents are not acidic and the solubilising effect is due to covalent bond formation between water and sulphonyl oxygen atoms or C=O groups, known as Irgalan dyes (GY), shown in Fig. 16.3 (Zollinger, 1961; Schetty, 1955).

Irgalan Brown Violet DL (1:2 metal complex dye), C I Acid Violet 78

16.3 Igralan dye.

16.5   References

ALLEN R L M (1971) *Colour Chemistry*, 1st Edition, London, Thomas Nelson & Sons Ltd.

Colour Index International (1987), 3rd Edition (3rd Revision), Bradford, UK, Society of Dyers & Colourists and AATCC.

GILLS C H (1944) 'The Mechanism of the Mordanting of Animal Fibres', *J. Soc. Dyers Color.*, **60**, 303–315.

HANNEMANN K (2001) 'Alternatives to after-chrome', *Int. Dyer*, **186**, 3, 16–19.

RACE E, BOYE F M and SPEAKMAN J B (1946) 'The Nature of the Dye-Chromium-Fibre Complex in the Case of Wool Dyed with Certain Chrome Mordant Dyes', *J. Soc. Dyers Color.*, **62**, 372–383.

SCHETTY G (1955) 'The Irgalan Dyes-Neutral-dyeing Metal-complex Dyes', *J. Soc. Dyers Color.*, **71**, 705–724.

SHENAI V A (1987) *Chemistry of dyes and principles of dyeing*, Mumbai, Sevak Publications.

SHENAI V A (2002) 'Dyeing of wool with chromium and metal-complex dyes', *Colourage*, **49**, 4, 25–28.

SZYMCZYK M and FREEMAN H S (2004) 'Metal-complexed dyes', *Rev. Prog. Color Related Topics*, **34**, 39–57.

TROTMAN E R (1994) *Dyeing and Chemical Technology of Textile Fibres*, 6th Edition (1st Indian Edition), Delhi, B I Publications Pvt Ltd.

ZOLLINGER H (1961) *Azo and Diazo Chemistry: Aliphatic and Aromatic compounds*, New York, Wiley Interscience.

Abstract: Basic dyes possess basic group in structures enabling to react with acidic group available in fibres like basic dyeable acrylic and cationic dyeable polyester in addition to dyeing of wool, silk and nylon in general forming electrovalent bond. Dyes possess high tinctorial value, soluble in mild acidic bath and shades are brilliant. Except limited cases, basic dyed textile shows poor light fastness with moderate wash fastness. Ability to react with anionic dyes extends use of basic dyes in topping on direct dyed textile.

Keywords: Cationic dye, ionic bond, mordanting, light fastness

17.1   Principle of application

Basic dyes are organic bases and retain free or substituted amino groups in structure, e.g. $-NH_2$, $-N(CH_3)_2$, $-N(C_2H_5)_2$, etc. These dyes are often referred as 'cationic dyes' as because in aqueous solution these release coloured cation; the latter is the chromophore itself possessing the empirical formula

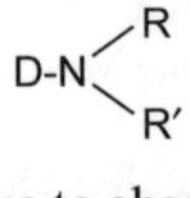

$$D\text{-}N\begin{matrix} R \\ R' \end{matrix}$$

Dyes do not show any colour due to absence of chromophore till converted to their respective salts, viz. hydrochlorides, oxalates, sulphates, nitrates and zinc chloride double salts, etc., which in turn are insoluble in water.

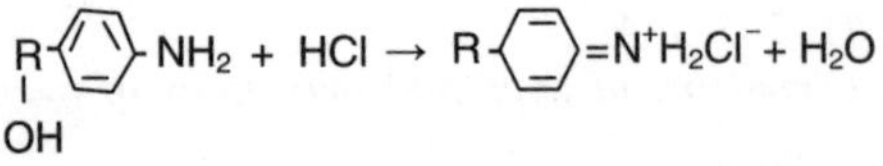

$$R\text{-}\langle\!\!\langle\ \rangle\!\!\rangle\text{-}NH_2 + HCl \rightarrow R\text{-}\langle\!\!\langle\ \rangle\!\!\rangle = N^+H_2Cl^- + H_2O$$
$$\underset{OH}{\mid}$$

Basic dyes do not possess systematic nomenclature – most of the dyes are still marketed and used under conventional names, e.g. Bismark Brown, Auromine O, Malachite Green, Methylene Blue, etc. However, latest dyes, specifically designed for dyeing of polyacrylonitrile are marketed under trade names, e.g. Astrazon (BFy), Basacryl (BASF), Maxilon (Gy), Symacryl (ICI), etc.

184

17.2   Properties

Basic dyes are cheaper, soluble in alcohol and methylated spirit but not easily soluble in water while a few are sparingly soluble. Shades are invariably brighter with high tinctorial value. Dyes form a sticky mass on storage posing difficulty in solubilisation and can be better solubilised in hot water alongwith little CH_3COOH or at cold with methylated spirit. Dyes are highly sensitive to alkali and on reaction liberate colourless basic dye (Trotman, 1994).

$$\begin{array}{c} NH_2C_6H_4 \\ \\ NH_2C_6H_4 \end{array}\!\!C = C_6H_4 = NH_2Cl \; + NaOH \rightarrow \begin{array}{c} NH_2C_6H_4 \\ \\ NH_2C_6H_4 \end{array}\!\!C\!\!\begin{array}{c} OH \\ \\ C_6H_4NH_2 \end{array} + NaCl$$

Use of soft water is a prerequisite in dyeing with basic dyes. Presence of calcium or magnesium bicarbonates in water causes precipitation of colourless dye, even presence of metal hydroxides or carbonates cause the same problem. Application of CH_3COOH in bath well before addition of dye reduces alkalinity and hardness both remarkably.

Due to cationic nature of dye, it can be precipitated by an anionic dye, e.g. direct or acid dye. The precipitation is due to formation of basic–direct dye or basic–acid dye complex which is insoluble in water. Basic dye shows no affinity for cellulose but can be applied on it if the latter has already been dyed with a direct dye or any other anionic mordant, e.g. tannic acid. Direct dyed cotton lacks brilliancy in shade and fastness – both these shortcomings can be invariably improved by topping with basic dye. All basic dyes can combine with tannic acid to form an insoluble compound in absence of mineral acid. When a solution of tannic acid and sodium acetate is added to dye, a precipitate is formed. The purpose of sodium acetate is to react with HCl liberated from the dyestuff and to maintain a pH favourable to formation of tannic acid–dye complex. Cellulose is dyed after mordanting while protein fibre is first dyed with basic dye followed by back tanning with tannic acid to improve wash fastness. $Na_2S_2O_4$ can decolourize non-azo basic dyes but treatment with CH_3COOH restores parent dye structure. Basic dyes containing azo groups are permanently discharged by $Na_2S_2O_4$.

17.3   Classification

Basic dyes belong to various chemical classes, viz. (i) Azo, e.g. Bismark brown, (ii) Di-phenylmethane, e.g. Auramine O, Auromine G, (iii) Tri-phenyl methane, e.g. Malachite Green, Magenta, (iv) Methane, e.g. Astrazon Yellow 3G, Orange G, (v) Acridine, e.g. Acridine Orange R, (vi) Xanthene, e.g. Rhodamine B, Rhodamine 6G, (vii) Azine, e.g. Safranine T, (viii) Oxazine, e.g. Meldola's Blue, Acronol Sky Blue 3G, (ix) Thiazine, e.g. Methylene Blue, Methylene Green, etc. (Allen, 1971). However, application technique remains almost same for all these various classes of basic dyes.

17.4   Specification of some important basic dyes

Structure and colour index specification of some important basic dyes belonging to various classes are shown in Scheme 17.1 (Colour Index, 1987).

CH$_3$COOH neutralizes alkalinity and controls pH of dye bath. It also helps in dissolving dye and acts as retarding agent by combining with cationic dye to reduce strike rate; adsorption of dye occurs steadily and level dyeing is obtained.

(i) Azo dyes

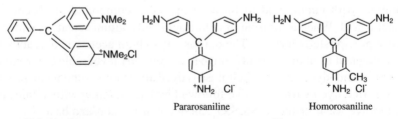

Bismark brown

C I Basic Brown 1, C I 21000

(ii) Diphenylmethane dyes

Auramine O
(C I Basic Yellow 2, CI 41000)

Auramine G
(C I Basic Yellow 3, CI 41005)

(iii) Triphenylmethane dyes

Pararosaniline   Homorosaniline

Malachite Green
(C I basic Green 4, CI 42000)

Magenta
(CI Violet 14, CI 42510 is a mixture of these two)

(iv) Methane dyes

Astrazon Yellow 3G
C I Basic Yellow 11, CI 48055)

Astrazon Orange G
C I Basic Orange 21, CI 48035

Scheme 17.1 Contd. ...

(v) Acridine dyes

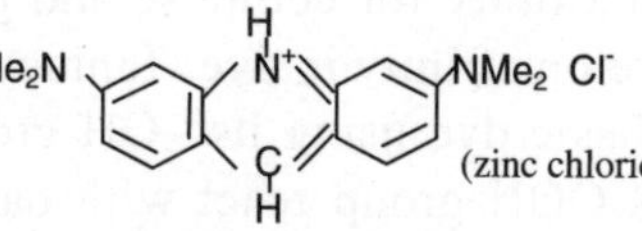

(zinc chloride double salt)

Acridine Orange R
(C I Basic Orange 14, CI46005)

(vi) Xanthene dyes

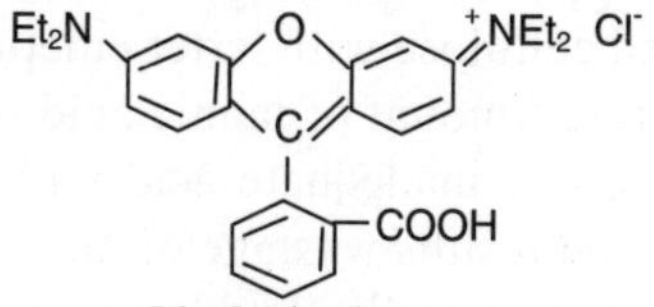

Rhodamine B
C I Basic Violet 10, CI 45170

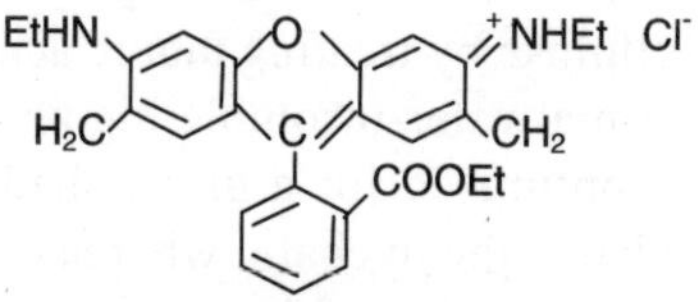

Rhodamine 6G
C I Basic Red 1, CI 45160

(vii) Azine dyes

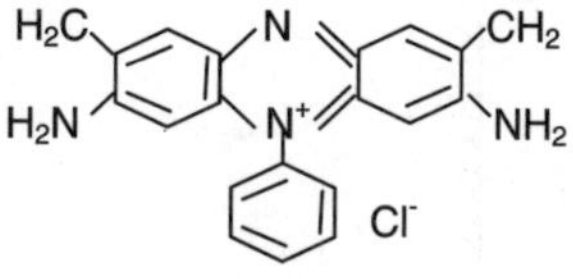

Safranine T

C I Basic Red 2, CI 50240

(viii) Oxazine dyes

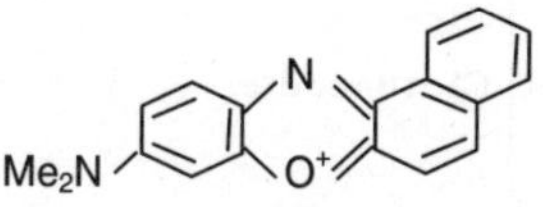

Meldola's Blue
(C I Basic Blue 6, CI 51175)

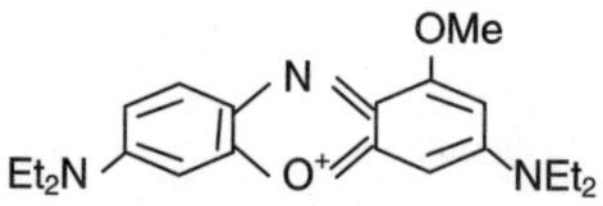

Acronol Sky Blue 3G
(C I Basic Blue 4, CI 51004)

(ix) Thiazine dyes

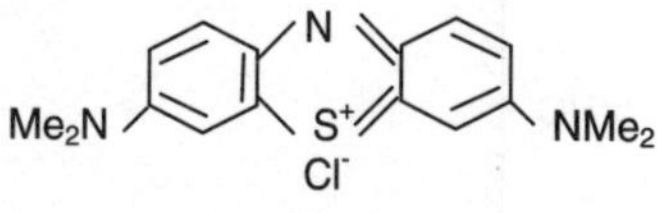

Methylene Blue

C I Basic Blue 9, CI 52015

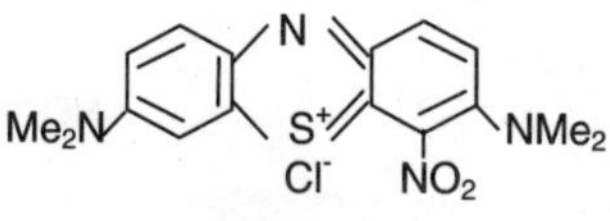

Methylene Green

C I Basic Green 5, CI 52020

Scheme 17.1 Structure and colour index specification of some important basic dyes.

17.5   Dyeing of cellulose

Basic dye possesses no affinity for cellulose and pre-mordanting of the latter is essential to develop affinity in dye. Tannic acid acts as a mordant in dyeing cotton with basic dye using its –OH group to make –H bond with cellulose while –COOH group react with basic dye, illustrated in Fig. 17.1 (Shenai, 1987). However, fastness is not good as neither dye–fibre reaction occurs nor tannic acid–basic dye complex is too bigger to develop excellent wash fastness. Even tannic acid being water soluble comes out in bath during dyeing and so applied tannic acid is fixed on cellulose by treating tannic acid mordanted cellulose with tarter emetic to form a water-insoluble big anionic complex. Amount of tannic acid is to be optimised for a given shade; application of inadequate acid will not exhaust the dyebath, whereas excess addition results wastage of material, causes rusting of dye on the mordanted fibre. Generally, double amount of acid is used in mordanting to that of weight of dye.

17.1 Sequence of fixation of basic dye on mordanted cellulose.

Cellulose adsorbs tannic acid rapidly at cold but very slowly at higher temperature. To avoid uneven mordanting, temperature of bath is raised to boil; tannic acid is added followed by cooling down the bath progressively. Absorption of tannic acid by cellulose is also retarded by acetic acid.

Cotton mordanted with tannic acid is treated with tartar emetic, chemically potassium antimony tartarate [$2\{K(SbO)C_4H_4O_6\}$] at cold for 30 min. Tartar emetic is soluble in water and it makes tannic acid insoluble by forming a big molecule of tannic acid – tarter emetic complex, which can not come out from cotton.

Mordanting by tannic acid and fixing by tartar emetic is a long process, synthetic mordants are used where no fixing is required, even a better wash fastness is obtained, e.g. katanol, resistone and tanninol – produced by heating phenol with sulphur alongwith a trace of iron compound. These mordants are pasted with a little Na_2CO_3 in cold water; little boiling water is added till completely dissolves. Cellulose is mordanted by immersing it in a solution of Katanol (3 parts) and Na_2CO_3 (2 parts) for 1–2 h at 60–90°C. The goods are then rinsed and dyed with basic dye at 40°C.

Practical application of basic dye onto cellulose includes three steps, viz. (i) mordanting of cotton with tannic acid, (ii) fixing with tarter emetic and (iii) dyeing with basic dye.

17.5.1 Mordanting

Tannic acid (double the weight of dye) is dissolved in water and heated upto boil. Cellulose is treated in this bath followed by cooling down in 2 h or overnight. During cooling, maximum absorption of tannic acid takes place.

17.5.2 Fixing

Tarter emetic (half of tannic acid) is dissolved in water and tannic acid treated cotton is passed through this bath for 10–15 min at cold followed by rinsing and hydro-extraction.

17.5.3 Dyeing

Dye is pasted with CH_3COOH (30%), boiling water is added and stirred for complete dissolution. Basic dye possesses higher affinity for tannic acid mordanted cellulose and so dye solution is added in three instalments.

Dyebath is prepared with water and CH_3COOH (pH~4.5–5.5) at room temperature. One-third of dye solution is added, mordanted cotton is entered at cold and dyed for 15 min; second one-third is added, temperature is raised to 40°C and dyed for further 20 min after which last one-third is added, temperature is raised to 70°C and dyed for desired time.

For better wash fastness, back tanning may be done by treating dyed cotton in a liquor of tannic acid (0.5 g/l) at cold for 30 min, squeezed and passed through a solution of tartar emetic (0.25 g/l). Back tanning is not as popular as it may alter the earlier shade.

17.6   Topping on direct dyeings

Cellulose is first dyed with a direct dye followed by redyeing in a basic dye bath at room temperature for 15–30 min in mild acidic pH, the process is called topping as the second dye remains at the top of first one. Bright shades with moderate wash fastness are produced. Direct dyes (DSO_3Na) form anion ready to react with basic dye cation as follows (Shenai, 1987):

$$DSO_3Na \rightarrow DSO_3^- + Na^+$$
$$R_4NCl \rightarrow R_4N^+ + Cl^-$$
$$DSO_3^- + Na^+ + R_4N^+ + Cl^- \rightarrow R_4N - DSO_3 + NaCl$$

17.7   Dyeing of wool and silk

Basic dyes show good affinity for protein fibres due to presence of –COOH group in polymer chain. Dye cation reacts with this –COOH site to form electrovalent bond between dye and fibre.

$$H_2N - W - COOH + RN_4^+Cl^- \rightarrow H_2N -W-COO^- RN_4^+ + HCl$$

When –COOH groups in fibre are ionised, negative sites are developed in fibre which reacts with cations of basic dye, HCl formed as by-product retards faster reaction resulting even dyeing. Basic dyed protein fibre shows good wash fastness but poor light fastness; exception is polyacrylonitrile, which is invariably dyed with basic dyes to develop brilliant shade with excellent fastness properties.

17.8   Fastness properties

Protein fibres are seldom dyed with basic dye show on fastness ground. When basic dyes are applied to PAN, it produces better wash and light fastness grades. Better wash fastness is due to higher binding force at the PAN-basic dye junction than protein fibre–basic dye junction. Furthermore, structure of protein fibres is loose one while PAN is compact and dyed above its glass transition temperature by means of opening the structure followed by collapsing in it at the end with better trapping of dye molecules. Incident light on basic dye when applied on protein fibre is not transmitted to the fibre causing degradation of dye. Another reason is that protein fibre retains moisture (H_2O) which on reaction with atmospheric oxygen produces H_2O_2 and this in turn oxidizes dye. In contrast, Incident light on basic dye when applied on PAN is transmitted to fibre and gets absorbed in fibre phase causing little or no damage to dye. Poor moisture regain of PAN does not allow converting moisture to H_2O_2. Except PAN, basic dyes are rarely used in dyeing of other fibres, rather used in preparation of various types of inks, cosmetics, crayons, food stuffs and other non-textile areas.

17.9   Cationic dyes for polyacrylonitrile fibres

Basic dyes are invariably used in dyeing of acrylic fibres. Though conventional basic dyes are cationic, the cation is not strong enough to make strong contact with PAN resulting in poor to moderate fastness properties. Introduction of anionic co-polymers makes PAN highly anionic and it is quite possible to make strong dye–fibre bond if cationic nature of dye can be enhanced and oriented in side structure properly. In advanced dyes synthesised for dyeing of acrylic fibre, the positive charge is supplied by a quaternary ammonium group (NH_4^+) group at one end of dye chain rather than a protonated amine (NH_3^+) at the centre. Fastness properties of latest cationic dyes on acrylics are excellent.

In one type, latest cationic dyes retain at least one quaternary nitrogen atom that is not a member of the aromatic ring and cationic dyes of these types are extensively used to produce red or orange shades. Arylaminoanthraquinones with externally bound quaternary ammonium group gives reddish blue shades on PAN. Azo derivatives of thiazole, benzothiazole, etc., are solubilised by dimethyl sulphate to get a wine red dye. A red dye based on hydrazinium has been found to be of paramount importance. Increase in length of chain is proportional to fall in light fastness; a shorter chain is the key of better light fastness. In earlier cationic dyes, the cation was with the chromophore whereas in latest dyes, it is isolated from the system by a saturated hydrocarbon link.

Light fastness and chemical constitution of dyes are also closely related. The lesser the basicity of dye, the better the light fastness; the more the dye–fibre bond is of electrostatic nature the lower the light fastness; increase in homopolarity decreases light fastness. In special cationic dyes, the cation is far away from rest part of dye fulfilling all these above requirements.

17.10   References

ALLEN R L M (1971) *Colour Chemistry*, 1st Edition, London, Thomas Nelson & Sons Ltd.
Colour Index International (1987), 3rd Edition (3rd Rev), Bradford, UK, Society of Dyers & Colourists and AATCC.
SHENAI V A (1987) *Chemistry of dyes and principals of dyeing*, 3rd Edition, Mumbai, Sevak Publications.
TROTMAN E R (1994) *Dyeing and Chemical Technology of Textile Fibres*, 6th Edition (1st Indian Edition), Delhi, B I Publications Pvt Ltd.

18

Dyeing with disperse dye

Abstract: Disperse dyes are water insoluble non-ionic dyes used to dye man made fibres from acidic bath, specially designed for dyeing of polyester and are applied alongwith dispersing agent to retain dyes in fine dispersion. Grinding is an important process to control particle size for ease in diffusion and migration. Due to their application invariably at higher temperature, a part of dye may get sublimised-off necessitating their classification based on stability against sublimation. Dyes are retained by fibre with physical forces. Fastness properties are excellent on polyester except certain cases.

Keywords: Dispersing agent, sublimation, high temperature, carrier, levelling agent

18.1 Characteristic property

Disperse dyes are electrically neutral, do not have affinity for any fibre and are mainly used for dyeing of non-ionic fibres, e.g. polyester, cellulose acetate, etc. Application of disperse dye on nylon, acrylic and other man made fibres is associated with technical problems, e.g. poor wash fast shades on nylon, only light shades on acrylic and so are not practiced. Dye particles have size ranging from 0.5–2.0 micron, generally contain $-NH_2$, substituted $-NH_2$ or $-OH$ groups in structure, get attached with fibre through H-bond and van der Waals force (Braun, 1983). Dyes possess higher saturation value, ranging from 30–250 mg/g of fibre with a corresponding value of 20% shades. Dyes are applied from homogeneous dispersion in bath with the help of a dispersing agent, which is either previously incorporated in dye mixture for efficient grinding of dye or during dyeing or both (Leadbetter and Leaver, 1989; Dawson, 1972, 1978). Dyes are transferred in molecular state from bath to fibre through a process of surface deposition, absorption and diffusion.

18.2 Classification of disperse dyes

18.2.1 Field of application

Based on criteria of critical dyeing temperature, migration, build up of shade, rate of diffusion and energy required for dye transfer from bath to fibre, all disperse dyes are categorised into four different groups A, B, C and D. In addition to, the relationship among structures, dyeing property and fastness characteristics are also considered for inclusion of a dye in a specific group; a dye showing best performance is included in D group, whereas the dye showing worst performance in A group, i.e. the performance gradually improves from A to D. Another group with 'no suffix' also exists, thus making total number of groups to five (Beffa and Steiner, 1973; Nunn, 1979). The characteristics of these groups may be summarised as follows:

(i)   'No suffix' dye – it is not suitable for polyester, but suitable for acetate and nylon.
(ii) Group A dye – it has poor sublimation fastness (~Grade 2), suitable for acetate and nylon but may be used for polyester.
(iii) Group B dye – it shows moderate sublimation fastness (~Grade 2–3), is suitable for acetate and nylon. Well-levelled shades on polyester, coverage on 'configurational barre' is excellent and so is invariably applied on textured polyester.
(iv) Group C dye – it has good sublimation fastness (~Grade 3–4), is suitable for all methods to dye polyester, e.g. carrier, HTHP, thermosol and other methods, where maximum sublimation fastness is not required.
(v)   Group D dye – it has maximum sublimation fastness (~Grade 5). Exclusively for dyeing of polyester in HTHP or thermosol technique but not in carrier method; few dyes produce wash fast shades on nylon.

The sublimation fastness grades shown are exclusively for polyester; the same for triacetate will be little better where as for acetate the highest. Sublimation fastness can be improved by selecting a dye of higher molecular weight.

18.2.2 Chemical structure

Almost 60% of all disperse dyes belong to azoic class, out of which only monazo dyes contribute for nearly 50%; anthraquinoids around 25% while rest from other miscellaneous classes, viz. methane (3%), styryl (3%), acrylenebenzimidazol (3%), ouinonapthalones (3%), amino napthyl amide (1%), napthoquinone–imine (1%) and nitro disperse dyes (1%). Disperse dyes belonging to these groups have also been documented in detail elsewhere (Nunn, 1979; Gulrajani, 1987).

18.3   Specification of few important disperse dyes

Colour Index specification and chemical structures of a few commonly used disperse dyes is shown in Scheme 18.1 (Colour Index, 1987).

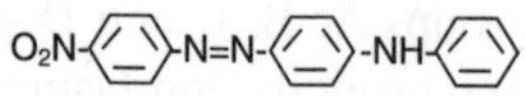

O₂N-⟨⟩-N=N-⟨⟩-NH-⟨⟩

Dispersol Orange B-A (ICI), Navilene Orange 5R(IDI)
(Redish Orange)
(C I Disperse Orange 1, CI 11080)

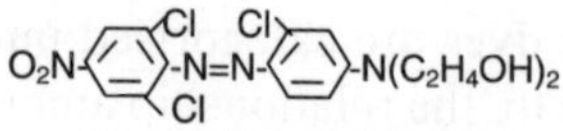

O₂N-⟨Cl Cl⟩-N=N-⟨⟩-N(C₂H₄OH)₂
⟨Cl⟩

Dispersol Bron D-3RA(ICI)
(Redish Brown)
(C I Disperse Brown 1, CI 11152)

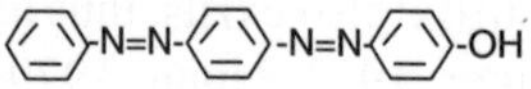

⟨⟩-N=N-⟨⟩-N=N⟨⟩-OH

Dispersol Yellow B-4R(ICI), Foron Yellow E-RGFL(Sandoz)
(Redish Yellow)
(C I Disperse Yellow 23, CI 26070)

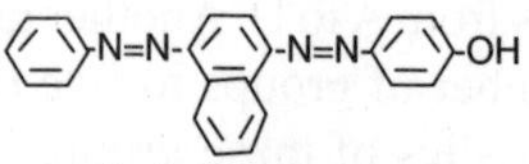

⟨⟩-N=N-⟨⟩-N=N⟨⟩-OH

Navilene Orange G (IDI)
(Yellowsh Orange)
(C I Disperse Orange 13, CI 26080)

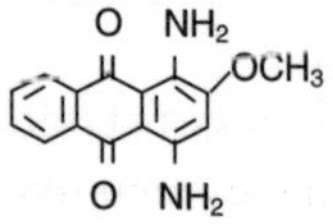

O   NH₂
      OCH₃

O   NH₂

Disperse Red B3B(Atic) / Navilene Violet 6R(IDI)
(Bright Bluish Pink)
(C I Disperse Red 11, CI 62015)

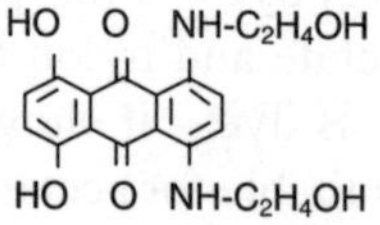

HO   O   NH-C₂H₄OH

HO   O   NH-C₂H₄OH

Disperse Blue 7G(ICI)
(Bright Greenish Blue)
(C I Disperse Blue 7, CI 62500)

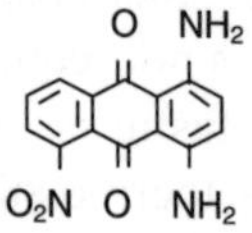

O   NH₂

O₂N   O   NH₂

Navilene Violet 3B (IDI), (Bright Bluish Violet)
(C I Disperse Violet 8, CI 62030)

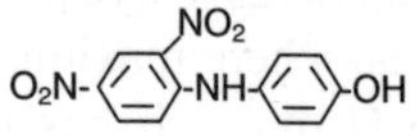

NO₂
O₂N-⟨⟩-NH-⟨⟩-OH

Navilene Yellow AL(IDI), (Bright Reddish Yellow)
(C I Disperse Yellow 1, CI 10345)

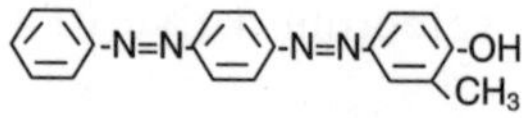

⟨⟩-N=N⟨⟩-N=N⟨⟩-OH
                        CH₃

Navilene Yellow 5RX (IDI), (Reddish Yellow)
(C I Disperse Yellow 7, CI 26090)

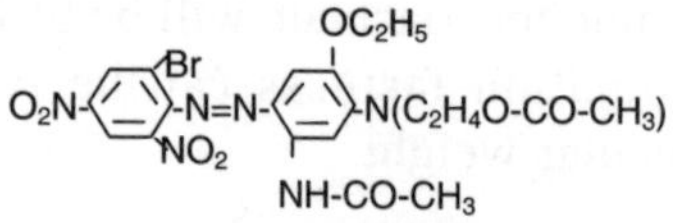

OC₂H₅
                    Br
O₂N-⟨⟩-N=N-⟨⟩-N(C₂H₄O-CO-CH₃)
           NO₂
                    NH-CO-CH₃

Foron Navy S-2GL(Clarient)
(C I Disperse Blue 79, C I 11345)

Scheme 18.1 C I specification and chemical structures of a few disperse dyes.

18.4   Technology of dyeing

Dyebath is prepared with dispersing agent (0.5–1.0 g/l), CH_3COOH (pH~5.5–6.5) and warm water. Well-scoured and bleached polyester is treated in it, temperature is raised to 50–60°C when dye is pumped in and agitated enough to make a fine stable dispersion. The bath is heated up to 90°C, beyond which rise in temperature is restricted to 1–1.5°C / min up to 120°C followed by rapid rise to 130±2°C. Dyeing is carried out at this temperature for 30–90 min after which the bath is cooled down to 80°C at a rate of 5°C/min and is dropped. For deep shades with better fastness, reduction cleaning is carried out at 50–60°C for 15–20 min with $Na_2S_2O_4$ and NaOH (3 g/l each); light shades require no reduction clearing. Soaping is imparted followed by washing, unloading and drying. A defoaming or deaerating agent (0.5–1.0 g/l) may be added for suppression of foam formed during agitation. Sodium hexa m-phosphate (0.5–1.0 g/l) is applied in bath if soft water is not used. When light shades are produced at boil or faulty dyeings are corrected at 135±2°C, carriers (0.5–1.0 g/l) find its way in dyeing recipe. A typical dyeing cycle of polyester with disperse dye is explained in Fig. 18.1 stepwise (Gulrajani, 1987).

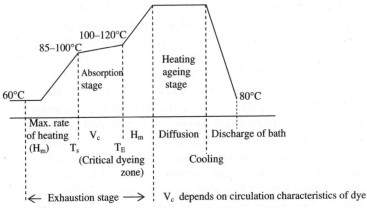

18.1 Dyeing cycle of polyester with disperse dye in HTHP method.

18.5   Temperature of dyeing

Nylon 66 and acrylic are dyed with disperse dye at boil due to adequate opening of fibre structure at this temperature. Nylon 66 has T_g around 71°C and beyond that opening is rapid offering adequate space to all disperse dyes at boil itself. Acrylic has T_g around 80°C and opening completes at around 85°C providing maximum opening just below boil; heating up beyond boil causes fibre degradation. In contrast, polyester has T_g at 80–85°C, but opening is too slow to affect any such opening at boil. Little opening may cause production of light shades with poor wash fastness. Dyeing temperature varies from dye to dye; the larger the structure of dye, more opening is needed necessitating a higher temperature of dyeing. However, all disperse dyes exhaust well at or below 117°C and a little higher temperature is maintained to compensate heat loss and better migration of dye molecules for levelled shades. In tropical areas and specifically in chilling winter, production of deep and levelled shades with better wash fastness is difficult to achieve due to condensation of steam during its long passage from boiler to the dyeing machine; the situation worsens if insulation of supply pipe is not proper.

18.6   Mechanism of dyeing

Disperse dyeing of man made fibres follows a solid solution theory where a solid dye is solubilised in another solid fibre phase and in all cases the dye forms no chemical bonds with fibre rather is retained by H-bonds and van der Waals forces, except reactive disperse dyes which reacts with only nylon. Solubility of disperse dyes is too feeble in water at room temperature (~0.2 mg/l) which is increased with heating of bath beyond 90°C or addition of dispersing agent or both. As disperse dyes are specifically applied at higher temperature, the dye when added, remains present in bath in two basic forms: (i) a very little part remains in completely soluble form and (ii) the rest in finely dispersed insoluble form; the later remains predominant and the ratio is being determined by solubility of dye in water at that temperature (Braun, 1983). When fibre is entered in dyebath, dissolved dye molecules slide past the narrow pores present in fibre and get attached with physical forces causing reduction in share of dissolved dye in bath which forces dispersed insoluble dye particles to break up and go into solution to maintain share of soluble dye in bath. As disperse dyes are hydrophobic in nature, soluble dye try to be separated out from the bath and are attracted by hydrophobic man made fibres; water acts as the vehicle for transfer of dye from bath to fibre. Though dispersing agents enhance solubility of dye for efficient dyeing and facilitate diffusion inside fibre, increase in temperature breaks down dye clusters improving solubility of dye and opens up fibre structure facilitating diffusion of dye inside, known as 'rate determining step'. The overall rate of dyeing is controlled by combined effect of temperature and type of dispersing agent used.

Dyeing takes place through (i) dispersion of dye in bath, (ii) dissolution of single dye molecule in water, (iii) deposition of dye on fibre surface, (iv) absorption of dye on fibre surface, and (v) diffusion of dye inside. Though rate of dyeing is determined by rate of diffusion, the latter is increased when more and more dye molecules are absorbed on surface. Rate of absorption, in turn, again depends on rate of deposition on surface and surface deposition is again influenced by extent of solubilisation of dye. The diffusion process takes more time as dye molecules are to be uniformly dispersed in fibre phase till a levelled shade is developed. Equilibrium is achieved when rate of dissolution, surface deposition, absorption and diffusion of dye become equals to each other with no further dye uptake. However, extent of solubilisation and diffusion of dye are the most two important criteria in disperse dyeing which affect extent of exhaustion as well as rate of dyeing. Dyeing time becomes more for deeper shades to position more dye molecules through simultaneous dyeing and stripping inside fibre, reduction in which may result insufficient diffusion with more and more dye molecules at the surface resulting poor rubbing and wash fastness.

18.7   Dispersing agent

Disperse dye is insoluble in water at room temperature though little solubility is achieved at higher temperature. Dye molecules after manufacture remain in clustered form and only agitation of bath is inadequate to retain dyes in dispersed state. It is essential to apply a little dispersing agent in dyebath for proper dispersion and to avoid settling of dye during dyeing. Dispersing agent imposes negative charge on dye to improve its aqueous solubility and inhibits formation of dye cluster through repulsion among negatively charged dye molecules; retains dye in consistent dispersion throughout dyeing by enclosing single dye particle with a protective film, as shown in Fig. 18.2, where 'D' denotes disperse dye and surrounding '⊖' ions are dispersing agents (Heimann, 1981).

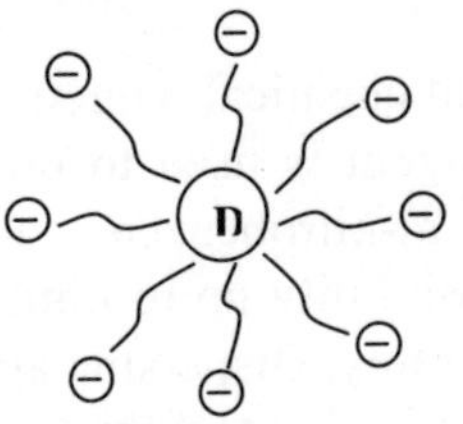

18.2 Interaction of dispersing agent with disperse dye.

The zeta potential of commonly used disperse dyes range from -30 to -80 mV. Dispersing agent, if in excess, enhances aqueous solubility of dye, reduces exhaustion and develops poor rubbing fastness. These are invariably mixed with disperse dye during grinding to keep control over particle size and do not allow to form cluster. While in bath, dispersing agents being highly soluble in water, surround a disperse dye molecule, keep it in dispersed state and release it slowly to get solubilised in bath for its acceptance by the fibre (Anon, 1981; Slack, 1979). Each disperse dye possesses varying solubility pattern at higher temperature and also in presence of dispersing agents due to difference in hydrophilicity of it which in turn is based on hydrophilic groups in its structure. Indeed, optimum concentration of dispersing agent required for each dye is different. As dispersion of dye is achieved through interaction between dyes and dispersing agent, concentration of the latter is based on amount of dye and do not have any relation with liquor ratio. Concentration of dispersing agent can further be reduced with increase in dyeing time to get a better levelled shade (Fig. 18.3).

Dispersing agents belong to mainly formaldehyde condensates of either of naphthalene sulphonic acid, cresol, 1-naphthol 6-sulphonic acids, fatty alcohol-ethylene oxide condensate, alkyl aryl sulphonates or lignin sulphonates; the condensate of naphthalene sulphonic acid is the most

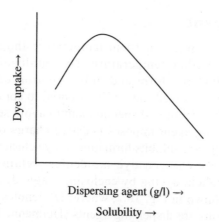

Dispersing agent (g/l) →

Solubility →

18.3 Influence of dispersing agent on solubility and uptake of disperse dye.

popular one. Lignin sulphonates brings about chemical reduction of some azo disperse dyes and so mild oxidizing agent is used to protect dyes (especially for polyester–wool blends) e.g. m-nitrobenzene sulphonates. While anionic dispersing agents increase solubility up to limited extent, other surfactants increase solubility considerably. Dispersing agents must have cloud point at least 5–10 °C higher over dyeing temperature to avoid precipitation of dye during dyeing.

18.8   Levelling agent

Levelling agents promote uniform distribution of dye throughout fibre structure by reducing the strike rate. These are non-ionic ethylene oxide condensates with aqueous solubility below cloud point; selected products should have cloud point beyond dyeing temperature otherwise may get precipitated in bath. Cloud point of levelling agents may also be enhanced with addition of anionic surface active agents. Levelling agents enhance solubility of dye and excess use reduce dye uptake. Some non-ionic levelling agents absorb dye causing loss in colour yield necessitating application of anionic products. Foam produced due to agitation in bath is suppressed by applying defoaming or deaerating agents.

18.9   Carriers

Carriers, when added to a disperse dye bath, interferes dispersion pattern of dye, physical characteristics of fibre, accelerate rate of dyeing to cause better dye up take at lower temperature. Chemically, these belong to hydrocarbons or substituted hydrocarbons, phenols, amides, alcohols, etc. Few popular carriers are monochloro-o-phenylphenol, o-phenyl phenol, diphenyl, monomethylnaphthalene, halogenated benzenes, methyl salicylate, butyl benzoate, etc., and are highly toxic. Several theories were put forward to emphasize mechanism of carrier action, e.g. increased swelling of fibre, increased water inhibition, transport theory, increased solubility of dye in bath, film theory, liquid fibre theory, loosening of fibre structure (lowering T_g), lubricity of fibre molecule and increased sites accessibility; exact mode of working of a carrier has not been established so far (Mittal and Trivedi, 1983). It is believed that these swell up fibre structure longitudinally, act as lubricant, cause adsorption and diffusion of dye rapidly at lower temperature. A high temperature method may be carried out in presence of carriers to get levelled results.

Selection of a carrier for a specific dyeing process is based on its toxicity, biodegradability, availability, cost, ability to show synergistic action on rate of dyeing, compatibility in bath, ease in removal, non-volatile nature, no adverse effect on light fastness and change in handle, non-irritate to skin, uniformity in absorption by the fibre and if insoluble in water, should have good emulsion stability; not a single carrier fulfils all these characteristics. Efficiency of a carrier is calculated based on its acceleration factor (a) which is defined as:

$$a = \frac{\text{amount of dye taken up by fibre in presence of carrier}}{\text{amount of dye taken up by fibre in absence of carrier}} = \frac{q_w}{q_0} \leq 1$$

Carriers are of two types: water soluble and insoluble. Water soluble type includes benzoic acid, salicylic acid and phenol; higher dose is required and are not used in dyeing. Insoluble types are mostly preferred, e.g. diphenol, benzene, naphthalene, o and p-phenyl phenol, o-chlorophenol, mono/di and trichlorobenzene, toluene, etc., which works better at 1–5 g/l but are highly toxic (Mittal and Trivedi, 1983). Tumescal OPE (ICI) is a popular carrier on these grounds.

18.10 Methods of application

Disperse dyes are applied in either of three different methods, viz. with carrier at boil, HTHP method or in 'Thermosol technique'. In the first method, dyebath is set at a pH~5.5–6.5 with CH_3COOH, carrier is added and the textile is treated in this bath up to 60°C followed by addition of dye. The bath is further heated up at boil and dyeing is continued for 1hr followed by soaping and washing. The method is restricted for production of only light shades as fibre opening as well as solubilisation of dye are not just enough for diffusion of dye inside causing poor wash fastness (Murray and Mortimer, 1971). In HTHP method, bath is set at 50–60°C with CH_3COOH and dye (pH~5.5–6.5); textile is placed in bath with agitation and temperature is raised to 90°C, succeeded by control over rate of heating at 1°C/min up to 120°C, beyond which the temperature is raised to 130±2°C and dyeing is continued for 1–2 h. For light shades, dyed textile is soaped and washed whereas for deep shades reduction cleaning is imparted to dyeings with $Na_2S_2O_4$ and Na_2CO_3 at 50–60°C for 15–20 min to remove superficial dyes for better fastness. Disperse dyes are insoluble in water up to 90°C, beyond which becomes soluble which increases with increase in temperature of bath, also opening of fibre (polyester) starts at 85–90°C and increases with increase in temperature of bath. To control diffusion of dye for a levelled shade, the temperature is controlled from 90–120°C, beyond which the bath is heated up to 130±2°C for more opening required to effective dye migration inside to produce levelled shades. The method is very popular due to (i) efficient levelling as well as covering of yarn irregularities with superior fastness through effective diffusion of dye and (ii) process efficiency is higher due to better exhaustion of bath. No auxiliary are required except little acid and dispersing agent. In thermosol method, the textile is padded with dye and thickener, dried and exposed to 190–210°C for 1 min when fibre structure is opened up and dye molecules diffuse inside. Efficiency of the method is based on dye particle size, extent of dispersion, sublimation property and solubility of dye. The method has restricted use due to poor migration of dye which is not desired for a levelled shade.

18.11   Gas fading of disperse dyes

Polyester, triacetate and acetate fibres dyed with anthraquinoid disperse dyes possessing primary or secondary $-NH_2$ groups in structure show fading effect on exposure to N_2O gases or other fumes. This happens when either primary $-NH_2$ groups are diazotised or secondary $-NH_2$ groups are converted to nitroso compounds. The problem can be solved by treating man mades with chemicals possessing affinity for fibre having general formula $R1R^2N(CH_2)_nN\ R1R^2$ during or before dyeing or by introducing electronegative groups in place of $-NH_2$ groups. Azo disperse dyes mostly remain stable against gas fumes except a blue dyes.

18.12   References

ANON (1981) 'High-temperature steam fixation of disperse dyes on polyester', *Textile Chem. Color.*, **13**, 4, 92–98.

BEFFA E and STEINER E (1973) 'Developments in metal-containing dyes', *Rev. Prog. Color Related Topics*, 4, 60–63.

BRAUN H (1983) 'Particle size and solubility of disperse dyes', *Rev. Prog. Color Related Topics*, **13**, 62–72.

Colour Index International (1987), 3rd Edition (3rd Revision), Bradford, UK, Society of Dyers & Colourists and AATCC.

DAWSON J F (1972) 'Developments in disperse dyes', *Rev. Prog. Color Related Topics*, **3**, 18–25.

DAWSON J F (1978) 'Developments in disperse dyes', *Rev. Prog. Color Related Topics*, **9**, 25–35.

GULRAJANI M L (1987) *Dyeing of Polyester and its blends*, Delhi, Indian Institute of Technology.

HEIMANN S.(1981) 'Textile auxiliaries: dispersing agents', *Rev. Prog. Color Related Topics*, **11**, 1–8.

LEADBETTER P W and LEAVER A T (1989), 'Disperse dyes - the challenge of the 1990's', *Rev. Prog. Color Related Topics*, **19**, 33–39.

MITTAL R M and TRIVEDI S S (1983) *Chemical Processing of polyester / cellulosic blends*, India, Ahmedabad Textile Industries Research Association.

MURRAY A and MORTIMER K (1971) 'Carrier dyeing', *Rev. Prog. Color Related Topics*, **2**, 67–72.

NUNN D M (1979) *The dyeing of synthetic-polymer and acetate fibres*, Bradford, UK, Dyers Company Publication Trust.

SLACK I (1979) 'Effects of auxiliaries on disperse dyes', *Canadian Textile J.*, **96**, 7, 46–51.

19

Colouration with pigments

Abstract: Pigments are water and solvent insoluble, possess no affinity for any fibre neither substantive and are not viable for exhaust dyeing. These are applied alongwith a binder when the latter reacts with fibre and traps pigment in between. Pigments are suitable for all fibres and blends. Absence of migration and diffusion develops one sided coloured effect when printed. Dyeing is carried out with a viscous pad liquor consisting of kerosene based emulsion thickener. Dyeings become stiffer due to presence of binder film; however no finishes or post-washes are imparted to cut down cost of dyeing.

Conventional thickening agents are not applied.

Keywords: Pigment, emulsion thickener, binder, catalyst, kerosene

19.1   Fundamentals of application

Pigments are colouring materials, often recommended for pastel to light shades and have the advantage, under certain conditions, of enabling dyeing and finishing to be combined in a single process (Carlier, 1993; Trivedi et al, 1976). Pigments do not possess affinity for nor react with textile fibres and are applied through a thickening system alongwith a binder and acidic catalyst. The highly viscous binder retains pigments and reacts with fibres during which the pigment get trapped in between the layers of binder and textile. Pigments remain embedded in transparent thermosetting binder film on the surface of fibre and do not diffuse at the interior. The binder–textile attachment occurs in the same way as that with an anti-crease precondensate and cellulose (Teli, 1985). Methylol precondensates are also used to form the film to retain pigments as well as to develop anti-crease finish simultaneously:

$$\text{Precondensate-CH}_2\text{OR} + \text{HO-Cell} \xrightarrow{\text{pH} \sim 3} \text{Precondensate-CH}_2\text{O-Cell} + \text{HOR}$$
[Methylol radical as cross-linking group] $[\because R = H, CH_3]$

202

When simple binders are used to form the film to retain pigments only:

pH ~ 3

Binder-COOH + HO-Cell   ⟶ Binder COO-Cell   + HOR

[Carboxy radical as film-forming group]   [∵ R = H, CH$_3$]

The complete attachment between binder and cotton seems to be as depicted in Fig. 19.1 while the location of fibre, film formed by binder and pigments are explained in Fig. 19.2. Needless to say, the longevity of coloured effect depends on force of chemical attachment between binder and fibre; the stringer the link the better the durability.

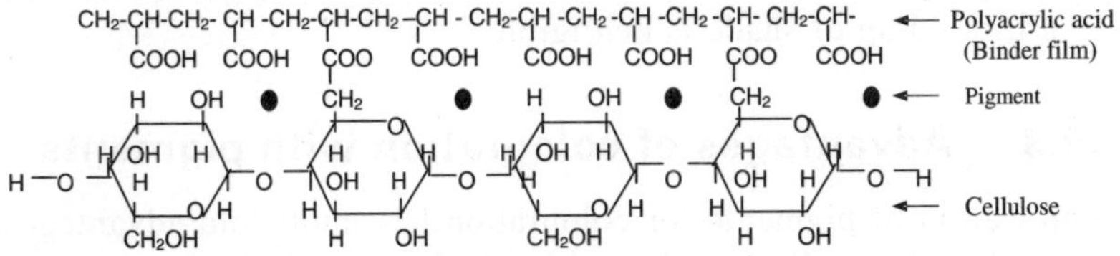

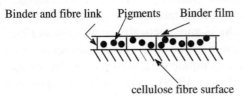

19.1 Attachment between binder and cellulose trapping pigment.

19.2 Location of fibre, film formed by binder and pigments.

The recipe for colouration with pigment essentially comprises a pigment, binder, thickener and an acidic catalyst – the catalyst develops required pH at higher temperature and initiates polymerization of binder.

19.2   Properties of pigments

Pigments belong to organic and inorganic origins, insoluble in water as well as most of organic solvents, viz. white spirit, chloroform, per and trichloroethylene and carbon tetrachloride. Due to lack of affinity for textiles, inorganic pigments are developed *in situ* fibre while organic pigments are invariably applied through concentrated padding liquor for dyeing or paste for printing (Kongdee and Bechtold, 2004; Kramrisch, 1986; Aspland, 1993). For efficient colouration, pigments should exhibit good fastness to light, wash, gas, chlorine, alkali, peroxide, perspiration, solvents and rubbing; emulsion thickenings used for dyeing liquor or print paste should ensure sharp impression, develop good brilliancy and should have good reproducibility. However, stability of paste or liquor depends on pigment particle size and its distribution, dispersibility, oil absorption rheology, electrical charge, specific gravity, crystal structure, hardness, solubility in solvent, purity in the intermediates used in pigment manufacture and flocculation of pigment. A particle size around 0.05– 2.0 μ ensures good colour yield with hiding power and brilliance. Coverage and light fastness increase with increase in particle size, but with a consequence of decrease in colour strength; particle size influences the processing characteristics too (Gunthert et al, 1989). Due to lack of affinity, 'build up' of shade is impossible rather a higher dose of pigment around 10 g/l is used – beyond which rate of build up is reduced and beyond 20 g/l, build up of shade is negligible.

19.3   Advantages of colouration with pigments

Application of pigments for colouration has multi-fold advantages, viz. (i) applicable to all natural, man made fibres and blends due to lack of affinity, (ii) full shade range is available, (iii) compound shades are produced with ease, (iv) shade matching is easier as no change in hue during colouration, (v) solid shade on fabric of cheaper quality, (vi) process is pad-dry-cure, no post-washing, and practically no waste-water load, (vii) cheaper and cost of production is too less, (viii) subsequent easy care finishing is omitted or dyeing liquor may incorporate anti-crease resin (Kruger and Thonig, 1972; Harper, 1989), (ix) colouration liquor and paste show good stability and (x) the process is simple and excludes complications (Schwindt and Faulhaber, 1984).

19.4   Related problems

Pigment dyeing is associated with a few problems, viz. (i) kerosene is generally used which raises chances of fire during polymerization of binder, (ii) shades become progressively dull with increase in number of washing cycles, (iii) screen may get choked giving light prints after working for some time, (iv) sharp impression is obtained with smooth paste or efficient liquor dispersion, (v) permanency of colouration depends on specific binder, (vi) printed wool becomes hard and hence not used, (vii) washing of the back-grey is troublesome, (viii) though colouration occurs on both the sides in dyeing, the interior of fabric remains undyed due to absence of diffusion, (ix) fastness to light, wet rubbing and dry cleaning is not so good, (x) binder stiffens handle – the problem gets worsened with deep shades due to use of more binder, (xi) auxiliaries added in recipe may show influence on binder to cause build up on padding rolls or sticking-in roller engraving, etc., (xii) deep shades can not be produced with good fastness properties and (xiii) brilliancy in shades like those produced with dyes is not available (Schwindt and Faulhaber, 1984).

19.5 Chemical specification of pigments

19.5.1 Inorganic pigments

These include white powders of ZnO, ZnS, TiO_2, white lead [basic lead carbonate, i.e. mixture of $2PbCO_3$. $Pb(OH)_2$], FeO (black), Fe_2O_3 (red brown), etc. Out of these, ZnO and TiO_2 have found their use in sizing and whitening of textiles. Combination of inorganic pigments is very useful too. Fe_2O_3 in combination with Cr_2O_3 produces Mineral khaki at different tones, silicates of copper and calcium produces Egyptian Blues (C I Pigment Blue 31, C I 77437), various Prussian Blues are produced through reaction of $FeSO_4$ with $K_4Fe(CN)_6$ (C I Pigment Blue 27, C I 77510, C I 77520, etc.), Ultramarine Blue (C I Pigment Blue 29, C I 77007) are the examples of combined inorganic pigments. The C I generic number from 77000 to 77999 has been assigned in colour index to all inorganic pigments, whether alone or in combination. For further details, readers are referred to more detailed literatures (Allen, 1971; Colour Index, 1987).

19.5.2 Organic pigments

Organic pigments are of three types:

(i)   Carbon Pigments, e.g. Carbon Black from natural gas, acetylene black from CaC_2, Lamp Black from soot of incomplete combustion of natural oils, etc. These have been assigned nomenclature in colour index as C I Pigment Blacks 6-10 with C I numbers 77265-68.
(ii) Lake type pigments, which are the metal salt derivative of anionic dyes; the final colour depends on metal salt used, and
(iii) Non-ionic organic pigments, which are mainly of azoic type.

C I specification and structure of few lake and non-ionic-type pigments are shown in Scheme 19.1(Allen, 1971; Colour Index, 1987).

19.6 Pigment

Due to lack of affinity, pigments are applied through padding along with a binder pre-condensate and catalyst solubilised in emulsion thickening followed by drying and curing at 140–150°C in polymerizer for 3–5 min, when the binder develops a thermosetting transparent polymer film entrapping pigment particles and also react with fibre. The more the reaction of binder with textile, the better is the stability of effect. Improper

Lake-type pigments

Al-salt of

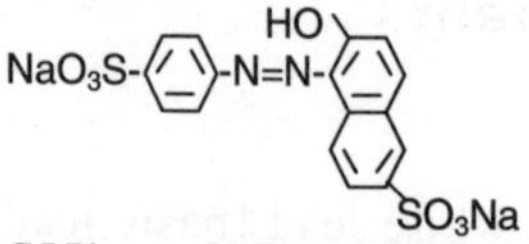

C I Pigment Yellow 104, C I 15985:1
(Bright Yellowish Orange)

Ba-salt of

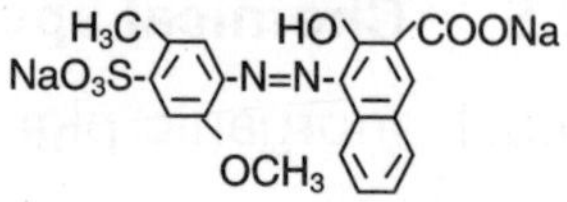

C I Pigment Red 56, C I 15870
(Bright Bluish Red)

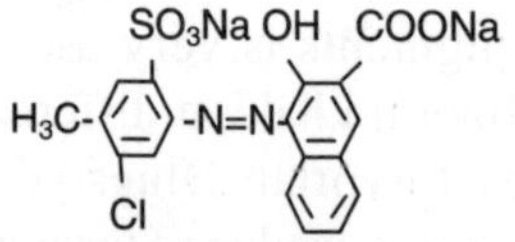

C I Pigment Red 48:1, C I 15865:1 (Bright Yel Red: Ba-salt)
C I Pigment Red 48:2, C I 15865:2 (Bright Red: Ca- salt)
(C I Pigment Red 48:3, C I 15865:3 (Red: St- salt)
(C I Pigment Red 48:4, C I 15865:4 (Bright Red: Mn- salt)

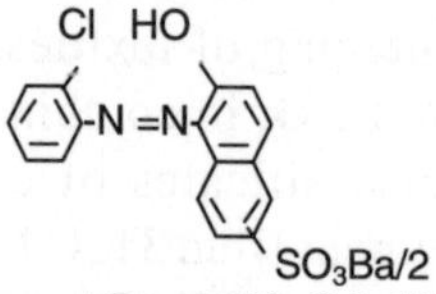

C I Pigment Orange 19, C I 15990
(Bright Reddish Orange)

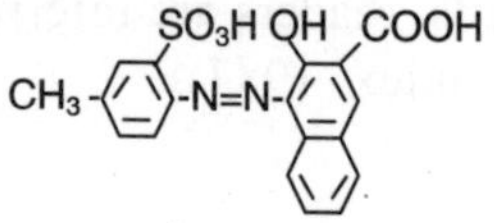

C I Pigment Red 57, C I 15850

(C I Pigment Red 57:1, C I 15850:1, Red, Ba-salt)

(C I Pigment Red 57:2, C I 15850:2, Red, Ca-salt)

Ba-salt of

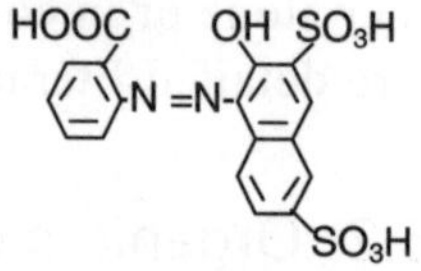

C I Pigment Red 60, C I 16105

(Bright Red)

Non-ionic pigments

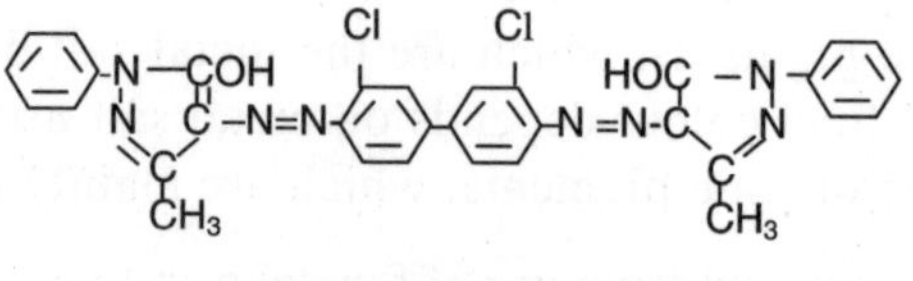

C I Pigment Orange 13, C I 21110
(Reddish Orange, fast to light and solvent)

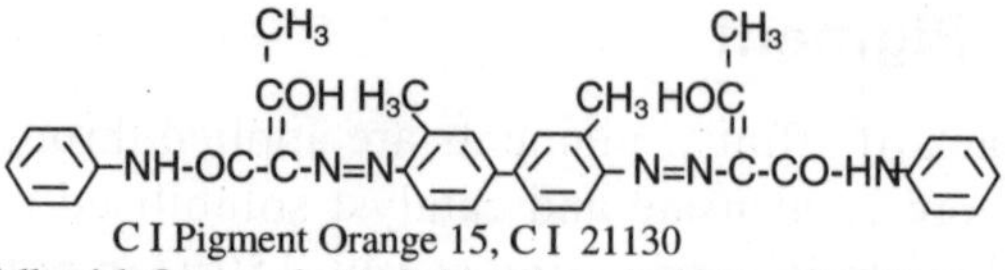

C I Pigment Orange 15, C I 21130
(Yellowish Orange, fast to light and poorly fast to solvent)

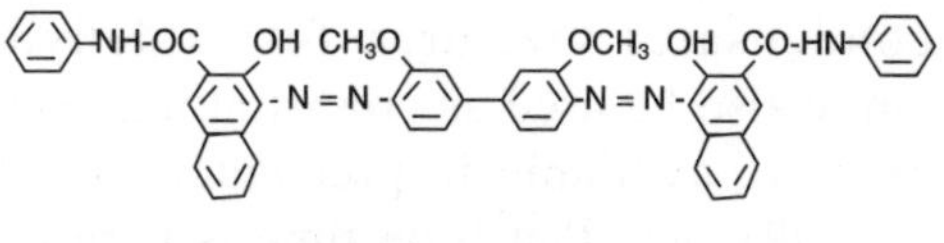

C I Pigment Blue 25, C I 21180
(Reddish Navy)

Scheme 19.1 Chemical specification of few lake and non-ionic type of pigments.

polymerization of binder implies its inadequate reaction with textile (determined by polymerization time, temperature, catalyst and type of binder) resulting poor wash fastness with stripping of colour during each progressive washing cycle. The size and shape of pigments are of due importance since small variations may seriously affect hue and tinctorial value; particle size should be as small as possible for sharp impression with better life and to achieve it, milling of pigment is done in presence of an emulsifying or dispersing agent; over-milling results loss of brilliancy and dulling of shade. Due to inherent tendency to absorb anions, pigments get negatively charged – anionic dispersing agents enhance negative charge. Free $-NH_2$ groups in pigment suppress negative charge and make it positively charged necessitating right selection of binders.

Pigment dyeing involves large amount of binder even to produce light shades which adversely affect handle. A new range of 'soft' pigment dyes from Ciba, known as Irgaphor SPD (Soft Pigment Dyeing), can be mixed with water to produce ready-to-use padding liquor. It involves a one-step process and does not from film on the fabric rather the pigment is fixed directly on the fabric (Anon, 2002).

19.7   Binder

These are simple chemicals alone or in combination, which forms high molecular weight transparent film through polymerization, establish chemical bonds with fibre and traps pigment molecules in between (Pfeiffer, 1975). Binders are applied by dissolving in solvents; during polymerization, solvent evaporates with formation of fine thermosetting film (Wiedemann et al, 1979a, 1979b). Fastness properties of coloured textile greatly vary with quality and stability of the binder film – incorporation of reactive groups in binder enable efficient linking within its molecules and that with fibre by a simple heat treatment with a consequence of enhancement in its chemical and physical resistance. Various olefinic unsaturated monomers, viz. acrylic acid and its derivatives (especially esters), butadiene, vinyl acetate, etc., are commonly used binders and are polymerized in emulsion polymerization technique, whereas few precondensates viz. urea formaldehyde, melamine formaldehyde, etc., may also be used as cross-linking agent cum binder. Commercially binders are available in precondensate state (partially polymerized with low DP) such that little time is required to complete polymerization. Binders generally form cross-link with each other – with hydroxy, amino and other groups of fibre and / or with emulsifying or thickening agents in paste or pad liquor during curing of prints. The emulsion polymerization is of condensation type which liberates water or methanol; rate of cross-linking with fibre and within binder are influenced by extraction of these by-products from treated textiles. In practice, curing is done at high temperature (140–150°C) at which either of water or methanol evaporates (methanol bp ~ 64°C) enhancing forward reaction.

Vinyl binders can be used alone or in combination to develop specific nature of film depending on end use, e.g. styrene, methyl methacrylate, vinyl cyanide and vinyl chloride develop hard and brittle film; butyl acetate and butadiene form tacky film. Components from these two groups can be mixed at right proportion to develop film with required characteristics in relation to textile monomers with reactive groups are also added to improve stability of film (Schwindt and Faulhaber, 1984).

Binder used for colouration should fulfil certain criteria's, viz. (i) it should not get coagulated due to shear forces operating during padding, (ii) the film must be colourless with desired handle, (iii) it should be pliable, (iv) should establish chemical linkage with fibre, (v) resistant to mechanical stresses, (vi) resistant to chemical and solvent action, and (vii) stable on exposure to light (Teli, 1985).

In case, a binder has no functional groups to activate self cross-linking, an other cross–linking agent such as urea formaldehyde or melamine formaldehyde condensate, etc., having at least two reactive groups per molecule is incorporated in the binder in small quantities even when the binder has self cross-linking groups too, since these help in better cross-linking and improve fastness properties.

Binders commonly used are of three types: (i) water soluble (i.e. resin binder), (ii) solvent soluble and (iii) emulsion miscible. Water-soluble binders are precondensates of urea formaldehyde, melamine formaldehyde, etc., (Lyofix A, Lyofix EH, etc.). Solvent soluble binders are fixed through drying (Sericose LC or IG); tricresyl phosphate may be used as a plasticizer to get sharp prints with bright shades and soft handle. Emulsion binders are applied through an emulsion and are of two types: (i) water in oil and (ii) oil in water (Humphries et al, 1985). In the first type, water is emulsified in oil medium with emulsifiers and requires organic solvent for cleaning of container, a serious problem why this system has been obsolete. The second type includes oil in dispersed state in water with emulsifier, which can be diluted with addition of excess water, cleaning is simple with water, use of plasticizers improve handle of textile; mostly kerosene (alternately turpentine) is used as oil.

19.8   Catalyst

Catalysts liberate acid at higher temperature to generate desired pH required for polymerization of binders or cross-linking agents to form the desired film (Modi et al, 1976; Parikh and Reinhardt, 1979). Numerous catalysts are invariably used for cross-linking, e.g. zinc nitrate, magnesium chloride, aluminium chloride, zinc chloride, ammonium salts, etc., out of which ammonium salts, viz. ammonium sulphates / chlorides / nitrates / phosphates are of importance while diammonium phosphate is invariably used in this context to initiate polymerization of binder at140–150°C for 4–5 min. Ammonium chloride is strongly acidic – stability of padding liquor gets reduced by starting polymerization in liquor itself, leaving lesser number of reactive sites in binder for reaction with fibre. This affects rubbing fastness due to inadequate reaction of binder with textile. A sulphamate based catalyst LCP developed by ATIRA is effective at a curing temperature of 110°C for 2–3 min, though diammonium phosphate (0.5–0.8%) is the commonly used catalyst in all the industries (Anon, 1973; Modi et al, 1978; Modi .et al, 1977; Trivedi, 1976; Shah, 1976).

19.9   Thickening system

Conventional thickeners, viz. hydroxyl ethyl cellulose, methylcellulose, PVA, alginate and gum tragacanth are not used in colouration with pigments except at very lower concentration just to balance the viscosity of padding liquor or print paste (optional); if used, get trapped between fibre and binder film that happens with pigments, react with binder and removal becomes impossible from coloured textile. As a result, the handle and related aesthetics get deteriorated. Conventional thickeners generally increase water content of liquor or paste and prevent drying up of the same – even problem related to removal of these in post-colouration stage has given birth of emulsion thickener concept in pigment processing for improved quality of coloured textile. In emulsion thickener, fine droplets of oil are dispersed and stabilized in water phase in presence of an emulsifying agent – stability of emulsion depends on extent of dispersion, type, quality and concentration of emulsifier used; both oil and water get evaporated during curing developing a soft handle. The most commonly used emulsions are made with kerosene in water along with an emulsifier. Emulsion thickeners leave no residue on textile and so are popular. It has been reported that out of a kerosene / water emulsion of 70:30 mixture that is commonly used, only 20–45% kerosene can be replaced with sodium alginate and primal ASE 60 without adverse effect on handle, while CMC, guar gum, etc., in emulsion produce low colour yield, poor scrubbing fastness and harsh feel. Very little deficiency in colour yield through partial replacement of kerosene with sodium alginate than that with emulsion alone could be enhanced by adding calcium chloride and paraffin oil in the liquor or paste.

19.10 Application

Padding liquor is prepared for dyeing, while a stock and print paste are separately prepared for printing. Padded or printed textile is dried and polymerized at 140–150°C for 4–5 min. Washing, soaping are optional and generally not imparted.

For dyeing, A 100 l padding liquor is prepared with Acramine SLN binder (10 kg), urea (2 kg), alginate paste (5%, 32 kg), kerosene (40 kg), diammonium phosphate (1 kg), pigment (x kg) and rest water.

For printing, the stock and print pastes are formulated separately as follows:

Stock paste			*Print paste*		
Emulsifier	–	2%	Pigment	–	x%
Kerosene	–	70%	DAP	–	3%
Binder SLN	–	15%	Urea	–	2–2.5%
Water	–	13%	Stock paste	–	rest
		100%			100%
(may or may not be boiled)					

Popular damask effect can be developed on white textile using titanium dioxide along with zinc oxide as pigment, when a white printed effect is obtained; for opaque colour effect along with TiO_2 coloured pigment is to be added.

Emulsion is prepared by dissolving Binder SLN in kerosene (70 parts) with high speed stirring followed by addition of urea, emulsifier and water (28 parts) and stirring is continued till a smooth white emulsion is obtained in which urea and excess water may be added with finally addition of pigment.

In printing, electrolytic effect of catalyst corrode doctor blade and engraved copper roller not due to its acidic nature; extent of corrosion with catalyst LCP is higher than that with DAP and is lower than with ammonium chloride and nitrate. Most efficient corrosion inhibitors, e.g. alkali phosphate, thiourea, formaldehyde, etc., are used with catalyst LCP to reduce the problem.

In dyeing, deposition of binder film on padding rolls occurs when catalyst LCP is used due to (i) higher acidic pH of padding liquor due to excessive acidity in the catalyst, which results in the premature polymerization of the binder, and (ii) reduced stability of bath, which results in separation of binder phase under the stresses arising at the nip of padding rolls. Most of kerosene is to be evaporated from padded or printed goods through drying to avoid explosion during curing.

Pigment printing over pigment dyed ground can be affected by padding with dyeing liquor followed by drying and printing with pigment, drying and curing. The problem of 'facing' (one sided print) is due to use of emulsion system, which does not diffuse through textile carrying pigment. This can be reduced by adjustment in viscosity of paste according to the depth of engraving – viscosity should be more with deep engraved rollers or any suitable emulsifier to the print paste. Fastness of prints can be improved by incorporating a fixing agent.

19.11   Garment dyeing with pigments

Garment dyeing is increasingly being adopted by manufacturers in order to offer unique colours and respond to changes in colour fashion. Differential construction of garments poses trouble in level dyeing with dyes due to affinity of the latter and differential diffusion in side fibre; dyeing of garments with pigments overcome this problem. The main usage of pigments on garments is to create a distressed or faded out look which is not possible with dyes (Anon, 2004) and fashionable ready-worn and wash-down effects can be achieved with greater ease (Lever, 1992). Pigments are feebly anionic, pre-treatment of garment with cationic reactant ensures even dyeing; cationization prior to dyeing also reduces waste-water load (Vishwanathan, 2004). Knitted garments, which can not be padded, may be dyed with pigment through exhaust technique (Shah, 1998; Chong et al, 1992).

19.12   Substitute for kerosene

Kerosene or turpentine oil develops sharp impression or bright shade with superior fastness while diesel oil impairs whiteness and creates dropping trouble in drying and polymerizing with poor fastness. When kerosene is substituted up to 40% with alginate, colour yield, handle and fastness properties remain excellent, but 20–40% substitution with Primal ASE 60 gives good colour yield, handle and wash fastness with little poor light fastness. Chemicals like Kerotex and Aquasynth XLT are useful to substitute 20% and 30% kerosene respectively with lower cost of processing as well as acceptable quality of processed textiles (Varghese, 1982). Attempts to substitute kerosene partly by using other thickeners normally reduce rubbing fastness and stiffer textile – the problem can be sorted out by evolving a water based emulsion binder system. Gum Indalca-U (1%) can replace maximum 20 kg of kerosene per 100 kg of paste with same handle and fastness properties, beyond which feel of fabric becomes harsh to an unacceptable level and fastness are also adversely affected.

Polyacrylic acid $(CH_2=CH–COOH)_n$, on neutralization with ammonia gives ammonium polyacrylate and develops viscosity (Bhagwat and Srivastava, 1984). However the thickened solutions become very 'long' or stringy and exhibits wrong rheology. To overcome this problem, a little cross-linking agent is added which allows polymer particles to swell (pH~4) but not to dissolve, a situation like kerosene in water emulsion – where oil droplets are dispersed in water phase and their relative movement get restricted resulting increased viscosity. In a true emulsion, globules or droplets can get deformed more easily under shear stress. In a paste containing ammonium polyacrylate, deformation of polymer particles occur in the direction of flow and the relative ease with which the particles can move past each other in the direction of flow. When the shear stress is removed, the globules return to their original shape and develop required viscosity. This pseudo-plastic behaviour can be employed to prepare paste of manageable viscosity. One of such commercial product is 'Alcoprint PTF', where the curing conditions remain unchanged, though uniformity and size of pigment particles contribute to the product stability (Humphries et al, 1985).

19.13 References

ALLEN R L M (1971) *Colour Chemistry*, London, Thomas Nelson and Sons Ltd.
ANON (1973), 'Development of a low temperature cure catalyst for pigment printing', *ATIRA Tech. Digest*, **7**, 3, 10–14.
ANON (2002) 'Soft pigments - A new dyeing concept', *Maschen Ind.*, **2**, 30–31.
ANON (2004) 'Garment pigment dyeing', *Int. Textile Bulletin*, **50**, 5, 72–77.
ASPLAND J R (1993) 'A series on dyeing - Chapter 14: pigments as textile colorants: pigmenting or pigmentation', *Textile Chem. Color.*, **25**, 10, 31–37.
BHAGWAT M M AND SRIVASTAVA H C(1984) 'Synthetic thickeners for pigment printing', *ATIRA Tech. Digest*, **18**, 3, September, 105–111.
CARLIER F (1993) 'Pigment dyeing of cotton', *Ind. Textile*, **1242**, April, 48–50.
CHONG C L, LI S Q AND YEUNG K W (1992) 'Pigment dyeing study of cotton garments by exhaustion', *Am. Dyestuff Rep.*, **81**, 5, 17–26, 63.
Colour Index International (1987), 3rd Edition (3rd Revision), Bradford, UK, Society of Dyers & Colourists and AATCC.
GUNTHERT P, HAUSER P AND RADTKE V (1989) 'Effect of pigment particle size on application properties', *Rev. Prog. Color Related Topics*, **19**, 41–48.
HARPER R J (1989) 'Cationic polyacrylates for garment dyeing', *J. Coated Fabrics*, **18**, April, 234–245.
HUMPHRIES A, MUFF J R AND SEDDON R (1985) 'Aqueous system for pigment printing', *Colourage*, **32**, 5, March 7, 15–27.
KONGDEE A AND BECHTOLD T (2004) 'In-fibre formation of $Fe(OH)_3$ - a new approach to pigment coloration of cellulose fibres', *Dyes & Pigments*, **60**, 2, 137–142.

KRAMRISCH B (1986) 'Pigment printing and dyeing of cotton', *Am. Dyestuff Rep.*, **75**, 2, 13 and 43.

KRUGER R AND THONIG W (1972) 'Combined pigment-dyeing and finishing of knitted goods', *Textil Praxis Int.*, **27**, 1, 43–44.

LEVER T (1992) 'Focus on cotton. Exhaust dyeing with pigments on cotton piece and garments', *J. Soc. Dyers Color.*, **108**, 11, 477–478.

MODI J R, PALEKAR AW AND TRIVEDI S S (1976) 'Catalyst systems for printing and dyeing with pigment colours', *Textile Dyer Print.*, **9**, April 8, 53–56.

MODI .J R, PALEKAR A W, MALI N C AND SHAH N C (1977) 'New catalyst systems for energy conservation in pigment printing and resin finishing', *Colourage*, **24**, 1, 23–28.

MODI J R, MITTAL R M AND SHAH N C (1978) 'Evaluation of catalyst LCP with different binders', *ATIRA Tech. Digest*, **12**, 3, September, 4–12.

PARIKH D V AND REINHARDT R M (1979) 'Low-energy curing catalytic systems for finishing treatments of easy-care, pigment padding and pigment printing', Proceedings: National Technical Conference, AATCC, NC, 238–252.

PFEIFFER G (1975) 'Pigment binders in continuous dyeing', *Chemiefasern / Textil Ind.*, **25/77**, 1, 62–67.

SCHWINDT W AND FAULHABER G (1984) 'Development of pigment printing over the last 50 years', *Rev. Prog. Color Related Topics*, **14**, 166–175.

SHAH D L (1998) 'Garment dyeing with pigment colours by exhaust process', *Man Made Textiles India*, **41**, 10, 445–446.

SHAH J K (1976) 'Practical experience with the use of pigments and catalyst LCP', *Textile Dyer Print.*, **9**, 8, April, 57–58.

TELI M D (1985) 'Choice of binder in pigment printing', *Textile Dyer Print.*, **18**, 18, Aug 28, 21–25.

TRIVEDI C G (1976) 'Practical experience with the use of pigment and catalyst LCP', *Textile Dyer Print.*, **9**, 8, April, 59–60.

TRIVEDI S S, GOKHALE S V AND MODI J R (1976) 'Theory and practice of pigment dyeing and printing', *Textile Dyer Print.*, **9**, 8, April, 50–52.

VARGHESE J (1982) 'Energy saving in pigment printing and resin finishing of textiles, *BTRA Scan*, **13**, 3, September, 16–22.

VISHWANATHAN N (2004) 'Pretreatment and cationization in garment pigment dyeing', *ATA J.*, **15**, 3, 64–66.

WIEDEMANN G, FRENZEL H AND MATZKE M (1979a) 'Crosslinking of polymeric binders in pigment dyeing and pigment printing of textiles', *Textiltechnik*, **29**, 2, 112–117.

WIEDEMANN G, FRENZEL H AND MATZKE M (1979b) 'Crosslinking of polymeric binders in pigment dyeing and pigment printing of textiles', *Textiltechnik*, **29**, 3 160–163.

Abstract: Silk retains both basic and acid groups on terminal ends
raising chances of dyeing with direct, acid, basic, metal complex,
mordant and reactive dyes when dye–fibre attachment takes place by
chemical means. Acid dyes produce bright light fast shades with
poor wash fastness and inadequate coverage of dye. Metal-complex
dyes produce dull shades with good fastness. Reactive dyes produce
bright shades with good wash, light and perspiration fastness.
Natural dyes produce only moderate bright light shades on mulberry
silk; dyes often require a mordant to make the colour permanent
while basic dyed silk suffers from poor light fastness.

Keywords: Silk, dye, fastness, coverage, brightness

20.1   Introduction

Raw silk is obtained as long filaments having around 275–900 m length;
a continuous protein filament spun by silk worm and has two layers. The
central layer known as fibroin constitutes the main fibre while the outer
layer is a gum, called sericin. Silk is available in different varieties, viz.
mulberry, wild, tussar, muga, eri, spider, etc. Silk industry is divided into
four sectors: sericulture, reeling, throwing and manufacturing. Sericulture
consists of production and culture of silk worms, their care, formation
and harvesting of cocoons. In reeling, long strands of silk fibres are
unwound from the cocoons, two or more filaments being combined to
make raw silk thread. When silk is to be reeled, the cocoons are softened
in hot water to loosen the outer mass, which are then removed. Reeled silk
is unsuitable for either weaving or knitting. It must be "thrown" or doubled
and the throwster takes the proper number of reeled threads together and
gives them a twist to obtain yarns suitable for weaving and knitting.

Raw silk possesses sericin as the outermost protective layer, also known
as gum, restricts penetration of chemicals in raw silk and is removed
through degumming while inherent yellowish colour necessitates
bleaching. Silk is easily attacked by UV rays of sunlight and is more
sensitive to light. Silk loses a part of its inherent lustre if treated in boiling
water.

20.2   Dyes for silk

Silk retains $-NH_2$ and $-COOH$ groups at either ends of its chemical structure and obviously it can be dyed with acid, basic, metal-complex, mordant, natural, reactive dyes. The dye–silk interaction occurs through ionic or covalent bonding. Application of vat colours is restricted due to very high alkaline pH of dyeibath, which may damage silk if the bath is heated up. Choice of colour mainly depends on shade, brightness and fastness grades (Gulrajani and Gupta, 1989). Acid and metal-complex dyes possess better affinity for fibre and are easily absorbed but have poor to moderate washing fastness. Reactive dyes offer good wash fastness and full range of dischargeable colours with high perspiration fastness while basic dyed silk possesses very poor light fastness with light shades due to formation of coloured cation which does not permit reaction with protonised amino groups when dyed from acidic bath.

20.3   Dyeing of silk with various dyes

20.3.1 Dyeing with acid dyes

Acid dyes possess lower affinity for silk as compared to that for wool and obviously lesser problem is encountered in level dyeing. Super milling dyes show very good fastness due to a long alkyl side chain in the molecule (Gulrajani, 1993).

Equilibrium uptake of acid dyes due to electrostatic or columbic interaction between dye anions and the fibre cations is known as sorption isotherm of 'Langmuir' type (Peters, 1975). Water soluble dyes have hydrophilic sites such as sulphonate groups (SO_3Na), but these also possess hydrophobic parts like benzenoid and aliphatic hydrocarbon chains. A hydrophobic dye–fibre bonding is also possible between the sparingly water soluble, large molecular weight dyes and fibres. Interaction between dye and silk varies from case to case as well as dye to dye and lies somewhere on line between the electrostatic (columbic) attraction of acid dyes (Langmuir type) and hydrophobic bonding of disperse dyes of Nernst type (Peters, 1975; Bird and Boston, 1975; Lewis, 1992).

Exhaust dyeing at neutral or acidic pH is the widely used method which starts at 40–45°C followed by addition of CH_3COOH (2–4%) and Na_2SO_4 (5–10%) in bath to maintain a pH~4–5. Dye is added and dyeing is continued for 10 min. Temperature is raised to 80–85°C in 45 min; dyeing at higher temperature beyond this reduces lustre. Dyeing is continued for further 30–90 min to promote diffusion of dye into fibre improving wash and rubbing fastness. A thorough wash removes superficial dyes. Strong acid dyes can also be applied at pH~3–4 using 1–3% HCOOH (85%) and 5% glauber's salt.

Acid dyes produce brilliant shades with good light fastness but poor to moderate wash fastness which can be improved by treatment with cationic dye-fixing agent (2–4 g/l) at 40–50°C for 20 min. Acid dyes generally show poor coverage on silk necessitating topping with basic dyes to produce levelled brilliant shades.

20.3.2 Dyeing with metal-complex dyes

Silk looses a part of its inherent lustre when treated in boiling water and to preserve aesthetic appeal, 1:2 metal-complex dye is the best choice for dyeing silk at 90°C in presence of either acetic acid alone or in combination with ammonium sulphate (5%); presence of ammonium sulphate as buffer increases dye uptake significantly (Muralidharan and Dharmaraj, 1993; Datye and Vaidya, 1984; Gulrajani and Balaji, 1992). After-treatment with cationic dye-fixing agent (2–4 g/l) at 40–50°C for 15–20 min indeed improves wash fastness (Gulrajani and Gupta, 1989). Metal-complex dyes adsorbed on silk inhibit marginal migration due to high stability of the dye–fibre linkages and produce invariably dull shades with good light and wash fastness.

20.3.3 Dyeing with mordant dyes

Mordant dyes are acid dyes having chelating sites to form stable coordination complex with metal ions from metal salts (mordants). Dyes can form chelates with different mordants to develop various shades with superior wash fastness. $Na_2Cr_2O_7$ and $K_2Cr_2O_7$ are mostly used mordants for wool because of presence of reducing thiol groups; absence of such reducing groups in silk can be overcome by using basic salts of chromium to form dye-chromium complex (Gulrajani et al, 1992). The dye–mordant reaction or chelation requires presence of salt forming groups, such as hydroxy or nitroso groups as well as presence of oxygen or nitrogen containing groups, such as carbonyl, carboxyl, azo, etc., so that a lone pair electron can be donated to the chromium atom.

Dye bath is set up with glauber's salt (10%) and acetic acid (4%). Temperature is raised to 50°C and silk is entered in the bath, temperature is raised to 90°C over 30 min after which formic acid (2%) is added for complete exhaustion of dye. Dyeing is continued for further 15 min and then dried without washing. Mordanting is done with mordant solution (3%) for 16 h at 30°C followed by washing and drying; amount of mordant exactly required depends on number of dye molecules on fibre as the chelation occurs at stiochiometric ratio. However, mordant dyes have restricted application on silk due to formation of dull shades and deposition of chromium on silk imparts rigidity to it with loss of its characteristic scroop.

20.3.4 Dyeing with natural dyes

Natural dyes produce moderate bright colours on mulberry silk but deep shades can be produced on tussar, spun and textured varieties only. These dyes often require a mordant to make the colour permanent. A few dyes have good light and wash fastness though some of the mordants are harmful to silk. Colouring matter is extracted from roots, stems, leaves, barriers and flowers of various plants as well as from certain insects and shell-fish by an elaborate series of processes used. Bright shades are produced with turmeric, berberis, dolu (yellow), annato (orange) and henna (brown). Substantivity of these dyes for silk could be partly due to presence of tannins which act as natural mordant.

Dye assistants include acetic acid to neutralize calcareous water, cream of tartar to brighten colours when used in conjunction with mordants; Na_2SO_4 to control level dyeing and use of boiled-off liquor causes the colouring matter to be attracted more slowly and evenly by silk and helps to preserve lustre.

Logwood black is applied on silk invariably. It is extracted from campeachy wood which contains a glucoside of haematoxylin. Fermentation of glucosides gives glucose and haematoxylin, the latter gets oxidized to form haematin which is the active dye and develops blue and black shades with chromium compounds. Dyeing of silk with this dye is popular due to beauty and bloom of the shade alongwith increased weighing, which results in heavier yarns with fuller handle. Wash fastness is very good which can further be improved by after-treatment of dyeings with $K_2Cr_2O_7$, $Na_2Cr_2O_7$ or $CuSO_4$ (Allen, 1971). Further details of natural dyes and their application on silk have been well documented in numerous literatures (Nishida and Kobayashi, 1992; Agarwal et al, 1992; Gulrajani et al, 1992a, 1992b; Siri, 2000).

20.3.5 Dyeing with reactive dyes

Reactive dyes produce bright shades with good wash, light and perspiration fastness properties due to its reaction with $-NH_2$ group of silk; a covalent bond is formed between reactive groups of dye and nucleophiles on fibre (Gulrajani, 1993). Similar reaction also takes place with sericine if present causing loss of colour value and so thorough degumming is a prerequisite for level dyeing. Weighted silk can also be dyed but build-up is poor.

Maximum fixation of dichlorotriazine dyes occur in weakly acidic medium at 60°C in 2 h, dye uptake depends on free $-NH_2$ and $-NH$ groups of histidine ; fixation can be increased by after-treating with 2% aqueous pyridine or 1% sodium bicarbonate to enhance reaction of dye with hydroxyl groups of tyrosine, serine and threonine. It has been reported that silk could be best dyed at 70°C for 1 hr at pH ~ 5.6 in presence of Na_2SO_4 (4 g/l). These dyes can also be applied at low temperature in pad-batch method, in which fabric is padded with dye solution at pH~8 and then treated for 5 h at 20–22°C, followed by cold and hot wash, soaping at 65°C for 15 min with a final cold wash. Impregnating fibres in diethyl phthalate at 90°C for 10 min while heating followed by treatment with base increases the percent fixation of dyes.

Monochlorotriazine dyes are generally used for printing because of their high stability and low substantivity. Reactivity of dye can be increased through treatment with tertiary amines like 1,4-diaza-2,2,2-bicyclo-octane (DABCO) which form quaternary ammonium salts with dye and these groups are faster leaving as compared to chlorine group (Bakker and Johnson, 1973). By using DABCO (0.3%) at pH~7.4, fixation of these dyes can be synergized. Dye–fibre bonds are relatively stable and dye concentration around the functional groups is also high when silk is dyed under natural conditions with maximum fixation. Addition of $Na_2S_2O_4$ (20 g/l) along with Na_2CO_3 (0.5 g/l) increases rate of fixation and by using 1.5% matexil DN-VL, level dyeing is achieved. Dyes can also be applied by a two step process in which dye is first exhausted from a neutral bath followed by fixation in alkali; the levelness depends on rate of dye sorption during fixation.

Vinyl sulphone dyes are marketed as sulphato-ethyl sulphone derivatives and are converted into vinyl-sulphone form by treating with alkali; fixation occurs by nucleophilic addition mechanism (Bhagwanth et al, 1970). Fastness properties are excellent, light fastness is comparable to that of dyed cellulose and provides good perspiration fastness; maximum fixation occurs at pH~7–8. Dye sorption is not dependent on number of protonated $-NH_2$ groups on silk. Addition of electrolyte improves both exhaustion and fixation of dye while addition of urea decreases rate of fixation. The hydrophobic dye having two methyl groups, has fastest dyeing rate while hydrophilic dye (having two hydroxyl ethyl groups) has the slowest. N-methyl taurine derivatives show maximum fixation at pH~6–7 at 95°C for 60–90 min in presence of NaCl (40–80 g/l).

Bromo-acrylamido dyes are known for brightness of shade, high reactivity and good all round fastness. These undergo reaction with polypeptides through both nucleophilic substitution mechanism and addition at the double bond (Shore, 1968; Lewis, 1982). In cold pad-batch method, silk fabric is padded with dye solution containing sodium alginate (10 g/l) and Na_2CO_3 (pH~10) and stored for 24 h at 25–30°C when an average fixation of 80–90% is achieved. In continuous dyeing process, padded fabric is steamed for 2 min at 100–102°C for fixation. In 4-difluoro-5-chloro pyridine dyes, bonds are formed between the reactive groups by reaction of two fluorine atoms in the dye with functional groups in the silk fibroin, mainly tyrosine OH groups (Yushu, 1983). The fixation ratio and dye fibre bond stabilizers are the two important factors to determine wash fastness of dyeings.

20.3.6 Dyeing with direct colours

Like acid dyes, direct dyes get attached to silk by electrostatic bonds between protonated amino groups (NH_3^+) of fibre and the dye anion (RSO_3^-) as well as by hydrogen bonds. Direct dyes produce better fast shades on silk than acid dyes; some direct dyed shades even possess excellent brightness even without subsequent after treatment.

Dyeing of silk fabric in light colours is started in a weakly alkaline dye bath containing soap (2–3%) and NaCl (5–10%, retarding agent). Dyeing in medium and deep shades is carried out in presence of Na_2SO_4 (10–20%). Dyeing is started at 40°C, slowly raised to 90–95°C and dyed at this temperature for 1 hr. Dyeings are rinsed in warm water and treated with CH_3COOH (5 g/l) at 30°C for 15 min. Direct dyes, which are poorly exhausted at neutral pH are dyed in presence of CH_3COOH (2–5%) and Na_2SO_4 (10–20%). Dyeing is started at 40°C, raised to 90°C and is continued for 45–60 min. Use of ammonium acetate or sulphates in place of acid produces well levelled shades with uniform distribution of dye.

20.4   Rapid dyeing of silk

It consists of (i) rapid diffusion of dye in solution, (ii) instantaneous sorption at the outer surface of fibre, (iii) faster diffusion at the interior of fibre and (iv) enhanced anchoring with fibre. Silk, when pre-esterified with methanol causes better dye uptake at 40°C, methanol creates more active sites increasing surface deposition and diffusion both (Vickerstaff, 1950). Esterification time as well as temperature of dyeing also influences exhaustion of dye ; rate of exhaustion increases marginally beyond a treatment time of 2 h (Mishra and Samy, 1992). An UV-radiation may also be used to enhance dye uptake at 45°C, which economizes heat energy and retains aesthetics of silk as dyeing at boil deteriorates lustre and strength both. Another development in dyeing technique used is use of ultrasonic energy for the purpose of chemical reactions, as powerful mechanical agitation is achieved and 'cavitation' effect of ultrasound provides tremendous energy to reaction at low temperatures (Saligram and Shukla, 1992).

20.5   Recent developments

National and international awareness about depletion of natural resources, ecological imbalance, pollution problems and over-disturbed environment due to ample usage of hazardous chemicals have forced to concentrate application of natural dyes which are non-toxic, non-allergic and non-carcinogenic.

20.5.1 Natural coloured silk

It has been proposed that in stead of dyeing silk, natural coloured silk can be developed in commercial scale. Although commercial silkworm strains are available which can produce greenish-yellow, yellow and golden yellow silk but after degumming, the colours are lost. Biotechnological methods have been introduced to produce genetically altered green fluorescent silk fibres. The cocoon colours are because of caretenoids, carotenes and xanthophylls derived from mulberry leaves besides the genetic constitution of silk glands.

20.6   References

AGARWAL A, GOEL A AND GUPTA K C (1992) 'Optimization of dyeing process for wool with natural dye obtained from turmeric (Curcuma longa)', *Textile Dyer Print.*, **25**, 22, October 28, 28–30.

ALLEN R L M (1971) *Colour Chemistry*, London, Thomas Nelson & Sons Ltd.

BAKKER P G H AND JOHNSON A (1973) 'The application of monochlorotriazinyl reactive dyes to silk', *J. Soc. Dyers Color.*, **89**, 6, 203–208.

BHAGWANTH M R R, DARUWALLA E H, SHARMA V N AND VENKATRAMAN K (1970) 'Mechanism of dyeing cellulose with sodium beta-arylsulphonylethyl sulphates (remazols)', *Textile Res. J.*, **40**, 4, 392–394.

BIRD C L AND BOSTON W S (1975) *The Theory of Coloration of Textiles*, Bradford, UK, Dyers Company Publication Trust.

DATYE K V AND VAIDYA A A (1984) *Chemical Processing of Synthetic Fibres & Blends*, New York, John Wiley & Sons.

GULRAJANI M L AND GUPTA S (1989) *Silk Dyeing, Printing & Finishing*, Delhi, Indian Institute of Technology.

GULRAJANI M L, SETHI B AND KAPUR V (1992) 'Application of chrome mordant dyes on silk', *Colourage*, **39**, 10, 38–42.

GULRAJANI M L AND BALAJI P (1992) 'Dyeing of silk with metal-complex and reactive dyes at low temperature', *Colourage*, **8**, 31–34.

GULRAJANI M L (1993) *Chemical Processing of Silk*, Delhi, Indian Institute of Technology. Gulrajani M L, Gupta D B, Agarwal V and Jain M (1992a) 'Some studies on natural yellow dyes-II: Flavonoids–Kapila / Onion / Tesu', *Indian Textile J.*, **102**, 5, February, 78–84.

GULRAJANI M L, GUPTA D B, AGARWAL V AND JAIN M (1992b) 'Some studies on natural yellow dyes-III: Quinones: henna dolu, *Indian Textile J.*, **102**, 6, March, 76–83.

LEWIS D M (1982) 'Dyeing of wool with reactive dyes', *J. Soc. Dyers Color.*, **98**, 5/6, 165–175.

LEWIS D M, (1992) *Wool Dyeing*, Bradford, UK, Society of Dyers & Colourists.

MISHRA S P AND SAMY P V (1992) 'Low-temperature dyeing of silk', *Textile Dyer Print.*, **25**, 4, February 19, 27–29.

MURALIDHARAN B AND DHARMARAJ V (1993) 'Low-temperature dyeing of silk using 1:2 metal-complex dyes', *Textile Dyer Print.*, **26**, 12, June 9, 25–26.

NISHIDA K AND KOBAYASHI K (1992) 'Dyeing properties of natural dyes under aftertreatment using metallic mordants', *Am. Dyestuff Rep.*, **81**, 5, 61–62.

PETERS R H (1975) *Textile Chemistry*, Volume-III, New York, Elsevier Scientific Publishing Company.

SALIGRAM A N AND SHUKLA S R (1992) 'Trends in silk processing' *Colourage*, **39**, 11, 39–41.

SHORE J (1968) 'Mechanism of Reaction of Proteins with Reactive Dyes: I-Literature survey', *J. Soc. Dyers Color.*, **84**, 408–412.

SIRI (2000) *Colours from Nature: Silk dyeing using Natural dyes*, New Delhi, Oxford & IBH Publishing Co.

VICKERSTAFF T (1950) *The Physical Chemistry of Dyeing*, London, Oliver & Boyd.

YUSHU X (1983) 'Formation of crosslinks in silk by dyeing with a difluoro chloropyrimidyl dye', *J. Soc. Dyers Color.*, **99**, 2, 56–59.

21

Dyeing of polyester

Abstract: Polyester is non-ionic, hydrophobic and thermoplastic in nature; highly compact structure permits efficient dyeing only at high temperature with disperse dyes, preferably in closed machines. Carriers swell up polyester at boil to facilitate dyeing in open baths that is in lieu of poor light fastness and light shades. Dispersing agent holds dye in dispersion, addition in excess reduces dye uptake enhancing solubility of dye, when excess dispersing agent helps to reduce barré in textured fibre. Complications associated with dyeing at high temperature have led to development of anionic as well as carrier-free dyeable polyester.

Keywords: Disperse dye, dispersing agent, carrier, high temperature, barré

21.1 Fundamentals of dyeing

Polyester is hydrophobic, non-ionic and dyeable with only disperse dye, though solubilised vat dyes are seldom used on polyester–cotton blends to dye both the components with light shades. Fibre structure is too compact to allow simple chemicals to enter inside below its glass transition temperature (T_g ~85°C); alkaline hydrolysis is an example where only the outer surface is scooped out by alkali with no change at the interior (Gorrafa, 1980; Shenai and Lokre, 1978). Beyond T_g, structure slowly opens up with corresponding increase in size of pores at the surface and interior; extent of opening is directly proportional to increase in temperature. For dyeing, required opening in structure is obtained just above 120°C for efficient diffusion of dye (IDI, 1988). Dyeing temperature is not constant, rather depends on required opening at specific temperature for dyes of various sizes – the smaller the size the lower the dyeing temperature and can be related as $T_d = T_g + \Delta T_{dye}$, where T_d is dyeing temperature and ΔT_{dye} is the rise in temperature beyond T_g to provide opening to accommodate a dye of specific size (Gulrajani and Saxena, 1979; Gulrajani, 1987). Under practical situations, most of the dye molecules possess the size to cause efficient dyeing at 120°C; however, for through distribution of dye through migration required to produce levelled shades, the dyeing temperature is maintained at $130 \pm 2°C$. Dyeing of polyester with disperse dye can be described as: (i) deposition of dye on surface at lower temperature, (ii) diffusion at the surface layers with increase in temperature and (iii) diffusion at the interior at higher temperature; the rate of diffusion at the interior is the rate determining step in dyeing. Once distribution of dye is over, the bath is cooled down to collapse fibre structure in which dye remains attached with fibre through H-bonds and van der Waals forces. In spite of weaker nature of these bonds, wash fastness seems to be excellent as coming out of dye from fibre is almost absent due to its compactness.

Non-disperse dyes (vat, sulphur, etc.) possess bigger structures and are not used in dyeing of polyester as these dyes can not penetrate fibre matrix even at higher temperature; also most of non-disperse dyes lack good thermal stability at very high temperature. A few dyes are water soluble and do not show any affinity for polyester, e.g. reactive, direct, acid and basic dyes.

The energy required for transfer of dye molecules from solution to fibre phase is termed as activation energy for a specific dye–fibre system and is expressed in kcal /mol. For polyester-disperse dye, it is ~30; nylon-disperse dye ~22; wool-acid dye ~22; cellulose acetate-disperse dye ~20–24 and viscose-direct dye ~14 and so on. The highest activation energy required for polyester-disperse dye system clearly reflects compactness of the fibre and why such a dyeing process can not be conducted at or below boil satisfactorily, even in presence of carriers.

21.2   Methods of dyeing polyester

There are four methods for dyeing polyester, viz. (i) dyeing at boil without carrier, (ii) dyeing at boil with carrier, (iii) high temperature high pressure (HTHP) dyeing and (iv) thermosol dyeing. The first three methods are batch methods whereas the thermosol method is a continuous one. Efficient pre-treatment of polyester is a pre-requisite to produce levelled and bright shades.

Molten beads formed during singeing develop spot effect and so singeing is done after dyeing when batch dyeing is followed. In contrast, continuous dyeing or thermosol process do not cause spot dyeing of molten beads as dyeing is affected not through exhaustion but through surface deposition and diffusion and so singeing may be included in pre-treatment sequence. Disperse dye manufacturers blend dispersing agent with dye during grinding, as application of dispersing agent is a must in dyeing with disperse dyes for any shade to avoid precipitation of dye in bath and this eliminates further use of dispersing agent in dyeing except for certain cases like dyeing of textured polyester.

21.2.1 Dyeing at boil without carrier

This method is used only for pale shades; fibre opening is too little at boil, necessitating use of only low molecular weight disperse dyes. Dyes do not penetrate due to restricted opening causing poor wash and sublimation fastness ratings; domestic pressing causes partial sublimation of dye too. Dye bath is prepared with dye, dispersing agent and acetic acid (pH ~5.5–6.5); dyeing is started at 50–60°C, raised slowly to boil and is carried out at this temperature for 2–3 h followed by soaping and washing. This method has least commercial importance except for very light shades on polyester–cotton or like blends.

21.2.2 Dyeing at boil with carrier

Carrier brings down T_g of polyester and swells up fibre structure for better diffusion of dye at boil. Dyebath is prepared with dye, dispersing agent, acetic acid (pH ~5.5–6.5). Dyeing is started at 50–60°C with slow rise in temperature to 80°C when required amount of carrier is added; temperature is raised to boil and dyeing is continued for 2–3 h. Dyebath is discharged and dyeings are washed with anionic or non-ionic detergent alongwith Na_2CO_3 (2–3 g/l each) to remove last trace of carrier. To enhance performance of the process, a closed jigger or winch may be swung into action to get uniform temperature throughout bath. This method is restricted to production of light shades with negative impact of carrier, e.g. irritation on skin, poor light fastness and spotting (due to volatile nature of carrier). Although no definite mechanism of carrier action has been devised, the more popular concept suggests swelling of fibre with ease in diffusion of dye inside. The swelled up structure subsequently collapses with cooling of bath followed by washing out of carrier (Mittal and Trivedi, 1983). Pattern of disperse dye uptake on polyester in presence and absence of carrier is shown in Fig. 21.1 (ICI, 1964).

21.2.3 High temperature high pressure dyeing

HTHP dyeing is the most popular commercial batch method of dyeing polyester to produce levelled shades. The process is based on opening of fibre structure at higher temperature, includes no carrier and generates lesser waste water load (Scott, 1982). Disperse dyes show feeble water solubility beyond 90°C; a part of dispersed dye goes into solution with simultaneous affinity for polyester: the higher the temperature the better the solubility and the higher the affinity due to more solubility of dye as well as opening of fibre. Equilibrium is only reached only when the subsidiary equilibria are achieved through a balance in all the dyeing steps, viz.

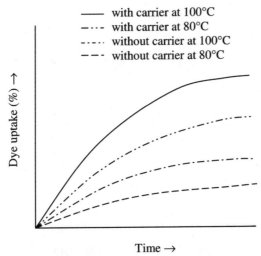

— with carrier at 100°C
—··· with carrier at 80°C
—····· without carrier at 100°C
——· without carrier at 80°C

Dye uptake (%) →

Time →

21.1 Disperse dye uptake on polyester in presence and absence of carrier.

(i)   dye dispersed in bath ↔ dye dissolved in bath
(ii) dye dissolved in bath ↔ dye adsorbed on fibre
(iii) dye adsorbed on fibre ↔ dye diffused in situ fibre

Strike rate is directly proportional to increase in temperature; the latter is increased from 90°C at 1°C/min to 117°C, the temperature at which all existing disperse dyes complete exhaustion due to sufficient opening in structure, seems to be right enough for levelled shades. However, another increase by 10–12°C is adopted in commercial application for uniform distribution and ease in migration of dye to compensate heat loss during dyeing, shown by the absorption curve in Fig. 21.2 (Vaidya, 1984). Dye uptake decreases initially with increase in heat setting temperature of polyester; remains approx. same from 160–200°C, again remarkably increased. If heat setting is carried out beyond 220°C, uneven dyeing may occur due to improper setting. Setting temperature must be within 180–200°C as explained in Fig. 21.3 (Marvin, 1954).

Dyebath is prepared with acetic acid (pH ~5.5–6.5) and dispersing agent. Temperature is raised to 60°C, dye is added and temperature is again raised to 90°C beyond which rise in temperature is restricted to 1–1.5°C / min up to 117°C, and beyond that temperature is raised quickly to 130±2°C at corresponding steam pressure around 3–3.5 kg / cm² and continued for 0.5–2 h depending on depth of shade. It is preferred to discharge the bath at 130±2°C to drain out liquefied low molecular weight oligomers which otherwise get precipitated on fibre surface during cooling below 120°C and causes harsh feeling as well as deteriorates rubbing fastness of dyeings.

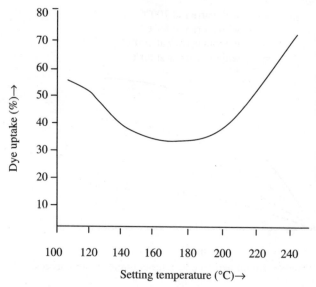

21.2 Effect of heat setting temperature on disperse dye uptake of polyester.

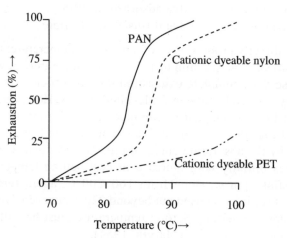

21.3 Dyeability of acrylic, cationic dyeable nylon and cationic dyeable polyester.

Alternately the machine may be cleaned after dyeing with NaOH (2 g/l) beyond boil. Dyeing follows reduction cleaning of dyed polyester with NaOH and $Na_2S_2O_4$ (3 g/l each) at 50–60°C for 15–20 min to remove superficial dye which otherwise inhibits wash fastness. HTHP dyeing is no doubt a unique method as all types of well levelled shades with superior fastness can be produced on polyester and its blends. Dyeings are generally performed either in jet dyeing machine in rope form or in open width in beam dyeing machines.

Levelling of shade crucially varies with control over temperature. Up to 100°C, as high as 10% of the bath is exhausted, 80% exhaustion tales place during 100–120°C and the rest 10% between 120–130°C. Neither the initial nor the final absorption stages affect levelling and so such control over heating of bath is necessary, but an efficient control is needed during heating in the range 100–120°C (Niwa, 1982).

21.2.4 Thermosol dyeing

This is a continuous method of dyeing polyester (Somm and Buser, 1984). In this process, dye is forced to get solubilised in solid fibre phase through partial loosening of intermolecular bonds under heat treatment when the internal structure of the fibre opens up to allow dye molecules to diffuse at the interior (Decheva et al, 1981a). The process is not viable for 100% polyester due to poor pick up of liquor during padding as the fibre is highly hydrophobic in nature. Though multiple dip and nip may improve pick up to some extent, chances of unleveled shade because of migration of dye poses a problem. Thermosol process is unique for continuous dyeing of polyester–cotton blends, where cotton picks up dye liquor from bath and transfer it to the polyester phase at high temperature during curing (Fulmer, 1983, 1984). The cotton part is subsequently dyed in usual ways (Bright, 1996; Adcock, 1974). The process is of non-migrating type i.e. no further scope of levelling once cured, a careful pre-treatment of fabric is a must to develop good and consistent wettability. Kinetics and thermodynamics of thermosol dyeing of polyester with disperse dyes has been detailed elsewhere (Decheva et al, 1981b, 1982a, 1982b).

Well pre-treated polyester–cotton fabric is padded with a liquor containing dye, acid and dispersing agent at 70–80% expression, followed by controlled drying at 90–100°C with uniform circulation of air to avoid migration (Greer, 1996; Lehmann and Somm, 1976). Dried fabric passes through stenter, curing chamber or loop ager at 180–210°C for 45–60 s for complete diffusion of dye in partially or fully sublimised state (Wersch, 1986; Oschatz and Somm, 1973) or passed through superheated steam (Von der Eltz, 1973). The coating of dye dispersion on fabric is very sensitive to touch and so the padded material be instantly passed through drying zone without any contact with guide rolls or other accessories. The best way is to place infra-red heaters just above the last nip for mild drying without migration of dye followed by drying in float drier at 90–100°C (Wersch, 1987).

Due to hydrophobic nature of polyester, dye concentration is kept on higher side. Liquid disperse dyes provide excellent result due to high rate of dispersion and diffusion; these diffuse more rapidly into fibre structure with saving of energy and short time of fixation. Dyes from B and C classes possessing moderate sublimation and wash fastness properties may be selected for levelled shades; dyes form D class produce excellent fast shades but are not recommended on grounds of having higher sublimation temperature and excess of dye is required to produce desired depth due to poor diffusion co-efficient. However, a reduction clearing definitely improves wash fastness and the latter can be verified by dipping dyed sample in dimethyl formamide at room temperature – stripping of colour in solvent confirms presence of dye on surface. Introduction of a migration inhibitor in padding liquor like sodium alginate or carboxymethylcellulose arrests migration of dye and produces levelled shades.

Increase in draw ratio during melt spinning promotes orientation and decreases rate of diffusion; even heat setting at 180–210°C causes surface melting of the polymer with less dye uptake. Uncontrolled drawing in filaments produce barré; stains on fabric must be removed prior to dyeing to produce levelled shades. Sometimes, a ring dyeing effect is produced due to insufficient diffusion of dye which may be rectified by exposing dyeings at 180–210°C for 30–45 s. Non-ionic surfactants, when applied as dyeing accelerants, promote rate of dye dissolution in auxiliary melt, thereby increasing the overall rate of thermosol dyeing. At high surfactant concentrations, dye retention in surfactant may adversely affect dye fixation (Manian and Etters, 2004; Kostadinov et al, 1978).

21.3　Dyeing of textured polyester

Polyester textiles are often textured to improve unpleasant handle, moisture regain, air retentivity, comfort and other aesthetic properties. Dyeing of textured polyester is bound to develop stripy dyeing or barré due to irregularities introduced in yarn or fabric during earlier processes viz. melt spinning, drawing, setting, winding or texturing itself. These irregularities cause variation in crystallinity of yarn throughout its length, which in turn imposes periodic stripiness or barré across the width or throughout the length of fabric. The problem is realized when such a textured yarn or fabric is dyed, the dye uptake varies form point to point, showing barré.

Problem of barré in textured polyester can be reduced by controlling dyeing process, selection of dyestuffs, compatibility of dyes in combination and dispersing agents (Vaidya, 1984a). Singeing is done at high speed in less intense flame to avoid formation of molten beads. Pre-treatment with Na_2CO_3 and non-ionic detergent (1–3 g/l each) at 60°C for 60–90 min removes coning oil completely, which otherwise solubilises a part of disperse dye during dyeing and activate post-thermo-migration; complete removal of detergent is a must to counteract post-migration of dye too.

Restoration of bulk and handle is affected using open width continuous machines. $NaClO_2$ bleaching is imparted at 80–85°C for 60–90 min with $NaClO_2$, $NaNO_3$ and HCOOH (1–2 g/l each) at pH ~3–3.5 succeeded by thorough washing; heat setting at 160–180°C for 30–90 s avoids stiffness.

Dispersing agent used in dyeing of polyester retains dye in dispersion through imposition of anionic charge such that repulsion occurs among dye molecules. This avoids formation of dye cluster and precipitation of dye. The solubility of dispersing agent, which is essential for dye dispersion, varies with its cloud point. The selective cloud point must be well beyond dyeing temperature to avoid precipitation of dispersing agent and the dye both. During periodic heating and subsequent cooling of dye solution while passes through heat exchanger, dye molecules of relatively larger size tend to crystallize – necessitating use of excess dispersing agent for consistency in dispersion pattern. Selected dye must be tested for its fineness and especially coverage power to suppress barriness – dyes from B and C classes show promising results. Non-carrier-type levelling agents belonging to non-ionic groups do not promote migration which is essential to cover barriness, increase dye solubility and are not recommended for textured polyester. Carrier accelerates rate of dyeing and migration both through swelling action and its application is a must for thorough distribution of dye; concentration and type of carrier should be optimised prior to application. Thorough washing of dyeings is also recommended to wash out last trace of residual carrier; otherwise it may affect light fastness.

Dyeing may be carried out in jet dyeing machine at 130±2°C for 1–2 h as usual followed by cooling at 2–5°C/min to 70–80°C when the bath is discharged. Thermosol process is generally not recommended for dyeing of textured polyester as it affects bulk and texture of fabric (Leube and Richter, 1973; Rau et al, 1975; Senner, 1977). If needed, padding may be carried out in a solution containing dye, anionic wetting agent (1 g/l), dispersing agent (1 g/l), anti-migrating agent (2–10 g/l) and CH_3COOH (pH ~4.5–5) at 90–100% expression with minimum nip pressure to avoid loss in bulk of fibre. A fixation accelerant, e.g. butyl benzoate, which lowers sublimation temperature of dye for better diffusion as well as opens up fibre structure effectively at lower temperature range, is recommended. Drying should be in two phases: pre-heating and final heating as stated in thermosol method for conventional polyester with low level air circulation. Thermo-fixation at 160–180°C for 45–60 s in tension free state preserves handle; fixation in superheated steam at 171°C is a better option to get excellent handle with wider choice of dye (Vaidya, 1988). Dyeing is followed by reduction clearing, washing and drying.

21.4 Dyeing of cationic and carrier-free dyeable polyesters

Compactness of normal polyester and related problems in its dyeing has led to development of carrier-free and cationic dyeable polyester. In the former, the compactness has been reduced by changing reactants and is dyed at boil for all shades without carrier; in the latter, an anion is introduced during polymerization to make the fibre reactive and to show response for cationic dyes.

21.4.1 Dyeing of carrier-free dyeable polyester

Carrier-free dyeable polyester is chemically of three types: (i) polybutylene terephthalate, (ii) polyethylene terephthalate – polyethylene oxide block copolymer and (iii) dicarboxylic acid modified terephthalate fibre. Few examples are Diolene 42 and 742, Trevira 210, 310, etc. This can be also produced by introducing some changes in spinning and post-spinning stages of unmodified polyethylene terephthalate to affect orientation and crystallinity. These have T_g around 75°C, paving way for dyeing at boil for all shades without application of carrier. Fibre compactness is slightly less, melting point is around 225–240°C, tensile strength and elongation are on little lower side, shrinkage at boil is quite high (2–3%) compared to that of normal polyester as 1%; wash and light fastness are little lesser than normal dyed polyester (Vaidya, 1988). Due to better fibre opening at boil, affinity of dye for fibre is higher; relative colour depth of carrier-free dyeable polyester dyed at boil is comparable to that of HTHP dyed polyester. Opening of fibre beyond 60°C is so rapid that chances of uneven shade formation is quite high due to localized absorption of dye at the boundary zone between the fibre surface and dye liquor (Hurten, 1982). Dyeings must be labeled for laundering below 50°C.

21.4.2 Dyeing of cationic dyeable polyester

Cationic dyeable polyester (CDP) is produced by introducing additives possessing sulphonic acid groups in the polymerization bath itself to produce anionic polyester, which can subsequently be dyed with a basic dye to produce unparallel bright shades. The additives may be either of 5-sulphoisophthalic acid, m- or p-phenylsulphonic acid, 2-naphthol-8-sulphonic acid, etc. (Vaidya, 1988). Additives act as plasticizer and bring down cohesive force in polymer chains thus reducing its tenacity (Rao and Datye, 1996). One popular brand is Decron T-64 (Du Pont).

Cationic polyester has a specific place in today's market, incomparable bright shades can be produced on it with greater ease which is not possible with disperse dyes; 'frosting' problem due to difference in abrasion resistance of components in polyester – cotton blend can be reduced by producing a balanced blend of cationic dyeable polyester, normal polyester and cotton (33:33:34) and dyeing these components with deeper, lighter and intermediate shades respectively. Fancy shades can also be produced on blends with greater ease. Dye up take pattern of cationic dyeable three main man made fibres, viz. polyester, nylon and acrylic is shown in Fig. 21.4 (Vaidya, 1988).

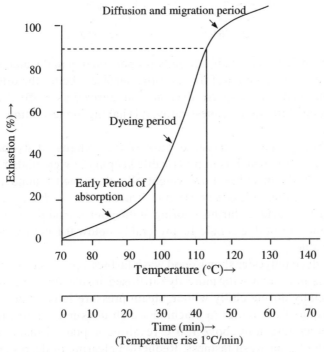

21.4 Typical absorption curve of polyester dyeing.

Cationic dyeable polyester is dyed at 110–115°C with basic dye. Dyeing beyond 120°C degrades fibre while dyeing below boil requires addition of carrier in bath; dyeings show excellent rubbing and wash fastness but light fastness lies in the range 3–7 (Renard, 1973). Heat setting must be done below 180°C. Disperse dyeing gives 10–20% higher dye uptake over that on normal polyester. Numerous works were carried out worldwide in regard to dyeing of cationic dyeable polyester due to its growing popularity (Karmakar et al, 1996; Teli and Prasad, 1989a, 1989b; Hashimoto and Tokuda, 1985; Oumae and Kinoshita, 1979; Satou, 1977; Teli and Vyas, 1990).

21.5   Fastness of dyeings

Disperse dyes generally exhibit overall good fastness properties. Light fastness depends on stability of dye structure with number of double bonds, substituent groups and their location in dye structure – lesser double bonds imparts better stability, the better the light fastness. Presence of $-NH_2$ groups in anthraquinone structure lowers light fastness through oxidation to $-NO_2$ on exposure to strong light, known as gas fading.

Sublimation fastness remains directly proportional to molecular weight of dye, polarity and amount of dye on fibre surface – better the diffusion the better the fastness. Disperse dyes are categorized as A, B, C and D based on their sublimation fastness ratings; D having highest ratings while A the least.

Poor rubbing fastness never occurs if the exhausted dyebath is discharged at 130°C which is impracticable keeping in view higher steam pressure inside the machine. Low weight oligomers melt around 120°C, accept more dye due to lesser crystalline structure during dyeing and get sticked to fibre surface during cooling with poor abrasion resistance; aqueous solubility of dye should be preferably below 0.1 mg/l with high saturation value.

Disperse dyed polyester shows good wash fastness, provided superficial deposition is minimum with more dye diffused inside fibre. In tropical areas, especially during chilly winter, steam pressure falls tremendously during transport from boiler to machine which does not support raising dyebath temperature at or above 120°C resulting in poor diffusion with a consequence of poor wash fastness. Reduction clearing of deep shades is a must for good wash fastness too. High temperature finishing also causes migration of a part of dye from interior to fibre surface, which is again enhanced in presence of silicone emulsions.

21.6   References

Adcock A (1974) 'Pretreatment and thermosol dyeing of woven polyester / cellulosic blends', *Textile J. Australia*, **49**, 9, 34–52.

Bright L (1996) 'Trouble shooting continuous thermosol dyeing of polyester fiber and blends, *Am. Dyestuff Rep.*, **85**, 8, 60–61.

Decheva R, Ilcheva R and Bechev K (1981a) 'Determination of the efficiency of disperse dyes in thermosol dyeing of polyester', *Textil Praxis Int.*, **36**, 6, 666–667.

DECHEVA R, VALCHEVA E AND ILCHEVA R (1981b) 'Kinetics and thermodynamics of continuous dyeing of polyester fibres with disperse dyes-III: The thermodynamics of thermosol dyeing', *Textilveredlung*, **16**, 12, 484–488.

DECHEVA R, VALCHEVA E AND ILCHEVA R (1982a) 'Kinetics and thermodynamics of continuous dyeing of polyester fibres with disperse dyes-IV: Influence of accelerators on the thermodynamics of thermosol dyeing', *Textilveredlung*, **17**, 3, 128–133.

DECHEVA R, VALCHEVA E AND ILCHEVA R (1982b) 'Kinetics and thermodynamics of continuous dyeing of polyester fibres with disperse dyes-V: Relationship between the kinetic and thermodynamic characteristics in the thermosol dyeing of polyester fibres with disperse dyes', *Textilveredlung*, **17**, 4, 166–168.

FULMER T D (1983) 'Continuous Thermosol dyeing- II', *Am. Textiles Int.*, **12**, 11 63–66

FULMER T D (1984) 'Continuous thermosol dyeing- IV'. Auxiliary equipment possibilities, *Am. Textiles Int.*, **13**, 3, 69–74.

GORRAFA A A M (1980) 'Caustic treatment of polyester filament fabrics', *Textile Chem. Color.*, **12**, 4, 83–87.

GREER J E (1996) 'Process for thermosol dyeing of polyester fabrics, USP 3 973 417.

GULRAJANI M L (1987) *Dyeing of Polyester and its blends*, New Delhi, Indian Institute of technology.

GULRAJANI M L AND SAXENA R K (1979) 'Studies of the glass transition temperature of polyester fibre by a dyeing method', *J. Soc. Dyers Color.*, **95**, 9, 330–333.

HASHIMOTO I AND TOKUDA S (1985) 'Dyeing behaviours of cationic dyes on cationic-dyeable polyester fibres, Journal of the Textile Machinery Society of Japan, 38, 12 (1985) T209–213.

HURTEN J (1982) 'Dyeing and finishing of woven fabrics containing polyester fibres that can be dyed without carrier', *Melliand Textilber.*, **63**, 4, 296–300.

ICI (1964) *'The dyeing of polyester fibres'*, UK, Imperial Colour Industries.

IDI (1988) *'Technical Manual on Navilene dyes'*, Mumbai, Indian Dyestuff Industries Ltd.

KARMAKAR S R, BHATTACHARYYA K AND DEB T (1996) 'Investigations into the dyeing behaviour of acrylic and CDPET fibres', *Colourage Ann.*, 47–58.

KOSTADINOV D, DECHEVA R, DUSHEVA M AND ILCHEVA R (1978) 'Effect of several dyeing accelerators during thermosol dyeing of polyester', *Melliand Textilber.*, **59**, 9, 745–747.

LEHMANN H AND SOMM F (1976) 'Optimal procedure in thermosol dyeing, *Chemiefasern / Textil Ind.*, **26 / 78**, 4, 320–321, 324.

LEUBE H AND RICHTER P (1973) 'Continuous dyeing of woven and knitted texturized polyester fabrics', *Textile Chem. Color.*, **5**, 3, 43–47.

MANIAN A P AND ETTERS J N (2004) Dye-surfactant interactions in thermosol dyeing *AATCC Rev.*, **4**, 9, 19–23.

MARVIN D N (1954) 'The Heat Setting of Terylene Polyester Filament Fabrics in Relation to Dyeing and Finishing', *J. Soc. Dyers Color.*, **70**, 16–21.

MITTAL R M AND TRIVEDI S S (1983) *Chemical Processing of polyester / cellulosic blends*, India, Ahmedabad Textile Industries Research Association.

NIWA T (1982) 'Rapid dyeing of polyester', *Am. Dyestuff Rep.*, **71**, 5, 25–30, 43.

OSCHATZ C AND SOMM F (1973) 'Optimum fixation times in thermosol dyeing', *Melliand Textilber.*, **54**, 2, 146–152.

OUMAE T AND KINOSHITA M (1979) 'Dyeing of cationic dyeable polyester', *JTN*, **297**, August, 96–102.

RAO B R AND DATYE K V (1996) 'Ionomeric polyester fiber', *Textile Chem. Color.*, **28**, 10, 17–24.

RAU R O, SELLO S B AND STEVENS C V (1975) 'Continuous dyeing of polyester fabric', *Textile Chem. Color.*, **7**, 2, 31–35.

RENARD C (1973) 'Dyeing properties of polyester fibres with modified affinity', *Melliand Textilber.*, **54**, 12, 1328–1335.

SATOU H (1977) 'Dyeing of regular and cationic dyeable polyester' *JTN*, December 101–102.

SCOTT M R (1982) 'Non-carrier concept of dyeing polyester', *Am. Dyestuff Rep.*, **71**, 11, 40–43.

SENNER P (1977) 'Methods for thermosol dyeing of textured polyester and high bulk acrylic yarns', *Chemiefasern / Textil Ind.*, **27 / 79**, 8, 696–698 and 700–703.

SHENAI V A AND LOKRE D B (1978) 'Action of alkali on polyester fibres', *Textile Dyer Print.*, **11**, 11, July, 27–28.

SOMM F AND BUSER R (1984) 'Present and future of thermosol dyeing', *Am. Dyestuff Rep.*, **73**, 12, 30–36.

TELI M D AND PRASAD N M (1989a) 'Influence of thermal modification of cationic-dyeable polyester on dyeability using cationic dyes', *Am. Dyestuff Rep.*, **78**, .9, 50–56,103.

TELI M D AND PRASAD N M (1989b) 'Influence of thermal modification of cationic-dyeable polyester on dyeability with disperse dyes', *Am. Dyestuff Rep.*,**78**, 10, 15–20, 40.

TELI M D AND VYAS U V (1990) 'Factors affecting dyeability of cationic dyeable polyester', *Man made Textiles India*, **33**, 11/12, 430 & 450–453.

VAIDYA A A (1984) 'Dyeing of textured polyester materials', *Col. Chronicle*, Sandoz, January to March, 6–13.

VAIDYA A A (1984a) 'Dyeing of textured polyester materials', *Textile Dyer Print.*, **17**, 8, April 11, 25–29.

VAIDYA A A (1988) *Production of Synthetic fibres*, New Delhi, Prentice-Hall of India Ltd.

VON DER ELTZ H U (1973) 'More reliability in the thermosol dyeing process-IV: Has high-temperature steam fixation any importance in thermosol dyeing, *Textil Praxis Int.*, **28**, 2, 105–108.

WERSCH K V (1986) 'Effect of fabric temperature in thermosol dyeing processes', *Textil Praxis Int.*, **41**, 1, V–VII.

WERSCH K V (1987) 'Economic aspects of drying and thermosol dyeing operations, *Textil Praxis Int.*, **42**, 1, 44–50; **42**, 2, 152–155.

22

Dyeing of nylon

Abstract: Nylon is moderately hydrophilic in nature. Presence of reactive acid and basic groups at terminal ends permits dyeing with acid, basic, metal complex, direct and reactive dyes at or below boil. Due to its thermoplastic nature, it can also be dyed with disperse dye in the same way as that is used for dyeing of polyester but with poor wash and light fastness and so not practiced. Metal complex and reactive dyes are mostly used for production of all shades. Absence of levelness due to barré may be handled with selection of right dye.

Keywords: Nylon, wash fastness, barré, differentially dyeable nylon

22.1 Introduction

Nylon 6 and 66 belong to polyamide group and resembles in physical as well as chemical nature; higher crystallinity, better molecular orientation and melting temperature of nylon 66 is higher than those of nylon 6. Being more crystalline, rate of dyeing of nylon 66 is relatively slower with reasonably better fastness of dyeings. Polymer chains of nylon are attached with H-bonds, which get detached when moisture surrounds this resulting higher moisture regain with drastic fall in its T_g and ease in dyeing compared to other man made fibres.

Nylon 66 is produced through reaction of hexamethylene diamine and adipic acid. It has the chemical formula

$$H [HN-(CH_2)_6-NH-CO-(CH_2)_4-CO]_nOH$$

While nylon 6 is produced from caprolactum and has the general formula

$$H [HN-(CH_2)_5-CO]_nOH$$

In short, H_2N-F-COOH, 'F' being the fibre part with two different terminal groups, viz. –COOH and –NH_2 and few intermediate –CONH groups; the –NH_2 group controls dyeing behaviour in dyeing with anionic dyes.

Nylon 66 is suitable for textile use while nylon 6 is not so much because of susceptibility to heat treatment of the later having a melting point around 215°C compared to that of nylon 66 around 260°C. Nylon 6 is more amorphous and rate of dyeing if obviously higher than that on nylon 66.

22.2   Dyes for nylon

Nylon can be dyed with numerous dyes in addition to disperse dye as it has relatively open structure with reactive end groups compared to acrylic and polyester (moisture regain of nylon, acrylic and polyester are 4%, 1% and 0.4%, respectively). The range of dyes suitable for nylon include direct, acid or metal complex, basic, reactive, disperse or reactive disperse and soluble vat dyes (Shenai and Parameswaran, 1980). Direct dyes generally produce dull shades and are used only on cheaper qualities; poor wash fastness is attributed to larger size of dye molecules and lack of diffusion inside. Acid dyes combined with ionic bonds and bright shades with moderate to poor wash fastness are produced – mainly used to produce medium and light shades, while metal complex dyes are used to produce deep shades only and these dyes form co-ordinate bonds with nylon; bright shades of superior fastness are produced with a possible problem of developing barré due to arrest in migration. Basic dyes are relatively costlier; produce both poor light and washfast shades. Cationic dyes can be applied to achieve excellent result, but are costly. Due to smaller size of basic dye molecules, overall fastness is poor in contrast to acid dyes of relatively larger structure. Selective reactive dyes produce bright shades with poor fastness while disperse reactive dyes produce bright fast shades. Disperse dyes are occasionally used on nylon as dyeings lack wash fastness due to its lower T_g in spite of good coverage of shades. Soluble vat dyes are too costlier and suitable for light dull shades only.

22.3   Dyeing of nylon

22.3.1 Direct dye

Affinity of direct dyes for nylon is directly proportional to number of – SO_3H groups in dye structure – higher the number, the higher the affinity and better the fastness. Dye with smaller size covers configurational irregularities (configurational barré) but develops poor fastness due to lack of affinity which can be improved through after-treatment with tannic acid–tartar emetic complex. Dyes are applied from an acidic bath in the same way as that used for acid dyes on nylon.

Dyeing of nylon 6 in presence of small amounts of benzyl chloride as dyeing accelerant leads to significant improvement in direct dye uptake;

the uptake varies with benzyl chloride concentration, time and temperature of dyeing. Half time dyeing and corresponding dyeing rate constant were smaller in the case of benzyl chloride / nylon fibre/ direct dye system where the apparent coefficient of diffusion was found to be three times higher than those obtained in dyeing without benzyl chloride (Abdou and Bendak, 1982).

22.3.2 Disperse dye

Disperse dye is non-ionic, water insoluble and does not react chemically with nylon; dyeing takes place irrespective of $-COOH$ or $-NH_2$ groups while dyes are retained by the fibre with van der Waals force and H-bonds. As a result, barriness of dyeing, which is a typical problem in dyed nylon, does not appear due to migration of dye; even irregularities in orientation and physical or chemical structures do not hinder formation of levelled shades.

Dye bath is prepared with dye, dispersing agent (1–2 g/l) and acetic acid at pH 4–5. Dyeing is started at 60°C, temperature is slowly raised to boil and dyeing is continued for further 1 h followed by soaping and washing. Disperse dyes are seldom applied on nylon due to poor wash fastness of dyeings except pale shades. Fading of disperse dyes on nylon on exposure to ozone and oxides of nitrogen is a great concern too (Aspland, 1993). Dyeing of nylon 66 with disperse dyes were also optimized (Carpignano et al, 1988).

22.3.3 Reactive dye

Reactive dyes belong to mainly two classes, e.g. substitution (chlorotriazinyl) and addition types (vinyl sulphone). Chemical attachment of nylon with reactive dyes occurs as follows:

$$D\!\!\triangleleft^{Cl}_{R} + H_2N\text{-}F\text{-}COOH \rightarrow D\!\!\triangleleft^{HN\text{-}F\text{-}COOH}_{R} + HCl$$

(triazine dye)   (nylon 66)

and $DSO_2 CH{=}CH_2 + H_2N\text{-}F\text{-}COOH \rightarrow DSO_2 CH_2\text{-}CH_2\text{-}HN\text{-}F\text{-}COOH$

Mode of attachment is either by ionic or covalent bond formation (Farias and Bryant, 2002; Gulrajani, 1974). For dyeing of nylon with Procion M and H dyes, dye bath is prepared with dispersing agent (1 g/l), CH_3COOH (pH~3.0–4.0) and dye. Dyeing is started at 40°C, slowly raised to 95°C and dyeing is continued for 1 h, when exhaustion is completed. Fixation is done by Na_2CO_3 (2–3 g/l) at pH~10–10.5 for 45–60 min followed by washing and soaping. For fixation, alkali is added at 95°C itself as rate of fixation is increased with increase in temperature of bath. However, it is beneficial to omit further heating and to allow the bath to cool down.

For dyeing with vinyl sulphone dyes, dyebath is prepared with trisodium phosphate (2%) and dye. Dyeing is started at 40°C, acetic acid is added, temperature is raised to 95°C and dyeing is continued for further for 45–60 min at this temperature. HCOOH helps to improve exhaustion if added in bath in place of CH_3COOH. Fixation is done in the same way as that is used with Procion M and H brand dyes.

Disperse reactive dye is non-ionic and insoluble in water and is applied in the same way as that used for other disperse dyes; labile halogen atom retained by these dyes can form covalent linkage with nylon in alkaline pH giving excellent wash fastness. Dyeing is performed in two phases: the first phase relates to exhaustion of bath in acidic pH with thorough distribution of dye throughout fibre structure followed by fixation in second phase in alkaline pH. A linear correlation exists between dyeing rate (diffusion) and thermodynamic affinity of disperse reactive dyes of monochlorotriazine type. The dye affinity and nature of heterocyclic ring were found to be the determining factors of the whole process (Giorgi, 1989). Dyeing is started with dye, non-ionic dispersing agent (1–2 g/l) and CH_3COOH (2–4 ml/l, pH 4–4.5) at 40°C and slowly raised to 85–90°C. Dyeing is continued for 30–60 min; fixation follows at pH 10–10.5 at boil with Na_2CO_3 (2–3 g/l) for 45–60 min with finally soaping, washing and drying. Once applied and fixed, migration of dyes cease and does not allow any post-dyeing shade correction – an issue why these dyes can not be successfully applied on textured nylon to cover barré. However, these dyes do not react with polyester and acetate fibres and can safely be used as normal disperse dye with post-dyeing corrections. Higher cost and incomplete shade range have restricted their use.

22.3.4 Acid dye

Rate of dyeing with acid dyes exclusively depends on pH – dye uptake slowly increases with fall in pH from 7 to 4, remains somehow unchanged from 4 to 2.7 while from pH 2.7 to 1, there is steep rise in dye uptake and the latter is related to charge on dye molecules. This analysis is based on titrating nylon with an acid or acid dye to obtain a titration curve as depicted in Fig. 22.1 (Rush and Miller, 1969).

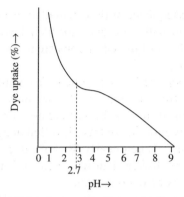

22.1 Influence of pH on acid dye uptake of nylon.

Positive charge on nylon indirectly varies with dye bath pH; the lower the pH the higher the charge and the higher the strike rate of dye as addition of more acid results protonation of more end $-NH_2$ groups. Around 90–95°C, dyeing occurs at much faster rate and dye–fibre attachment is formed through formation of ionic bond between acid end group of dye and protonated $-NH_2$ group of fibre.

$$H_2N\text{- F- COOH} \xrightarrow{\text{acid } (H^+)} H_3N^+\text{- F- COOH} \xrightarrow{\text{dye } (D^+SO_3^-)} D^+SO_3^- \ H_3N^+\text{- F-COOH}$$

pH of dyeing is always maintained above 3 to avoid fibre degradation due to over-dyeing, otherwise at pH below 3 the internal $-CONH$ groups also form ionic bond with dye. Dye saturation value on fibre depends on the number of $-SO_3H$ groups in dye molecules.

At pH ~5–6, sorption is slow but increases as the pH is lowered down progressively; all the $-COOH$ groups are neutralized as the pH moves below 2.7 with maximum cationicity of fibre, called 'isoelectric point' of nylon. Excess acid absorbed below pH 2.7 attacks $-NH_2$ groups forming more salt sites on fibre. On completion of attacking end $-NH_2$ groups, excess acid attack $-CONH$ sites making these cationic too. The net result is that if starting pH of dyeing is maintained below 2.7, all $-NH_2$ as well as $-CONH$ groups remain cationic and dye anions attack all these sites

$$\begin{array}{c} \text{O} \\ \parallel \\ DSO_3^- \ NH_3^+ \ F\text{-C-NH-COO}^-H^+ \\ \mid \\ H^+ \\ DSO_3^- \end{array}$$

Over-dyed nylon

causing over-dyeing with a consequence of fibre degradation and unleveled shade formation. To dye nylon with acid dyes for well levelled shades, only the $-NH_2$ groups are to be protonized at consistent and slower rate leaving $-CONH$ groups intact.

In dyeing light shades, pH should be maintained throughout on higher side for lesser protonation of fibre as limited sites are only required for dye–fibre attachment; addition of more acid will create more protonated sites with chances of uneven dyeing. In contrast, levelled deep shades can be formed at a pH around 4–4.5. When nearly 80–85% dye is exhausted from bath, pH of bath can be lowered down even at 2 to complete exhaustion without any chances of over-dyeing.

Acid dyes are classified based on their structure and pH of dyeing. Class 1 dyes belong to strong or levelling or equalizing dyes possessing low affinity and are applied at pH 1–2 using H_2SO_4 (2%) at 95°C in presence of Na_2SO_4 (10%) for 1 h; class 2 is milling dye having moderate affinity and is applied at pH 4–5 with CH_3COOH (3–4%) at 95°C for 60–75 min while class 3 is super-milling dye possessing very high affinity and is applied at neutral pH using $(NH_4)_2SO_4$ or CH_3COONH_4 (3%) at 95°C for 1 h. The strike rate is controlled by pH, temperature and presence of retarding agents (electrolyte).

Stiochiometrically, a nylon polymer chain should retain one $-NH_2$ and one $-COOH$ group at terminal ends but practically the weightage of $-COOH$ group is higher compared to that of $-NH_2$ groups as CH_3COOH is used as chain stopper in synthesis of nylon.

$$NH_2\text{--- F – C-NH}\sim\sim\text{COOH} \xrightarrow{CH_3COOH} CH_3\text{-C-NH--- F – C-NH}\sim\sim\text{COOH}$$

In practice, weightage of $-NH_2$, $-COOH$ and $-CONH$ groups are 36, 90 and 0.885×10^4 m equiv / mg respectively.

Nylon when immersed in aqueous bath causes protonation of $-NH_2$ end groups through transfer of the proton from $-COOH$ end group producing a fibre chain with zweater ions.

$$NH_2\text{--- F–C-NH}\sim\sim\text{COOH} \xrightarrow{H_2O} NH_3^+\text{-- F– C-NH}\sim\sim\text{COO}^- \xrightarrow{acid[H^+]} NH_3^+\text{--F– C-NH}\sim\sim\text{COOH}$$
(zweater ion)

Basic dyes can also be successfully applied on nylon due to more $-COOH$ end groups, but poor fastness of shades has restricted their use. Acid dyes produce cheaper bright shades – fastness varies from dye to dye. Modified acid dyes, viz. metal complex dyes (1:1, 1:2) produce relatively dull shades and covering power is not good due to insufficient affinity of these dyes for nylon.

Fastness of acid dyed nylon can be improved by post-treatment with tannic acid (2%) and tartar emetic (potassium antimony tartarate, 2%) at 60–90°C for 30 min. Dyeings are treated with tannic acid at boil with progressive cooling down the bath, unexhausted tannic acid is drained out followed by treatment with tartarate with a consequence of dulling of shades.

The technique of dyeing nylon with levelled shades (6, 66, 11 and aramids) with acid dyes has been reviewed critically (Farias and Bryant, 2002; Perkins, 1996; Zahn, 1972). Role of dyeing auxiliaries has been studied in detail (Richardson, 1972; Fiebig et al, 1993) and a mathematical model of dyeing nylon fibres with acid dyes has also been proposed (Broda, 1988).

Thermal setting of nylon prior to dyeing plays crucial role on dye uptake. With increase in heat-setting temperature, physical structure and amine end-group content of nylon are changed; acid dye uptake beyond 2 h dyeing decreases, but not monotonically (Seu, 1983).

22.3.5 Metal-complex dye

There are two types of metal-complex dyes: first type is 1:1 metal complex dye, in which one mole of dye is attached with one atom of metal; and 1:2 metal complex dye, in which two moles of dye are attached with one atom of metal. The size of 1:2 type is obviously bigger than 1:1 type.

1:1 metal complex dye

Dyebath is prepared with dye, H_2SO_4 (8%) or HCOOH (3–4%) and electrolyte (0.1–3%) at pH~1.5. Dyeing is started at 40–45°C, slowly raised to boil and is continued for 1.5–2 h followed by soaping and washing.

1:2 metal complex dye

Dyebath is prepared with dye, disodium phosphate (0.5 g/l) and non-ionic detergent (2%) at pH ~6–6.5. Dyeing is started at 45–50°C and raised to boil. Acid liberating agents like $(NH_4)_2SO_4$ may be added to improve exhaustion. Migration of dye is poor due to larger structure of dye molecule causing poor coverage of shades; compound shades are produced with least effort and light as well as wash fastness properties are good.

22.4   Barré

Dyed nylon shows continuous visual stripy pattern parallel to yarn direction in knitted or woven fabrics, which is basically due to physical, optical or dye differences in yarns or geometrical differences in fabric structure or poor coverage power of dye, acting either singly or in combination to produce a barré pattern (Razafimahefa et al, 2004; Davis et al, 1996; Dusheva et al, 1989; Holfeld and Shepard, 1977; Willingham, 1973).

22.4.1 Rating of barré

Depending on intensity of the defect, barré is rated from 1–5; very objectionable barré is rated as 1, objectionable barré – 2, slight barré – 3, very slight barré – 4 while absence of barré is ratted as 5 (Pratt, 1984). All these ratings are assessed with the help of a grey scale or more precisely with a 'Hatra Barre'ness Scale', which consists of five black frames with a grey card, graded from 5 to 1; each having horizontal lines at varying frequency intervals and the principle is to match fabric bars with standard frequency interval lines (Zaukelies, 1981; Jaeckel, 1975; Kok et al, 1975, 1977; Wright and Hager, 1975; Knoll, 1977).

22.4.2 Types of barré

Barré produced due to variation in physical structure includes orientation of polymer chains and their packing, crystallinity as well as the difference in morphology (Pratt, 1977). Geometrical or optical differences in fabric arise due to (i) irregular spacing between two successive yarns caused by mal-adjustment of weaving and knitting machine (Makansi, 1987a), (ii) bulk difference between yarns (Makansi, 1987b; Hes, 1986; Vaidya and Nigam, 1977), (iii) difference in lustre due to irregular distribution of TiO_2 in manufacturing (dull looks lighter), (iv) difference in filament cross-sections may arise due to partial choking of spinneret (trilobal produces brightest shade), (v) differences in denier of filaments (microfibres produce lighter shades) (Texas, 1990; Clapp, 1995), (vi) differences in twist of yarns (higher twist produces dull shades) (Makansi, 1987a) and (vii) common woven fabric defects, e.g. weft crack on a warp, crammed pick, starting mark, some defective textured weft, missing end, wrong drawing, weft bar, imperfections in take-up motion of loom (Bhattacharyya and Dweltz, 1986; Jhala and Solanki, 1985) or defects acquired through faulty knitting (Wiume, 1983; Anon, 1977; Holfeld and Nash, 1976). These defects are purely physical in nature and are acquired through improper control over various stages of formation of yarn and fabric; they cannot be corrected during dyeing and it is better to sell-off the fabric in white state itself as full bleached product. These defects due to physical or optical variations are called configurational barré. Barré produced due to poor coverage power of dye is known as orientation barré and it arises from

(i) differences in chemical structure with respect to terminal $-NH_2$ groups, viz. (a) varying amount of dye absorbed by different polymer chains in regard to $-NH_2$ groups due to difference in degree of polymerization in chains, (b) difference in merge number results differential dye uptake (Holfeld and Shepard, 1977) while in capacity barré, a few $-NH_2$ groups are oxidized during heat setting as well as texturing and do not take part in dyeing; variation of 1°C in heat setting produces barré after dyeing.

Type of barré can be diagonised in various ways. Configurational barré is diagonised by dyeing of nylon sample with a small molecular weight disperse dye which is insensitive to crystallinity, orientation and end – NH_2 groups; non-uniform dyeing confirms presence of configurational barré. In 'plastic replica test', fabric sample is sandwiched between two plastic sheets and heated up when the sandwiched dyed fabric leaves impression on plastic sheet; raised points on fabric surface produce dots, if distance is uniform then it is due to fault in fibre – if non-uniform then due to weaving. Capacity barré is diagonised through dyeing nylon sample with a small molecular weight acid dye when dyeing occurs in proportion to end $-NH_2$ groups; level dyeing confirms absence of capacity barré. Dyeing with large size acid dye or spraying of acid dye solubilised in a solvent on nylon produces unleveled dyeing effect, called orientation barré (Bello, 1985).

22.4.3 Remedies to barré

Problem of barré can be minimized by dyeing nylon with lower molecular weight disperse dye, which shows better migration and dyeing occurs irrespective of end $-NH_2$ groups (Greider, 1974; Hendrix et al, 1974). Shades formed are levelled but brightness not so good; wash fastness is also poor. Another approach is dyeing with equalizing acid dyes which possess lower affinity for nylon. During absorption and desorption of dyes, more levelled shade is expected to be formed through even distribution of dye and uniform reaction with end $-NH_2$ groups throughout fabric; bright shades are produced with poor wash fastness – the latter can be improved through back tanning or treatment with nylofixan PI (HCHO condensed products of dyesyn diamide). A third remedy is to use levelling agents during dyeing. Anionic levelling agents, e.g. turkey red oil, carboxylic soaps or calsoline oil HS (with multiple $-OSO_3$ groups) competes with dye and gets absorbed on positively charged fibre. As a consequence, fibre becomes temporarily neutral and dye is not attracted towards fibre – strike rate is reduced with chances of levelled shade. As dyeing proceeds, dye replaces levelling agents.

Cationic levelling agents form complex with anionic dye, become electrically neutral and strike rate is obviously reduced due to bigger size of dye–levelling agent complex to produce a levelled shade. At high temperature, complex breaks down and releases dye. Choice of surfactant is limited to avoid precipitation of anionic dye-cationic levelling agent complex. Higher cost and yellowing of fibre during storage limits the application of cationic levelling agents. Non-ionic levelling agents form loose complex with dye with slow release of dye from complex, dye–surfactant combination changes λ_{max}. An efficient method to solve barre' problem is to use swelling agents. Benzyl alcohol, n-propanol can swell up fibre effectively, help internal migration of dye, but are costly. Dyeing of nylon for longer time may reduce barré problem due to extensive migration. Even high temperature dyeing with acid dyes causes better migration for high affinity dyes producing a levelled shade.

Combination of fibre-reactive anionic and surface-active agent based on sulphonic acid and a dye-reactive cationic and surface-active agent based on an aliphatic diamine can also remove barriness from textured nylon filament (Wassileva et al, 1991).

22.5   Dyeing of differentially dyeable nylon

Structural characteristics as well as dyeability of nylon can be varied through structural modification (Partlett and Whiteside, 1969; Vaidya et al, 1977a, 1977b; Dawson, 1971; Vaidya, 1988). Deep acid dyeable nylon is produced by introducing additional $-NH_2$ groups in structure to enhance basicity of fibre. This is achieved by means of adding N:N'-bis (3-aminophenyl) piperazine to nylon salt before spinning to improve its dyeability with anionic (acid) dyes and deeper shades can be produced with greater ease due to high proportion of end $-NH_2$ groups (100 meq/kg) and poor $-COOH$ groups (7–8 meq/kg) compared to those in normal nylon with 40 meq/kg $-NH_2$ and 65 meq/kg $-COOH$ groups respectively.

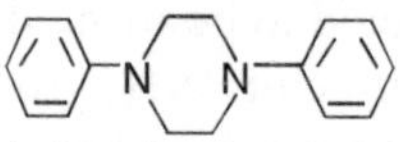

N:N'-bis (3-aminophenyl) piperazine

In cationic dyeable nylon, dyeability with basic dyes can be synergized if additional anionic groups are introduced through incorporation of monomers containing $-SO_3H$ groups. Addition of 5-sulphoisophthalic acid, sulphanilic acid, etc., before polymerization adds extra anionic nature to fibre, in which end $-NH_2$ groups has a poor share of only 7–8 meq/kg with respect to that of $-COOH$ groups with 100 meq/kg and the related reaction is shown in Fig. 22.2 (Vaidya et al, 1977b).

HOOC—⟨ ⟩—COOH

SO$_3$H

(5-Sulphoisophthalic acid)

+   H-[HN-(CH$_2$)$_6$-NH-CO-(CH$_2$)$_4$-CO-]$_n$H

(Nylon 66)

↓

H$_2$N – (CH$_2$)$_6$-NH-CO—⟨ ⟩—CO-NH-(CH$_2$)$_6$-NH

SO$_3$H

22.2 Synthesis of basic dyeable nylon

Regular or normal nylon is dyed with acid dye with a consequence of lesser dye uptake. In deep dyeable nylon, the weightage of –COOH and – NH$_2$ groups are of opposite order to that of cationic dyeable nylon and obviously the fibre is dyeable with almost all shades with acid dye only. Making required blending of various types of these, fancy shades can be developed with various dyes or dye combinations, e.g. disperse dye will produce levelled shades on any such blend as it works irrespective of end groups, but mixture of disperse dye and acid dye or only acid or basic dye depending on type of blend will give multi-coloured effects. Differential dyeable nylon has little poor wash fastness and so is exclusively used in carpet industry.

22.6   References

ABDOU L A and BENDAK A (1982) 'Accelerated dyeing of nylon 6 with direct dyes', *Am. Dyestuff Rep.*, **71**, 11, 35–39.

ANON (1977) 'Diagnosing knit barré', *Textile Ind.*, **141**, November, 77 & 79.

ASPLAND J R (1993) 'A series on dyeing: Chapter 9 – The structure and properties of disperse dyes and related topics', *Textile Chem. Color.*, **25**, 1, 21–25.

BELLO K A (1985) 'Reactions of coloured carboxylates with polyamide fibres', Ph D Thesis, Leeds University, UK.

BHATTACHARYYA S and DWELTZ N E (1986) 'Objective detection of various fabric defects', *ATIRA Tech. Digest*, **20**, 2, 60–71.

BRODA J (1988) 'Mathematical model of dyeing nylon fibres with acid dyes', *Przeglad Wlokienniczy*, **42**, 8, 329–331.

CARPIGNANO R, SAVARINO P, BARNI E, VISCARDI G and BARACCO A (1988) 'Dyeing of nylon 66 with disperse dyes: an optimization study, *Dyes & Pigments*, **10**, 1, 23–31.

CLAPP D M (1995) 'Influence of varying micronaire values on barré in knitted fabrics', *Int. Textile Bulletin*, **41**, 2, 72–74.

DAVIS H, McGREGOR R, PASTORE C and TIMBLE N (1996) 'Human perception and fabric streakiness', *Textile Res. J.*, **66**, 8, 533–544.

DAWSON T L (1971) 'Developments in coloration for the carpet industry (June 1967 to September 1970)', *Rev. Prog. Color Related Topics*, 2, 3–10.

Dusheva M, Icherenska M and Gavrilova E (1989) 'The phenomenon of barriness in the dyeing of polyamide fibre material - I: What is known about the theory of barriness', *Melliand Textilber.*, **70**, 5, 360–364.

Farias L T and Bryant R (2002) 'One bath exhaust dyeing of cotton nylon woven fabric with reactive and acid dyes', *AATCC Rev.*, **2**, 9, 30–32.

Fiebig D, Herlinger H and Brennich W (1993) 'Role of nonionic auxiliaries in the dyeing of polyamide with acid dyestuffs', *Melliand Textilber*, **74**, 11,1168–1172.

Giorgi M R D (1989) 'Correlation between dyeing rate and thermodynamic affinity of disperse reactive dyes', *Dyes & Pigments*, **11**, 4, 293–302.

Greider K (1974) 'Experimental study of the dyeability of barré nylon fibre with acid dyes', *J. Soc. Dyers Color.*, **90**, 12, 435–441.

Gulrajani M L (1974) 'Dyeing of nylon with reactive dyes', *Colourage*, *21*, 26, December 26, 25–32.

Hendrix J, Farmer L B and Kuhn H H (1974) 'Barré control', *Textile Ind.*,**138**, 9,168–169.

Hes L (1986) 'Reasons for barriness in woven and knitted fabrics from textured yarns', *Textil*, **41**, 4, 126–129.

Holfeld W T and Nash J L (1976) 'Knitting machine barré: a multi-fibre problem', *Canadian Textile J.*, **93**, 1, 67–73.

Holfeld W T and Shepard M S (1977a) 'Water as a carrier in dyeing and processing nylon'- III: Barré as an indicator of fibre structural changes', *Canadian Textile J.*, **94**, 8, 65–75.

Holfeld W T and Shepard M S (1977b) 'Water as a carrier in dyeing and processing nylon IV: Causes of orientation barré with rate-sensitive acid dyes', *Canadian Textile J.*, **94**, 9, 82–85.

Jaeckel S M (1975) 'Precision and applicability of the Hatra Barriness Scale', *Textile Inst. Ind.*, **13**, 8, 247–249.

Jhala P B and Solanki M G (1985) 'How to control barré due to imperfections in take-up motion of loom', *ATIRA Tech Digest*, **19**, 1, March, 1–3.

Knoll A L (1977) 'Some calculations related to the instrumental assessment of barré and streakiness, *J. Textile Inst.*, **68**, 8, 260–262.

Kok C, Kruger P J, Lake R, Turner R and Vlist N V D (1975) 'Photoelectric instrument for the assessment of fabric appearance and structure', *J. Textile Inst.*, **66**, 5, 186–192.

Kok C; Kruger P J, Lake R, Turner R and Vlist N V D (1977) 'Some calculations related to the instrumental assessment of barré and streakiness', *J. Textile Inst.*, **68**, 8, 260–262.

Makansi M (1987a) 'Perception and control of fabric streaks-I: Theory', *Textile Res. J.*, **57**, 8, 463–472.

Makansi M (1987b) 'Perception and control of fabric streaks-II: Experimental', *Textile Res. J.*, **57**, 9, 495–502.

Partlett G A and Whiteside J A B (1969) 'Prospects and problems in the dyeing and finishing of differential dyeing nylon 66 yarns', *J. Soc. Dyers Color.*, 85, 12, 621–629.

Perkins W S (1996) 'Today's principles of dyeing nylon', *Am. Textiles Int.*, **25/10**, 60–64.

Pratt H T (1977) 'Some causes of barré, *Knitting Times*, **46**, 25, June 13, 18–22.

Pratt H T (1984) 'A descriptive terminology for barré, *Textile Chem. Color.*, **16**, 3, 61–62.

RAZAFIMAHEFA I, VROMAN I and VIALLIER P (2004) 'Dye diffusion and the phenomenon of barriness on polyamide fabrics, *Coloration Tech.*, **120**, 2, 61–65.

RICHARDSON G M (1972) 'Auxiliaries in the dyeing of nylon', *Modern Textiles*, **53**, 3, 34 & 36–37; **53**, 4, 21, 26 and 32.

RUSH J L and MILLER J C (1969) 'Nylon dyeing: Its relation to physical and chemical properties of yarn', *Am. Dyestuff Rep.*, **58**, February, 37–44.

SEU G (1983) 'Effect of heat-setting on dyeing of nylon 66 fibres', Ann. Chim. (Rome), **73**, 5–6, 333–339.

SHENAI V A and PARAMESWARAN R (1980) 'Dyeing of nylon - a review', *Textile Dyer Print.*, **13**, 18, October 1, 21–26 and 42.

TEXAS TECH UNIVERSITY (1990) 'Effect of micronaire on fabric barré, *Textile Topics*, **19**, 4, December, 1–3.

VAIDYA A A and NIGAM J K (1977) 'Minimizing barriness in textured nylons', *Textile Inst. Ind.*, **15**, 3, 96–101.

VAIDYA A A, CHATTOPADHYAY S and RAVISHANKAR S (1977a) 'Flame-resistant textile fibres-II: Flame-resistant polyester and polyester/cotton blends', *Textile Dyer Print.*, **10**, 6, 49–52.

VAIDYA A A, CHATTOPADHYAY S and RAVISHANKAR S (1977b) 'Flame-resistant textile fibres. III. Flame-resistant nylon, acrylic and other specially developed synthetic fibres', *Textile Dyer Print.*, **10**, 8, 37–41.

VAIDYA A A (1988) *Production of Synthetic fibres*, New Delhi, Prentice-Hall of India Ltd

WASSILEVA V, STOYANOV K and DUSCHEWA M (1991) 'The 'barriness' phenomenon in the dyeing of nylon-II: Methods for solving this problem in practice', *Melliand Textilber.*, **72**, 3, 200–202.

WILLINGHAM W L (1973) 'Barré problem', *Textile Ind.*, **137**, 9, 193–212.

WIUME W (1983) 'Systematic approach to assessing faults in fabric appearance, *Textil Praxis Int.*, **38**, 10, 1071–1075.

WRIGHT W D and HAGER L W (1975) 'Tool for assessing barre', *Textile Chem. Color.*, **7**, 1, 17–20.

ZAHN C (1972) 'Level dyeing of polyamide fabrics', *Int. Textile Bulletin*, **18**, 1, 66, 71.

ZAUKELIES D A (1981) 'Instrument for measurement of barré in knitted fabrics', *Textile Res. J.*, **51**, 1, 1–4.

23

Dyeing of acrylic

Abstract: Pure acrylic is brittle, tough and undyeable with any dye due to absence of reactive sites. Being a thermosetting man made fibre, it can not be dyed with disperse dye at high temperature to avoid formation of only light shades and tendering of fibre. Addition of co-monomers in polymerization bath develops amorphousness, flexibility and introduces reactive site enabling it to make dyeable at room temperature. Generally acidic co-monomers are added due to which all acrylic fibres are popularly dyed with basic dyes. Rapid opening of fibre throughout its T_g my be handled by either adding retarding agent in bath or controlling rate of heating.

Keywords: Co-monomer, basic dye, glass transition temperature, retarder

23.1   Problem in dyeing pure acrylic fibre

Acrylic fibre is synthesized through polymerization of acrylonitrile (vinyl cyanide) and has the chemical name polyacrylonitrile (PAN) with the empirical formula

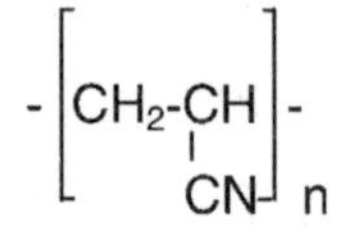

$$-\left[\begin{array}{c} CH_2\text{-}CH \\ | \\ CN \end{array} \right]_n -$$

Where 'n' is the degree of polymerization.

Melt spinning of acrylic is not possible as the polymer is thermosetting in nature and does not melt at higher temperature rather degraded forming infusible mass. Wet or dry spun acrylic is produced by dissolving polyacrylonitrile in HNO_3 (65%) or dimethyl formamide (DMF) respectively. Pure acrylic is tuff and compact rod shaped, brittle, possesses poor abrasion resistance, lacks textile properties, T_g is very high (105°C), electrically inert and difficult to chemically process. Fibre was chemically modified prior to dyeing, through 'cuprous ion technique' to introduce cation within fibre with $CuSO_4$ and hydroxylamine sulphates, when copper in cuprous state remained attached to the fibre. The modified fibre was thus dyeable with acid dye.

$$-CH_2-CH-CH_2-CH-CH_2-CH-CH_2-CH-$$

The process has been obsolete on grounds of its complexity, lack of reproducibility, temperature sensitivity, difficulty in stabilising pH, liquor ratio, dosing of $CuSO_4$ and hydroxylamine sulphates (Chakraborty, 1996).

23.2 Modification of pure acrylic

To impart textile value, a copolymer of PAN has been produced by adding 5–10% neutral co-monomers, e.g. methyl acrylate, methyl methacrylate or vinyl acetate to reduce its T_g and to enable it to process in open bath. However, electrical inertness impaired level dyeing and was inferior in aesthetics. Modern PAN is a copolymer and is produced by adding electrically active ethylenic co-monomers in polymerization bath. Thus produced PAN has two nomenclatures: if share of co-monomer is $\leq 15\%$, it is named acrylic and if it is $\geq 15\%$, then it is called modacrylic. This modern concept of introducing co-monomers in single or in combination at varying extents during polymerization has given birth of varieties of PAN to cut demands of present day market (Vaidya, 1988).

Based on electrical nature of co-monomer, fibre acquires electrical charge, e.g. allyl sulphuric acid, acrylic acid, methacrylic acid, itaconic acid, sodium allyl sulphonate, etc., carry negative charge while vinyl pyrazine, vinyl pyridine, ethylene imine, allylamine, etc., carry positive charge. Basic dye is preferred in the first case and acid dye in second case.

$$CH_2 = CH \qquad CH_2 = CH$$
$$CH_2OSO_2H \qquad COOH$$

Allyl sulphuric acid   Acrylic acid

The co-monomer may be either from vinyl chloride, vinyl bromide, vinylidene chloride (a halogen derivative), etc., if flame retardancy of PAN is to be improved; while co-monomers like 2-hydroxymethyl methacrylate, acrylamide, etc., are used to improve hydrophilicity and related anti-static properties (Vaidya, 1988).

The negative charge is also introduced by activator or initiator like peroxysulphate or thiosulphate during polymerisation.and these remain as end groups at a concentration of 20–25 meq/g. Both these anions effectively take part in dyeing with basic dye.

$$^-O_3SC—CH—CH_2—CH—CH_2—CH—CSO_3^-$$

$$\begin{array}{ccc} | & | & | \\ CN & CN & CN \end{array}$$

Cationic co-monomers are expensive, after-treated acrylic turns yellow on storage and that is why anionics are preferred. Introduction of co-monomers show plasticizing effect on PAN and lowers down its T_g from 105°C to 85°C.

The COO⁻ (weak acidic group) or SO_3^-, SO_4^- (strong acidic groups) introduced in the polymer during polymerization using acrylic acid co-monomer and peroxysulphate or thiosulphate as initiator remain as integral part of fibre in ionized state at their own pk_a values (a pH at which 100% ionization of end groups take place) which is the pH value for different end groups and if fibre is dyed at this pH, the corresponding end groups will only take part in dyeing, e.g. pk_a value of dyeing for COO⁻ ion is 6.5 while that with either of SO_3^- or SO_4^- is 1.65. This implies that if pH of dyeing is maintained at 1.65, amount of dye absorbed will be proportional to SO_3^-, SO_4^- ions and if at 6.5, in proportion to COO⁻ ions. Number of end groups and their types on fibre can easily be assessed from optimum pH of dyeing (either 6.5 or 1.65) which will be directly proportional to number of basic dye molecules react on these sites. Even, when dyeing is carried out at pH 4.5 and 5 separately, the latter gives deep shade. Various types of PAN retain COO⁻, SO_3^- and SO_4^- ions at varying ratios; as a result, the optimum pH of dyeing is to be adjusted somewhere between 1.65 and 6.5 according to the ratio of these ions to get maximum dye uptake, e.g. Orlon 42, Acrylon 16, Courtelle E and Dralon contain 46 and 17, 31 and 21, 0 and 154, 48 and 53 strong and weak acid ions (meq/g) respectively. Selection of spinning process also has tremendous impact on processability of PAN: dry spun fibers possess lesser voids and picks up 60–70% more dye compared to wet spun fibres due to difference in available sites in fibre.

Dyeing is to be kept limited to fibre saturation value to retain excellent fastness properties. From given fibre saturation factor, the maximum depth of shade on fibre can be calculated by multiplying dye saturation factor of a dye with percent of dye with respect to fibre in recipe and summing all the data. If the sum remains below fibre saturation factor, the shade can be easily developed. Say, if a shade is to be produced in a recipe with three different dyes and the fibre has dye saturation values as X, Y, Z and concentrations (%) as a, b, and c in recipe in regard to fibre, then aX + bY + cZ must remain below fibre saturation value A for successful dyeing. Hence it is essential to have knowledge of dye and fibre saturation values to produce fast shades. If retarder is used in dyeing, its concentration and saturation factor are also to be included to calculate saturation value of fibre and obviously not used in dyeing deep shades. Cationic dyes show poor migration below 100°C – a reason why it is difficult to produce consistent well levelled shades on PAN in commercial dyeing; but migrate well above 110°C and higher the temperature the better the migration. As degradation of acrylic starts from 110°C onwards, dyeing temperature is kept little below or at 110°C. Cationic dyes are well stable at 110°C and reproduction of shades is excellent.

Special cationic dyes have come up and colour index specification of some of these is still awaited. These dyes are modified basic dyes, costlier but dyeing process is too simple; bigger in structure and do not diffuse so easily. Rise in dyebeth temperature is not important as these dyes do not react with PAN below boil because the cationic site of dye remains balanced with some other anionic molecules and the dye do not show affinity for fibre rather only gets deposited on surface. At boil, anionic halogen atom splits up, exposes cationic site of dye which then reacts with fibre. Retarders are not used, at least for medium and deep shades. With special cationic dyes, dyeing starts at 80°C with NaCl (5–10%), CH_3COONa (0.5–1%) and CH_3COOH (0.5–1%) at pH 4.5 for 5 min followed by addition of dye, raised to 95°C in 15 min and dyeing is carried out for 1–2 h. However, their higher cost has forced many dyers to apply conventional basic dyes in commercial dyeing.

23.3   Dyeing with basic dye

Basic dyes are invariably applied on acrylics. It has some advantages over other dyeing methods, viz. (i) all deep shades can be developed with greater ease, (ii) shades are brighter and (iii) fastness properties are excellent when special cationic dyes are used. Maximum dye uptake is governed by number of acid groups in fibre though a small amount of dye is also held by dissolution process. In producing composite and light shades, dyes with same K value, called compatibility index, are mixed to get similar rate of adsorption; dye with higher K takes more time for exhaustion (Kamat et al, 1991; Popescu et al, 2001a, 2001b; Bhattacharya et al, 1997; Anon, 1972). Efficient dyeing of acrylic in yarn, package as well as hank form is also carried out extensively (Park and Shore, 1981; Park, 1979; Heane et al, 1979; Peretto, 1973).

23.3.1 Troubleshoots in dyeing

PAN starts and completes opening at 80°C and 85°C respectively. Below 80°C, no such dyeing or diffusion of basic dye occurs except some surface deposition whereas beyond 80°C due to fast opening of fibre structure dye molecules rush towards fibre, react where it is adsorbed, block passage for migration of other dye molecules causing unlevel dyeing (Emsermann, 1996; Vaidya, 1984; Nunn, 1979; Grosso, 1973; Jowett and Cobb, 1972). To produce levelled shades, either slow heating of bath in this range is preferred for controlled opening as well as diffusion of dye (non-retarder method) or retarding agents are used to get level dyeing (retarder method); related problems are that while non-retarder method lengthens time of dyeing the retarder method shortens dyeing time but adds to dyeing cost.

23.3.2 Practical application

In retarder as well as non-retarder methods, dye is first pasted with CH_3COOH (0.5 g/l) and dissolved by adding boiling water. It is the dyeing temperature and concentration of retarder those decide maximum dye uptake but not the pH; change in pH only causes change in colour.

Non-retarder method

In non-retarder method, the bath is heated up in a controlled manner to cope up with opening of fibre structure. Dyeing is started at 40°C, quickly raised to 70°C beyond which, rise in temperature is restricted to 1°C / min up to 85°C, raised to 95°C rapidly with further dyeing for 1–2 h followed by soaping and washing (Fig. 23.1); a better control over temperature as shown in Fig. 23.2 gives best levelled shades.

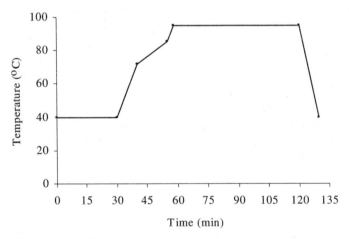

23.1 Dyeing cycle of acrylic in non-retarder method.

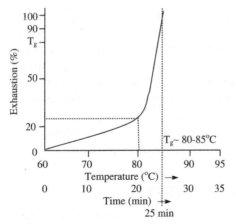

[Rate of heating-1°C/ min beyond 70°C up to 85°C, Tg : Temperature at and beyond which diffusion starts.

75% of dye exhausted in only 5 min (80–85°C) and related Rate of dyeing = 75/5 = 15%/ min or 15%/°C.

Strike rate may be reduced by controlled heating or retarder may be added in bath. Fast heating beyond 85°C is a must to reduce time of dyeing and for better migration of dye].

23.2 Absorption of basic dye at Tg of PAN (non-retarder method).

Retarder method

In cationic retarder method, the bath is set at pH 4.5, retarder (2%) is added at 60°C, PAN is treated in this bath for 10 min followed by addition of basic dye and dyeing is carried out in this bath at 95°C for 1–2 h.

In anionic retarder method, a bath is set at 60°C with retarder (0.5–2%), non-ionic dispersing agent (2–3%) and CH_3COOH in combination with CH_3COONa to get a pH around 4.5. PAN is treated in this bath for some time followed by addition of dye. Bath is heated up to 85°C at a faster rate, then slowly to boil for steady acceptance of complex and dyeing is further done for 1–2 h followed by soaping and washing. This method has a unique feature that, as not the dye but the complex positively takes part in dyeing, rate of dyeing with any basic dye is somewhat same.

Polymeric retarders, due to bigger in size, are absorbed on surface and can not diffuse inside; presence of multiple charges on these retarders increase their affinity for PAN many-folds – resulting a better controlled process with levelled shade at even lesser dose (0.1–0.2%).

Basic dye has limited tendency to migrate throughout acrylic below 95°C. Absorption is slow up to 80°C but above 90°C, rate of absorption increases suddenly and in spite of taking all precautions to control rate of dyeing, uneven dyeing is invariably obtained in case of medium and pale shades due to problem of migration (Rohner and Zollinger, 1986; Carbonell et al, 1974; Harwood et al, 1972).

According to BASF, sources of unevenness in dyeing PAN may be better demonstrated as (i) yarn differences, i.e. faults in PAN yarn in manufacturing stage, which can't be rectified, (ii) temperature difference at different points of bath (it is 30°C between bottom and top) which causes bottom part to be dyed 30 times faster necessitating thermostatic control with maximum agitation (constant temperature method) and (iii) difference in dye concentration, i.e. at the time of addition top part gets maximum dye while bottom part gets less dye necessitating pumping of dye solution at multiple points with efficient circulation system.

23.4   Retarder and its types

Retarders are colourless chemicals and slow down rate of dyeing. Cationic retarder is smaller in structure and possesses cation like that on cationic dye but its mobility is faster than dye while anionic retarders possess opposite electrical charge to that of dye. Polymeric retarders are polyquaternary ammonium compounds with high molecular weight of 1000–2000 compared to conventional cationic retarders 300–500 and carry multi-positive charges on structure. All retarders, irrespective of their electrical nature, reduce strike rate of dye to ensure level dyeing (Clarient, 1985; Galafassi and Scheidegger, 1976; ICI, 1972; Puhlmann, 1971; Parham, 1993; Shukla and Mathur, 1993; Gawish and Shakra, 1983).

23.5   Mechanism of retarder action

Cationic retarders possess smaller molecular size and its mobility is faster than dye. During dyeing, these compete with dye, reaches to fibre surface first, occupy sites and neutralizes fibre charge temporarily. As a result, dye molecules do not feel attraction for fibre causing slow and even deposition on fibre and displace retarders slowly as final affinity of dye for fibre is much higher than that of retarder. In this way, cationic retarders help positioning of dye molecules uniformly to produce level dyeings. Anionic retarder forms complex with dye, increases molecular weight of dye, reduces its mobility and hence the strike rate. Precipitation of such a complex during dyeing is avoided by applying a non-ionic dispersing agent. Dye-retarder complex from different parts of bath strikes fibre surface at different times. With heating up of bath, the complex breaks, dye is released from complex and diffuses inside fibre.

Polymeric retarders are larger in size, possess multiple cationic charges and are attracted by fibre. These can not diffuse inside fibre, rather remain at the surface and controls rate of surface deposition and diffusion of dye on fibre.

23.6   Matching of shades

In case a specific dye is to be added to bath to match shades, dyebath is to be cooled down below 80°C otherwise there is every possibility of rushing of dye molecules towards fibre at complete opening stage. After required addition, the temperature is again raised and dyeing is continued.

23.7   Dyeing with disperse dye

Mechanism of fixation of dye is similar to that with polyester but production of heavy shade is restricted as solubility of dye in polymer is low and fibre has lower saturation value; fastness properties are acceptable. Dye is pasted with dispersing agent (1–2 g/l), water is added followed by CH_3COOH (0.5–1 ml/l) to maintain a pH of 4.5–5.5. PAN is entered in bath at 40°C, temperature is raised slowly to boil and dyeing is continued at boil for 60–90 min. Dyeing may also be carried out in HTHP method as that used for polyester with pressure 3.5 kg/cm² at 125–130°C for 30 min. After dyeing, bath is cooled down and discharged followed by soaping and washing. Under practical situations, disperse dyes are never applied on PAN as medium and deep shades can never be built-up as well as excessive shrinkage occurs beyond 110°C. Shades are dull with poor sublimation but good light fastness, wash fastness and gas fading properties.

Other dyeing methods for PAN include constant temperature dyeing method and rapid dyeing method (Makhija, 1973; Siepmann and Bruns, 1969). In constant temperature dyeing method (Defitherm process, BASF), dyeing is started at 85°C (not at 60°C as followed in conventional method) and maintained constant. Dyeing temperature is to be reduced by 1°C for 1–1.5 denier fibre as these finer fibres have large surface area and so rate of dyeing will be much higher. In contrast, dyeing temperature is to be increased by 1°C for 15 denier fibers. The problem related to this method is that special auxiliary Defithermol TR is to be essentially used with equal weight of dye.

In Astrazon rapid dyeing technique (Bayer), dyeing is started at constant temperature depending on dye–fibre system succeeded by rising in it by 2°C/5min to 100°C or beyond that. Rate of exhaustion of bath is controlled by adding suitable retarder. Amount of retarder required for a given quantity of dye to produce a 'relative saturation' of 90% for wet spun fibers and 60–70% for dry spun fibers is calculated precisely and used, e.g. if accessible site is 100 for PAN, 90 sites is to be dyed for wet spun PAN while 60–70 sites is to be dyed for dry spun PAN to reach to relative saturation value of fibres. By dyeing 1% shade, if 25% sites are occupied and 65% sites are left; this causes unlevelling of shades. In such a case, 1% dye is mixed with retarder (x%) to occupy 90 sites for levelled shades, this x% is to be calculated (sufficient for all the sites). For light shades, maximum retarder is to be used and vice-versa. The dyes are classified by K values ranging from 1 to 5, which are based on compatibility index, migration capacity, affinity and rate of diffusion of dyes (Elsner, 1969; Beckmann and Jakobs, 1969).

23.8   Various types of basic dyes for acrylic fibre

Conventional basic dyes show uneven results in dyeing and poor fastness properties. In special cationic dyes, the cation has been positioned in dye structure in such a way that the dye can not react with fibre immediately after reaching to fibre surface, rather distributed thoroughly unlike that occurs with normal basic dyes.

Group I dye (conventional basic dye)

The positive charge resonate within the dye structure and remains as part of chromophoric system. These are concentrated basic dye with high tinctorial value and can produce brilliant shades, cheaper, light fastness is around 1 for cotton and 3 for PAN, e.g. Malachite green. The chromophore is TPM, i.e. triphenylmethane.

Group II dye (modified basic dye)

Cost is around ten times higher and less bright than Group I dye. Light fastness is around 6. The positive charge is not a part of chromophoric system rather is positioned at some distance from chromophore. Light fastness is determined by chromophoric system, e.g. anthraquinone, azo, etc.

Group III dye (special cationic dye)

These are also called migrating cationic dyes and costs at per with Group II dye. Cationic group exists in dye but in balance with halogen to start with, migrate for uniform distribution on PAN and the cationic group is generated during dyeing. The dye is adsorbed like dispersed dye and later on generate the positive group followed by reaction with PAN.

Azo dye

Halogen (X) is attached to central N and so positive charge of N is in balance with X. At high temperature, X gets detached, N becomes positive and migration becomes excellent.

23.9   Stripping of basic dyes

Partial stripping is possible with detergent (0.5–1 g/l) and CH_3COOH at pH~4 for 1–2 h. Reducing agents partially strip out dye and flatten the shade. Partial stripping with $NaClO_2$ (0.5 g/l) and HNO_3 (1 g/l) at pH~4 for 30–60 min at boil gives good result, but this can not be applied on acrylic–wool blends as HNO_3 damages wool. Complete stripping is not so easily possible.

23.10   Some special cationic dye types for PAN

(i)   A quaternary ammonium group is carried on the pendent alkyl chain in the coupling component.

(ii) The basic site is located in the diazo component.

(iii) Cationic site at one end and far away from chromophore

23.11 Differentially dyeable PAN

By selecting type of co-monomer during polymerization based on its electrical nature, acid or basic or disperse dyeable PAN can be produced and blended to finally give differential dyeable PAN (Vaidya, 1988; Balasubramanian, 1977, 1979).

23.12 References

ANON (1972) 'Compatability test for combinations of basic dyes in the dyeing of acrylic fibres' *J. Soc. Dyers Color.*, **88**, 6, 220–222.

BALASUBRAMANIAN S (1977) 'Acrylic fibres- I, II', *Man Made Textiles India*, **20**, 5, 221–229 & 231; **20**, 6, 277–290.

BALASUBRAMANIAN S (1979) 'Recent developments in acrylic fibres', *Man Made Textiles India*, **22**, 4, 191–202.

BECKMANN W AND JAKOBS K (1969) 'Strazon rapid dyeing method for acrylics, *Bayer Farben Revue*, **16**, 1–16.

BHATTACHARYA S D, AGARWAL B J, PATEL N I AND SHAH V R (1997) 'Dyeing and dye-induced changes in the physico-mechanical properties of polyacrylonitrile fibres', *Indian J. Fibre Textile Res.*, **22**, 1, 38–43.

CHAKRABORTY J N (1996) 'Dyeing of Acrylics', *Indian Textile J.*, **106**, 1, 24–27.

ELSNER W D (1969) 'The dyeing of acrylic fibres with cationic dyes by the Astrazon rapid dyeing method', *Int. Textile Bulletin*, **3**, 250–259.

EMSERMANN H (1996) 'Dyeing processes of acrylic fibers-I & II', *Am. Dyestuff Rep.*, **85**, 8, 48–54; **85**, 10, 15–22.

GAWISH S M AND SHAKRA S (1983) 'Cationic surfactants to level basic dyes on acrylics and wool', *Am. Dyestuff Rep.*, **72**, 2, 20–23.

GROSSO E N (1973) 'Continuous dyeing of acrylic fibre', *Tinctoria*, **70**, 9, 321–324.

HEANE D G, HILL T C, PARK J AND SHORE J (1979) 'Package dyeing of acrylic fibre yarn: important parameters which influence level dyeing', *J. Soc. Dyers Color.*, **95**, 4, 125–142.

JOWETT A M AND COBB A S (1972) 'Dyeing of acrylic fibres', *Rev. Prog. Color Related Topics*, **3**, 81–89.

KAMAT S Y, ALAT D V AND MOORTHY P K (1991) 'Processing aspects for dyeing of acrylic fibres with Sandocryl B dyes', *Colourage*, **38**, 3, 21–30.

MAKHIJA R K (1973) 'Acrylic fibre dyeing by Defitherm process', *Silk and Rayon Ind. India*, **16**, 2/3, 78–89.

NUNN D M (1979) *The dyeing of synthetic polymer and acetate fibres*, Bradford, UK, Dyers Company Publications Trust.

PARK J (1979) 'Package dyeing of acrylic yarns', *Textile Chem. Color.*, **11**, 8, 156–157.

PARK J AND SHORE J (1981) 'Bulk-scale dyeing of acrylic yarn packages', *J. Soc. Dyers Color.*, **97**, .5, 223–226.

PARHAM R (1993) 'New dyeing system for acrylic/cationic dyeable polyester', *Am. Dyestuff Rep.*, **82**, 9, 79.

PERETTO C (1973) 'Dyeing of high bulk acrylic yarns in hank form', *Textilia*, **49**, 10, 71–74.

POPESCU V, BUTNARU R, POPESCU G L, PENCIUC M AND DAN D (2001a) 'Statistical analysis of acrylic fibres dyeing levelness- Part I: Temperature and dye concentration influence, *Revista Romana Textile Pielarie*, **4**, 77–82.

POPESCU V, BUTNARU R AND POPESCU G L (2001b) 'The analysis of variance applied in case of tinctorial system acrylic fiber-cationic dye', *Revista Romana Textile Pielarie*, **1**, 85–90.

PUHLMANN G (1971) 'Retarder systems in the dyeing of polyacrylonitrile fibres', *Deutsche Textiltechnik*, **21**, 9, 577–583.

ROHNER R M AND ZOLLINGER H (1986) 'Porosity versus segment mobility in dye diffusion kinetics – a differential treatment: dyeing of acrylic fibres', *Textile Res. J.*, **56**, 1, 1–13.

SHUKLA S R AND MATHUR M (1993) 'The retarding action of polyacrylamide in acrylic dyeing', *J. Soc. Dyers Color.*, **109**, 10, 330–333.

SIEPMANN E AND BRUNS H J (1969) 'First anniversary of the Defitherm process for dyeing polyacrylic fibres', *Int. Textile Bulletin*, **2**, 147–156.

VAIDYA A A (1984) 'Dyeing of acrylic fibre and its blends', *Indian Textile J.*, **94**, 8, May, 115–119.

VAIDYA A A (1988) *Production of Synthetic fibres*, New Delhi, Prentice-Hall of India Ltd.

24

Influence of fibre and dye structures in dyeing

Abstract: Irrespective of dyeing method introduced for a specific dye–fibre system, both dye and fibre must possess either suitable free volume inside with canals to diffuse and / or reactive site to establish dye–fibre linkage which is essential to achieve desired shade with promising fastness. If required, fibre structure may be modified to make it efficient for the purpose. There are various parameters in dyeing which control the dyeing phenomenon and decides type of shade to be produced.

Keywords: Fibre structure, dye structure, free volume, reactivity, modification

24.1   Basic concept of dyeing

Dyeing of textiles is exercised through either of exhaust dyeing and padding techniques. In exhaust dyeing, selected dye must have affinity for fibre to develop desired shade with ease. In contrast, affinity does not play an important role in padding as the colouring material is mechanically applied on textile using a highly concentrated solution or paste followed by drying and fixation. A padding technique whether continuous or semi-continuous in nature, is economical only for processing longer length of fabric and reasonably the technique is not frequently recommended in industrial application. This has paved the way for batch dyeing using jiggers, winches, etc.

Exhaust or batch dyeing of textiles consists of four basic steps: dissolution of dye in bath, transport of dye from solution to surface of fibre, adsorption at the surface followed by diffusion at the interior; these steps are irrespective of whether dye is retained by fibre through chemical bonds or physical forces. Dye uptake, i.e. amount of dye taken up by fibre can be attributed to the final step of dyeing, i.e. rate of diffusion and is dependent on (i) fibre structure, (ii) dye structure, and (iii) dyeing parameters.

24.2   Fibre structure

Salient features of fibre which govern its dyeability are (i) permeability – ease with which dye molecules diffuse into fibre structure, which in turn is influenced by its physical structure, (ii) weightage of reactive functional groups influenced by its chemical structure, and (iii) nature and extent of modification imparted to fibre at the manufacturing and post-spinning stages, during its pretreatment and in various other ways (Bird and Boston, 1975; Holme, 1970).

24.2.1 Physical structure

Influence of physical structure in dyeing is governed by two factors: (i) volume fraction, shape, size and configuration of voids or disordered regions accessible to dye (amorphous regions), and (ii) volume fraction, shape, size, configuration and distribution of ordered regions, which resist dye penetration (crystalline regions) (McGregor and Peters, 1968). These two factors, in turn, depend on ratio of amorphous and crystalline regions for a given fibre. The micro-physical state of fibre influences its dyeability by extending access to dye and water molecules to internal functional groups. In general, amorphous regions are more absorbent and accessible to dye molecules as compared to crystalline regions and are responsible for higher dye uptake. Extent of crystalline and amorphous regions in natural fibres is inherent characteristics and dye uptake can be enhanced only up to some extent through efficient pre-treatment or various other modifications of fibre structure, while dye uptake on manmade fibres can be varied up to remarkable extent with greater ease.

Shape of free volume or voids in a fibre is of utmost importance to give passage to the dye molecules; the latter possessing remarkably smaller volume may not diffuse inside fibre due to incompatible shape of voids.

Regularity of polymer chains

The degree of regularity in polymer chains is reflected in two different but related forms of physical order, viz. orientation and crystallinity. Orientation refers to the extent at which the molecular chains are directionally correlated over the entire fibre. Better the orientation along the axis of fibre, lesser will be dye uptake and vice versa, shown in Fig. 24.1 (Bird and Boston, 1975). Orientation is unequally distributed between crystalline and non crystalline regions and the latter is more significant to affect dye uptake.

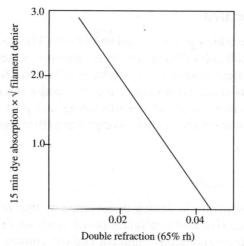

24.1 Impact of increasing draw ratio on rate of dyeing.

Within the family of cellulose origin, ramie and jute have extremely high degree of orientation while cotton has much lower degree of orientation with a consequence of higher dye up take. In case of man made fibres, degree of orientation can be controlled at the initial extrusion and subsequent drawing stages as described by effect of double refraction on rate of dyeing of filament viscose when dyed with direct dye (Bird and Boston, 1975).

Segments of polymer chains tend to remain not only parallel to fibre axis but are also arranged side by side in full 3D order. All natural and man made textile fibres show good crystallinity, but a high degree of crystallinity is detrimental for efficient dye up take.

Rate of dyeing with disperse dye on secondary cellulose acetate (CA), cellulose triacetate (CTA) and polyester (PET) are not equal due to differences in degree of crystallinity in respective fibres as explained in Fig. 24.2 (Bird and Boston, 1975). It is evident that rate of dyeing is bound to be more for less crystalline cellulose acetate.

Mobility of molecular chains

It comes into effect at and beyond glass transition temperature (T_g) of polymers. Changes in properties of man made fibres those occur at and above T_g affect its dyeability. As temperature increases beyond T_g, mobility of molecular chains increase in non-crystalline region until adequate space is created for migration and to accommodate dye molecules with increase in accessible free volume for diffusion of dye. Glass transition temperature (T_g) is a function of crystallinity, e.g. T_g of polyester increases with increase

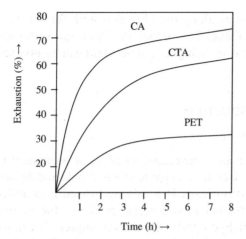

24.2 Effect of crystallinity on rate of dyeing.

in crystallinity and so polyester with required T_g can be produced by keeping control over process conditions during its manufacture as well as spinning. Lower the T_g, faster the mobility of the molecular chains at lower temperature with a consequence of increase in rate of dyeing.

Rate of diffusion varies with mobility of polymer chain segments (Bell and Murayama, 1969). Polymer chains may be assumed having unsymmetrical folded configuration and reasonably at very low draw ratios, rate of diffusion of dye remains on higher side, marked in Fig. 24.3 (McGregor and Peters, 1968). Slow increase in draw ratio ruptures

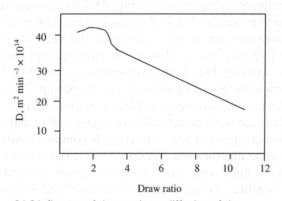

24.3 Influence of draw ratio on diffusion of dye.

crystallinity present in the fibre with simultaneous creation of more disordered regions by opening folds in chains. This increases mobility of polymer chain segments, when the chains tend to lie more or less parallel to each other reducing voids through which dye can diffuse with consequence of lowering in dye up take. Further increase in draw ratio stabilizes molecular chains along the fibre axis with further decrease in diffusion coefficient of dye. The higher the draw ratio, the lower the diffusion of dye and obviously higher colour yield can be predicted with lesser dye uptake.

24.2.2 Chemical structure

Interaction of fibre with solvent

Ability to absorb water and consequent swelling is prominent in fibres containing polar groups which are capable of ion–dipole and dipole–dipole interactions, but more specifically H-bonds are formed with molecules of water when no such attachment is possible in fibres due to absence of polar groups. In general, hydrophilic fibres swell appreciably in water and are able to accept water soluble ionic dyes, while hydrophobic fibres show little or negligible swelling in water and are permeable to non-ionic dyes possessing low aqueous solubility (Bird and Boston, 1975).

Moisture regain and degree of swelling

Ability of fibres to absorb water is a clear indication of its dye accessibility. Degree of swelling in water exhibited by textile fibres reflects the chemical composition and dyeing behaviour. Also, the degree of swelling in water and moisture regain is closely related to the dyeability of a fibre. Polar nature of fibre to absorb water and presence of disorder regions in the polymer matrix influences degree of swelling. PAN has a moisture regain of around 1% due to presence of anionic co-monomers with polarity in it; otherwise pure fibre does not absorb water at all. Polyester has very low moisture regain (~0.4%) due to absence of polar groups and highly crystalline nature of fibre. Due to its hydrophobicity, swelling in water remains almost absent necessitating the only way to open up fibre structure using swelling agents called carriers or expansion in fibre structure through heating. Nylon has relatively better moisture regain (~4%) and swelling in spite of its higher crystallinity and orientation. In contrast, wool contains polar groups as well as disorder arrangement of molecular chains implying higher dye uptake. Fibres with accessible polar groups generally show high degree of swelling during dyeing (McGregor and Peters, 1968; Campbell, 1966).

Functional groups

Fibres which contain polar groups can be dyed with greater ease. Concentration of polar functional groups in fibre determines absorptivity of dye on fibre for any given dye–fibre system. Polar groups in fibre are capable of ionising under favourable conditions to interact with dye ions, e.g. due to presence of less polar acetyl groups in secondary cellulose acetate, absorption of water at saturation is only 8%, although molecular chains are less oriented (McGregor, 1972). Fibre swelling is therefore limited and accessibility of less ordered regions of fibre is insufficient for penetration of water soluble direct dyes.

Surface potential/zeta potential

Man made fibres possess relatively high negative potential and the resulting potential barrier at the surface is sufficient to prevent adsorption of dye anions. In case of natural fibres, this high negative potential is of relatively little importance, since this potential is reduced or reversed by increasing the ionic strength of aqueous dye solution, e.g. protonation of amino groups in wool give rise to net positive charge on it thus producing favourable electrostatic environment for adsorption of acid dye anions.

24.2.3 Modification of fibre structure

Change in orientation as well as modification of chemical structure by changing or modifying functional groups is another unique approach to initiate higher dye up take. It must be emphasised that changes in chemical structure in general be accomplished by changes in physical structure too. However, both these changes can enhance rate of dyeing as well as extent of dye uptake and vice-versa (Bird and Boston, 1975).

Modification of physical structure

(a) Thermal treatment/heat setting

Structural changes can be introduced in a polymer through heat treatment to influence the dye uptake pattern (Bird and Boston, 1975; Kloss, 1972; Zimmerman, 1972), e.g. receptivity of disperse dyes by polyester is dependent on temperature at which dry heat setting is imparted before dyeing. With increase in heat setting temperature, dye uptake decrease initially, remain on lower side from 140–180°C, beyond which increases rapidly up to 200°C followed by a drastic increase beyond that, the trend has been marked in Fig. 24.4 (Marvin, 1954). This change in pattern of dye uptake is due to more liberalisation of chains with least potential energy inside along with least resistance for penetration and diffusion of dye at very higher temperature.

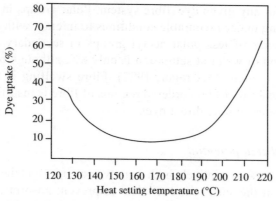

24.4 Influence of setting temperature on dye uptake.

(b) Drawing

Drawing results in reduction of dyeability of fibre. Changes in properties of filaments those occur on drawing are caused by change in physical structure of polymer. There is increase in degree of crystallinity and orientation both in molecular chains, the relative importance of each type of changes depends on fibre type; increase in both reduce dye uptake as shown in Fig. 24.5 (Bird and Boston, 1975).

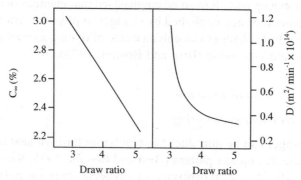

24.5 Effect of draw ratio on equilibrium sorption (C∞) and diffusion coefficient (D) of disperse dye on polyester.

(c) Reshaping of fibres

Cross section of a fibre can be modified to increase absorbency. Treatment of cellulose in concentrated NaOH under tension causes fibre cross-section to change from oval to spherical one enabling it to be more absorbent with more uptake of water-soluble dyes. This principle is used in mercerisation of cotton.

(d) Creating voids on surface of fibre

Heating of fibres beyond certain temperature will cause expansion of pores on the surface giving rise to increased absorbency. Synthesis of polyester through micro-emulsion polymerisation technique produces voids in the structure.

(e) Enhancement in thermo plasticity

Reduction in compactness of polymers without introducing functional groups can be brought about by introducing plasticizers in polymer matrix which bring down T_g through reduction in cohesive force (Bird and Boston, 1975). Plasticizers may be added either internally during manufacturing or externally as a preparatory treatment before dyeing. Introduction of methyl methacrylate or vinyl acetate as co-monomer results PAN with increased permeability for disperse dyes. T_g of PAN is lowered down from 105°C to 80–85°C through addition of co-monomers. Incorporation of non-isomorphic co-monomer into a homopolymer reduces crystallinity as reflected by decrease in T_g of the resultant polymer. Melting point of co-polyester (Toyobo) produced by introducing isophthalic acid is found to be 233–240°C while that of polyester lies in the 260–265°C range.

Modification of chemical structure

(a) Introduction of functional groups

(i) use of ionic co-monomers

Various ionic co-monomers when added to monomers during polymerisation make fibres electrically active and change affinity of dye for that fibre (Hobson, 1972). PAN is thermosetting in nature and resemble rod like structure, making it undyeable with almost any dye. Addition of ionic co-monomers develops affinity of different dyes – while acidic co-monomers develop negative charge on fibre making it dyeable with cationic dyes, basic co-monomers impart cationic charge making the fibre dyeable with acid dyes. Cation active chemical treatment on polyester produces anionic dyeable polyester which can be dyed with cotton dyes too.

(ii) use of additives

Incorporation of functional groups before extrusion can change dyeing pattern of fibres. Absence of reactive sites and high crystalline nature makes polypropylene difficult to dye. Organo-metallic compounds consisting of polyvalent transition metals are capable of forming chelates with selected dyes and polymers possessing high electron donating groups and are added to the polypropylene melt before extrusion to result enhanced substantivity for disperse dyes. Incorporation of suitable components in viscose dope before extrusion results in differentially dyeable viscose (Levine and Weimer, 1970; Fox, 1973).

(b) Modification of functional groups

(i) Intensifying reactive sites

Amino end groups in nylon are responsible for adsorption of acid dyes and rate of dye uptake depends on extent of cationization of end amino groups, which in turn, depends on acidity of bath – the more the acid the higher is the extent of cationization. By increasing effective amino end group content, rate of dyeing as well as overall dye uptake can be intensified, e.g. deep dyeing nylon yarns are produced in this way which is due to more dye uptake than that on regular nylon yarns (Vaidye, 1988).

(ii) Modifying functional groups

Structure of polymers can be modified to improve dye uptake through its pre-treatment with some specific chemicals. In acetylation of cellulose, many of the –OH groups in non-crystalline regions are converted into acetyl groups, causing it to become more hydrophobic and dyeable with disperse dye rather than with a water soluble dye. In this case, dye uptake increases at initial stage due to disruption in the physical orientation of polymer chains and then gradually decreases as the degree of acetylation increases.

24.3   Dye structure

Dyeing is a function of fibre structure as well as dye structure and the dyeing behaviour must be related to the detailed structure of both the components (Bird and Boston, 1975).

24.3.1 Size, shape and molecular weight of dye

Size of a dye molecule influences the dyeing pattern for a given fibre. In general, dyes of smaller size require lesser activation energy to diffuse easily whereas dyes of bigger size cause poor diffusion and remain mostly on the surface of fibre resulting poor wash fastness (Fig. 24.6). With a smaller dye, accessibility of the amorphous region will be a less important factor but increase in size of dye plays more and more decisive role. For example, the relationship between molecular weight of dyes and their diffusion coefficients for PET from perchloroethylene at 121°C has a coefficient of correlation of 0.94. Size and shape of a dye molecule in relation to those of pores of a fibre structure play a decisive role in diffusion of dye inside fibre.

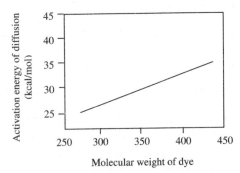

24.6 Activation energy vs molecular weight.

24.3.2 Plane of dye structure

The geometrical structure of dye in relation to the fibre structure influences dye uptake as it can experience spatial hindrance. For efficient dye uptake, molecular structure of dye should be coplanar, i.e. all the side groups of the dye molecule should lie on the same plane.

24.3.3 State of the dye

Many dyes exist in aggregates with weaker bond among themselves. Being bigger in size, these aggregates have lesser accessibility of fibre pores. Heating of dyebath breaks down these aggregates, facilitating smooth entry of dye molecules into fibre structure.

24.3.4 Electrical nature of dye molecule

Dye should have opposite electrical charge to that of fibre for efficient dye uptake. Chances of uneven shade formation may be reduced by applying retarding agents to control initial strike rate during dyeing. In contrast, if dye and fibre both possess identical electrical charge, electrolyte is to be added to promote exhaustion, e.g. dyeing of cotton with direct dyes.

24.3.5 Chemical groups attached to dye

The nature, size and number of chemical groups present in a dye molecule influence dye uptake. Generally, polar groups improve solubility of dyes and penetrate into fibre structure more effectively. These groups also decide substantivity of dyes for different classes of fibres.

24.3.6 Stability of dye molecule

The dye should be stable under dyeing conditions like temperature, pH and auxiliaries used, etc.

24.4   Dyeing parameters

It is a well established fact that various parameters in a specific dyeing process play key role to form levelled shades as well as extent of dye uptake. Inadequate reduction of vat dye due to lesser temperature of vat or dyebath, lesser alkali or lesser $Na_2S_2O_4$ does not retain reduced form of dye resulting lesser and unlevelled dye uptake. Concentration of acid primarily controls acid dye uptake as degree of cationization of protein fibres directly varies with it and hence the rate of dyeing; pH of bath is to be kept on very weaker acid side for levelled dyeing. Dyeing of wool with equalising acid dyes require salt as retarding agent from the very beginning of dyeing to reduce strike rate, otherwise an unlevelled shade is invariably formed. Similarly all dyeing processes are run following some technical guidelines; the latter is to be strictly followed if efficient as well as levelled dye uptake is to be expected.

24.5   References

BELL J P and MURAYAMA T (1969) 'Relations between dynamic mechanical properties and melting behaviour of nylon 66 and polyethylene terephthalate', *J. Polymer Sci.*, part A2, **7**, 6, 1059–1073.

BIRD C L and BOSTON W S (1975) *The Theory of Coloration of Textiles*, Bradford, UK, Dyers Company Publications Trust.

CAMPBELL B (1966) 'The Dyeing of Secondary Cellulose Acetate', *J. Soc. Dyers Color.*, **82**, 303–313.

FOX M R (1973) 'Fixation processes in dyeing', *Rev. Prog. Color Related Topics*, **4**, 18–37.

HOBSON P H (1972) 'Acrylic and modacrylic fibres: development, use and potential', *Textile Chem. Color.*, **4**, 10, 232–237.

HOLME I (1970) 'Fibre physics and chemistry in relation to coloration', *Rev. Prog. Color Related Topics*, **1**, 31–43.

KLOSS E (1972) 'Dyeing and finishing of texturized yarn and piece goods', *Canadian Textile J.*, **89**, 9, 97–103.

LEVINE M and WEIMER R P (1970) 'Acid-dyeable polypropylene', *Textile Chem. Color.*, **2**, 15, 269–272.

MARVIN D N (1954) 'The heat setting of terylene polyester filament fabrics in relation to dyeing and finishing', *J. Soc. Dyers Color.*, **70**, 16–21.

McGregor R and Peters R H (1968) 'Some Observations on the Relation between Dyeing Properties and Fibre Structure', *J. Soc. Dyers Color.*, **84**, 267–276.

McGregor R (1972) 'Ionizable groups in fibres and their role in dyeing: An examination of the sorption of anionic dyes by cellulosic fibres and films', *Textile Res. J.*, **42**, 9, 536–552.

Vaidye A A (1988) *'Production of synthetic fibres'*, New Delhi, Prentice-Hall India Ltd.

Zimmerman J (1972) 'What's the outlook for nylon', *Textile Chem. Color.*, **4**, 8, 212–215.

25
Processing of micro-fibre

Abstract: Textile fibre fineness below one dtex has been considered micro-fibre which possesses quite higher surface area when compared to conventional textile fibres and are getting popularity due to their good tensile strength, dimensional stability and stable to wear and tear, etc. Micro-fibres essentially belong to man made class as such a high draw ratio can not be imparted to natural one. Due to exceptional fineness and high surface area, chemical processing becomes difficult and excess dye is required to produce a specific shade. Dyeings lack required fastness properties too.

Keywords: Dtex, surface area, dyeing, fastness properties

25.1 Definition and importance of micro-fibre

Micro-fibres belong to man made origin and fabrics made from these possess soft handle with excellent drape, wear comfort, lighter and easy care properties. Garments made from micro-fibre fabrics exhibit excellent tensile strength, bulkiness, durability, dimensional stability, weather resistant and are less vulnerable to wear and tear; a reason why these are finding increased use in functional and fashion clothing (Partin, 1991a). Application of amino modified silicone polymers and epoxy functional silicones on polyester microfiber–cotton blended wovens can further enhance their tactile properties, i.e. tearing strength, wettability and recovery angle, etc. (Sabia, 1995).

For a given count, micro-fibres have at least four times higher surface area due to four times more filaments than that in conventional fibre and as a result, more light is reflected from surface appearing lighter in colour (Fig. 25.1). Dyeings lack required wash, light and rubbing fastness. Woven micro-fibre fabric is too light, densed and finer one. Absolute count range has not been defined till date, though fibres less than 1 dtex (10000 m/g) or 1 dpf (9000 m/g) has been considered as micro-fibres (Table 25.1) (Partin, 1991a; Hilden, 1991; Parton, 1996; Fong, 1995).

(a)	(b)
Conventional fibre with lower total surface area	Micro-fibre with higher total surface area
Total surface area = 7 × surface area of each fibril	Total surface area = 30 × surface area of each fibril

Total surface area of (a) < Total surface area of (b)

25.1 Comparison of surface area in conventional and micro-fibres.

Table 25.1 Fineness nomenclature of textile fibres

Tex system*	Fibre status	Denier system
7.0 dtex ≤	coarser fibres	=7.0 dpf
7.0 – 2.4 dtex	fine/normal fibres (cotton, wool, polyester)	7.0 – 2.4 dpf
2.4 – 1.0 dtex	extremely fine fibres (silk, acrylic)	2.4 – 1.0 dpf
1.0 – 0.3 dtex	micro-fibres	1.0 – 0.3 dpf
0.3 dtex > 0.0001 dtex	super ultrafine micro-fibres latest development	< 0.3 dpf –

* Denier (d) = 0.9 × decitex (dtex)

The count of conventional polyester fibre ranges between 2.0 and 5.0 denier/filament (dpf), whereas a polyester micro-fibre is 1.0 dpf or even less. Polyester and polyamide micro-fibres have secured commercial importance, though polyacrylonitrile and viscose rayon micro-fibres are also of specific interest. Micro-fibres and derived fabrics are susceptible to develop crease marks badly. Diameter and surface area of some important fibres are also shown in Table 25.2 (Fong, 1995). Surface area of a fibre (F) and its diameter (d) are approximately related as $F = \pi d \times 10^7$ dtex, where $d = 3.6 \times 10^{-4} \sqrt{(dtex / \rho)}$, ρ is the density of fibre in g/m³. Total surface area is a vital issue since rate of surface deposition of dye is directly proportional to it (Fig. 25.1).

Table 25.2 Diameter and surface area of various fibres

Parameter	Fibre type					
	Cotton	Wool	Silk	Polyester		
				Normal	Micro-filament	
dtex	2	11.5	1.3	5.55	0.7	0.5
Kg/m³	1500	1360	1370	1380	1380	1380
d, 10⁻⁶m	13.1	34	11	22.8	8.1	6.8
F, m²/Kg	206	93	290	129	363	430

25.2   Nomenclature of micro-fibre yarns and fabrics

Yarns made of micro-fibres is coded as X f N, where 'X' denotes count of yarn, 'f' the micro-fibre filament and 'N' is the number of filaments. Accordingly, a 90 f 150 yarn consists of 150 filaments whose total fineness is 90 dtex, with an individual filament count of 90 / 150, i.e. 0.6 dtex. Woven fabrics made of such yarns are named as polyester or nylon micro-fibre yarn or fabrics (Parton, 1996).

25.3   Application of micro-fibres

These are mainly used as warp or weft and to make blends with natural fibres. Polyester micro-fibre fabrics are in greater demand and are used in manufacture of wind proof jackets and waterproof fabrics, viz. hydrophobically laminated film finished with water vapour permeable but water droplet impermeable. Due to superior handle, polyester micro-fibres are also used for production of silk like woven and knitted fabrics with excellent drape. Other application of polyester micro-fibres include sophisticated special fabrics with attractive design through emery treatment or crashing (Partin, 1991a).

Nylon micro-fibre fabrics are coated with breathable waterproof membrane, e.g. Cyclone, Gore-Tex; knitted Tactel as swim wear, body apparel, socks, tights and sports wear (Parton, 1996). Flexibility is further improved by introducing various elastomeric materials, e.g. polyurethane (lycra). Nylon micro-fibre is mainly blended with wool to manufacture socks while that with cotton for sports wears, swim wears, leisure wears, ladies fashion outerwear, fashion wears and space wears. Garments made of micro-fibres are very densed so that these are by themselves water vapour and air permeable, windproof; films or coatings are just a protection.

Myoliss, a high performance acrylic micro-fibre is produced by Enichem through co-polymerisation of acrylonitrile with vinyl acetate as co-monomer, having count range around 0.8 dtex and is used mainly in knitwear production (Spaggiari and Trevisan, 1993). Other application of micro-fibres includes medical, household and miscellaneous fields (Partin, 1991a; Mukhopadhyay, 2002). Make and brand of a few commercial micro-fibres is described in Table 25.3 (Partin, 1991a; Hilden, 1991; Spaggiari and Trevisan, 1993).

25.4   Typical micro-fibre characteristics

Micro-fibres are made of three regions, viz. amorphous domains of micro-fibril, crystalline domains of micro-fibril and inter-micro-fibril regions. Crystalline and amorphous regions are evenly distributed and the folded

Table 25.3 Some important commercial micro-fibre types

Fibre type	Brand name	Manufacturer
PET	Diolen Micro	Akzo / Enka
PET	Trevira-Finesse	Hoechst AG
PET	Trevira-Micronesse	Hoechst AG
PET	Terital Microspun	Montefibre
PET	Setila Micro	Rhone-Poulenc Fibres
PET	Tergal Micro	Rhodia AG
Nylon 6.6	Meryl Micro	Rhodia AG
Nylon 6.6	Tactel Micro	ICI
Nylon 6	Lilion	Snia Fibre
Acrylic	Myoliss	Enichem

chains forming crystalline regions have a well defined element structure with a width of 60–200Å through interlacement, branching and fusion without an abrupt end. In the highly extended inter-micro-fibrillar regions, the chains are oriented with absence of crystallinity, is called 'oriented amorphous regions', which causes fibre shrinkage below T_m. The inter-fibrillar regions do not contribute to fibre strength, rather the extended chain inter-fibrillar domains are the strongest elements of fibre structure and controls fibre strength. The increase in fibre strength on drawing is attributed to an increase in volume fraction of extended chain molecules those are formed as a result of the relative displacement of micro-fibrils; the latter imparts dimensional stability at elevated temperature (Datye and Vaidye, 1984).

Micro-fibrils can be defined as another element of fibre structure whose shape is ribbon like, about 300–400 Å thick and are presumed to consist of several micro-fibrils and inter-micro-fibril regions. The size of micro-fibrils is highly dependent on thermo-mechanical treatment (Hongu and Phillips, 1997).

25.5   Processing of polyester micro-fibre

25.5.1 Pre-treatment

The pre-treatment processes resemble those of conventional polyester with certain exceptions. Highly densed with higher surface area and negligible gap among fibrils retain chemicals and dyes firmly making removal difficult (Partin, 1991a). To protect distortion of fibre, sizing is done with polyester or polyacrylate sizes due to extreme fineness, where polyvinyl alcohol is used for conventional polyester. Sizing material to be used is in far excess with a pre-condition of ease in complete washing out. However, size removal becomes costlier. All pre-treatments and washings should be imparted in open width with high rate of liquor flow under least tension to avoid distortion. Jet dyeing machine may be swung into action if open width processing is not possible due to machinery constraints. To attain efficient desizing, polyester sizes are best removed at pH 8.0–8.5 and for polyacrylates at pH 10.5–11.0. Before desizing, identification of size is a must (Hilden, 1991). Desizing may be imparted with H_2O_2 (3 g/l), Na_2CO_3 (4 g/l), defoamer (0.5 g/l), detergent (2 g/l) and Na_2SiO_3 (0.05 g/l) (Fong, 1995). Wash out effect is to be tested with a preliminary sample dyeing test after each pre-treatment. Chlorite bleach may be carried out with $NaClO_2$ (1–3 g/l), $NaNO_3$ (1–3 g/l), HCOOH (1–3 g/l to get pH~4) at 90°C for 1 hr followed by hot wash, neutralization with Na_2CO_3 (2–5 g/l) for 5–10 min at 90–95°C, thorough cold wash, anti-chloring and final washes (Menezes, 2001).

Heat setting is carried out before dyeing at 180°C for 30–45s for 100% polyester micro-fibre plain warps; in case no plain warps are available, heat setting is omitted but dried at 140–150° to moderately set the fabric and to prevent shrinkage. Pre-setting is required to hinder formation of crease marks during dyeing at high temperature followed by final setting after dyeing or during finishing (Goldstein, 1993). Adequate care must be ensured to avoid sublimation of dye during post-setting as most of dye molecules remain embedded at the surface unlike that with conventional dyed polyester where dye diffuses *in situ*. Emery treatment facilitates sueding, reduces creasing tendency and is imparted to develop slightly napped, peach-like surface and a pleasant soft luxurious handle in pre-dyeing stage; if imparted in post-dyeing stage develops unevenness and stripiness in shade (Hilden, 1991). Micro-fibre fabrics are emerised in a sueding machine at a speed of 15–20 m/min on specialized rollers coated with 30–500 grade emery paper followed by heat setting at 180°C and final washing in jet dyeing machine to remove fibre dust; poor sueding reduces tensile strength of filling yarns up to 60% (Partin, 1991a).

25.5.2 Dyeing

With the same overall count, micro-fibres have many times greater total fibre surface, leading to greater area for acceptance of dye – nearly four times higher when compared to conventional fibres. Obviously, dyeing rate is also very high (almost five times higher) which frequently leads to uneven dyeing, shown in Table 25.4 (Fong, 1995). Dyeing rate for different types of micro-fibres is to be pre-evaluated with 'Resolin S process' (Bayer AG). Due to more surface area, shade depth becomes lighter with identical dye concentration and excess dye is needed to get desired depth of shade. Another reason may be the greater degree of light reflection which lightens the shade. Penetration of dye molecule is faster raising chances of uneven dyeing. Ring dyeing occurs in flat polyester whereas thorough dyeing occurs in micro-polyester fibre – a reason why micro-fibres develop lighter shades with given concentration of dye. Jet dyeing or soft-flow dyeing machines are only suitable to preserve bulk; package or HTHP beam dyeing machine are unsuitable keeping in view the high density of micro-fibres, which severely impairs the flow necessary for level dyeing (Partin, 1991a). For levelness and reproducibility, the characteristics associated with dye exhaustion, heating gradient, colour depth, dye liquor flow rate and related dyeing parameters are also of paramount importance (Partin, 1991b; Griesser and Tiefenbacher, 1993).

Table 25.4 Dye fixation on polyester micro-fibre at lower temperature

Fibre type/Temp (°C)	40	50	60	70	80	90
Conventional	15.58	21.75	31.17	35.51	50.65	68.18
Micro-fibre	8.72	9.49	9.23	9.23	10.26	11.50

Continuous dyeing technology is yet to be established. To secure level dyeing in batch process, following guidelines are quite important (Hilden, 1991; Menezes, 2001).

(i)   Jet flow is adjusted to affect higher fabric speed (>250 m/min).
(ii)   Fabric rope is to be plaited down evenly.
(iii)   Crease inhibitors, e.g. lubricants are to be used to prevent creasing.
(iv)   Sticking to machine walls during dyeing is to be restricted through smooth and consistent flow of liquor.
(v)   Slow rate of heating ($\leq 1°C$/min) is a must for on-tone exhaustion and migration; cooling rate is as high as 1°C/min.
(vi)   Optimum compatible dyestuff selection and metered addition, i.e. injection of dye solution in few installments are useful.
(vii)   Addition of excess dispersing agent is required to disperse large number of dye molecules required for given depth of shade.
(viii) Deaerating agent is to be applied to suppress foam and to visualize status of fabric in side the machine.
(ix)   Levelling agents may be applied to suppress rate of dyeing in order to build up levelled shades.
(x)   For dyeing of blends of different counts, e.g. standard warp and micro weft, special care should be taken to ensure level dyeing.
(xi)   Reductive after-washing improves wash, light and rubbing fastness.

Dyeing is started at 60°C with deaerating agent (0.25%), levelling agent (0.50%), lubricant (1%) and CH_3COOH (pH~4.5–5.0); fabric is loaded and run for 10 min succeeded by addition of dye and the fabric is again run for 10 min. Dyebath is then heated up @ $\leq 1°C$/ min up to 96°C and dyeing is continued at this temperature for 20 min (hold step) followed by further heating at the same rate up to $130\pm2°C$ and dyeing is carried out for 1 hr. Cooling is affected @ 1°C/ min up to 70°C, the bath is drained, dyeings are washed thoroughly. Reduction clearing is imparted at 70°C with $Na_2S_2O_4$ and NaOH (4.0 g/l each) for 15 min, rinsed, neutralized with CH_3COOH (0.5 g/l) to improve fastness properties. Dyeings are finally rewashed and unloaded (Fong, 1995).

Alternately, the fabric is loaded at room temperature, run in blank bath for 5 min after which dye, defoaming agent (0.5 g/l), levelling agent (1 g/l), dispersing agent (0.5 g/l) and CH_3COOH (pH~4.5–5.0) are added. The bath is heated up to 45°C and run for 5 min, further heating is carried out up to 98°C @3°C/min and run for 30 min, further heated up at 130±2°C @1°C /min and dyeing is carried out for 45 min after which the bath is cooled down @1.5°C/min to 80°C and discharged, shown in Fig. 25.2 (Fong, 1995). Reduction clearing is a must for improved fastness properties.

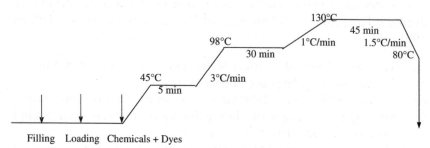

25.2 Dyeing of polyester micro-fibre in jet dyeing machine.

Since primary outlet of micro-fibres is apparel, i.e. fashionable outerwear, light fastness is not a major obstacle; wash fastness can also be improved through efficient reduction cleaning. To improve handle as well as functional properties, sueding, softening and fluorochemical finishes are the pre-requisites (Datye and Vaidye, 1984).

25.6   Assessment of dyestuff requirement

It is learnt that for identical shade, micro-fibre fabric will need more dye than that required on conventional polyester. No scientific approach has been made so far in this regard. As a rule of thumb, the following equation may be used to calculate the requirement (Fong, 1995; Partin, 1991b).

$$C_1 = \sqrt{(W_2/W_1)} \times C_2$$

C_1, C_2, W_1 and W_2 are % dye on micro-fibre, % dye on conventional fibre, denier per fibre (dpf) of micro-fibre and dpf of conventional fibres respectively. For example, if a 1.6% shade is to be developed on 1.5 dpf conventional polyester fibre, dye requirement to produce the same shade on 0.75 dpf micro-fibre will be

$$C_1 = \sqrt{(1.5 / 0.75)} \times 1.6 = 2.26\%$$

Textured polyester, irrespective of conventional or micro-fibre, will required little more dye over respective flat counterpart polyester.

25.7   Processing of nylon micro-fibre

25.7.1 Pre-treatment

These have count range of 0.7–1.2 dtex and are on the borderline of micro-fibres. Processing of nylon micro-fibre has typical problems in all the steps, viz. fabric preparation and level dyeing with build up of desired shade having satisfactory fastness. Silicone oil used to lubricate lycra during knitting has a tendency to redeposit onto the fabric during rinsing, producing random dye resist marks. Efficient scouring is to be imparted followed by over-flow rinsing. A controlled double scouring also improves performance with over-flow rinsing above 80°C after each scouring and reduces deposition of lubricant (Parton, 1996).

25.7.2 Dyeing

Two to three times more dye is required for a specific shade in comparison to that with conventional nylon. Other influencing factors which control dye uptake are fibre cross-sectional shape, degree of texturising and delustring; the greater the degree of delustering, more will be dyestuff requirement which may increase upto four times than that used for flat nylon (Parton, 1996). However, nylon micro-fibre delustered with TiO_2 produces matt effect and attain handle as well as appearance of cotton. Rate of dyeing is little higher compared to that in dyeing of conventional nylon. Higher dyestuff requirement leads to dyeing at or near saturation point of fibre lowering wash fastness grade; back tanning of dyeings may improve the situation depending on dye structure; a syntan (Nylofixan PM, Clariant) after-treatment (6–10%) can improve wash fastness effectively (Blackburn and Burkinshaw, 1998, 1999). Back tanning doubles time of production and partly gets precipitated causing harsh handle. Dyes of higher molecular weight with moderate migration and slower levelling properties result better wash fastness, especially for dyeing of sports wear, hosiery and body apparel, where required fastness can not be achieved with conventional dyestuff. Metal complex dyes comparatively give better results, with an obvious lesser dye uptake; light fastness is also severely affected due to more precipitation of dye on surface. Selection of dyes of lower molecular weight which shows higher affinity and lesser levelling property increases formation of barré but develops a well levelled shade in presence of a cationic levelling agent through formation of dye-levelling agent complex thereby slowing down rate of migration; at higher temperature, the complex breaks down and release dye for fibre. Anionic levelling agents are also important to prevent barré and working mechanism is based on blocking fibre sites to reduce strike rate of acid dyes. Dyeing at higher temperature should be the preference to develop levelled shades through efficient migration in more opened up fibre structure, but nylon may get oxidized and at 105–120°C; lycra if present, too gets adversely affected.

25.8   Fastness of dyed micro-fibres

Dyed polyester micro-fibre fabrics lack wash, light and rubbing fastness than those of conventional one. This is because of larger effective surface for colouration with congested interstices and densed structure which hinders efficient removal of dyes and chemicals during washing. As a result, both wash and rubbing fastness suffer (Hilden, 1991). Light fastness is too adversely affected because of (i) higher surface area absorbs more UV-radiation to affect fading at faster rate and (ii) most of dye molecules remain at the surface to cause ring dyeing or surface dyeing, which is attacked by heat and moisture within a lesser period of exposure.

25.9   References

BLACKBURN R S and BURKINSHAW S M (1998) 'Aftertreatment of 1:2 metal complex acid dyes on conventional and micro-fibre nylon 6.6 with a commercial syntan/cation system', *J. Soc. Dyers Color.*, **114**, 3, 96–100.

BLACKBURN R S and BURKINSHAW S M (1999) 'Aftertreatment of 1:2 metal-complex acid dyes on conventional and micro-fibre nylon 6.6 with a commercial syntan/cation system-Part 2: Repeated washing', *J. Soc. Dyers Color.*, **115**, 3, 102–105.

DATYE K V and VAIDYE A A (1984) *Chemical processing of synthetic fibres and blends*, New York, John Wiley & Sons.

GOLDSTEIN H B (1993) 'Mechanical and chemical finishing of microfabrics', *Textile Chem. Color.*, **25**, 2, 16–21.

GRIESSER W. and TIEFENBACHER H. (1993) 'Polyester micro-fibres', *Textilveredlung*, **28**, 4, 88–96.

HILDEN J (1991) 'Effect of fibre properties on the dyeing of micro-fibres', *Int. Textile Bulletin*, **37**, 3, 19–26.

HONGU T and PHILLIPS G O (1997) *New Fibres*, Cambridge, Woodhead Publishing Ltd

FONG'S NATIONAL ENGG CO LTD (1995) 'Processing of microfiber in jet dyeing applications', *Am. Dyestuff Rep.*, **84**, 3, 46–51.

MENEZES E.(2001) 'A practical guide to processing polyester micro-fibres', *Colourage*, **48**, 7, 31–32.

MUKHOPADHYAY S (2002) 'Micro-fibres - An overview', *Indian J. Fibre Textile Res.*, **27**, 3, 307–314.

PARTIN A R (1991a) 'Wet processing of polyester micro-fibre fabrics', *Am. Dyestuff Rep.*, **80**, 11, 45–49.

PARTIN A R (1991b) 'Wet processing of polyester micro-fibre fabrics–II', *Am. Dyestuff Rep.*, **80**, 12, 43–44 and 56.

PARTON K (1996) 'Dyeing nylon micro-fibre, *Int. Dyer*, **181**, 6, 14–21.

SABIA A J (1995) 'Modification of the tactile and physical properties of microfiber fabric blends with silicone polymers', *Textile Chem. Color.*, **27**, 9, 79–81.

SPAGGIARI W and TREVISAN E (1993) 'A new high performance acrylic micro-fibre', *Asian Textile J.*, **1**, 7, May, 29–33.

26

Dyeing of blend

Abstract: Deficiency in a specific fibre is compensated by means of making blends to improve overall performance. Technology of dyeing of a fibre is that fibre specific but when in blend, required adjustments become mandatory to dye component fibres as dyeing of one component may harm the second fibre if dyed in single or multiple baths. Chances of formation of differential fancy shades on the blends make the situation further critical.

Keywords: Blend, shade, method, dyeing

26.1 Reasons for blending

All textile fibres have their own inherent characteristics in terms of market value, look, feel, physical as well as chemical properties, etc., due to which products made of any one among these is not capable of meeting all the end use requirements. This has prompted to blend fibres to reduce cost of product and to balance physical prospects like look, tensile properties, moisture regain, comfort, crease recovery, etc., finally paving way for overall improvement in aesthetic properties of textiles. Fibres can be blended in two major ways, e.g. intimate and physical blending. Intimate blending is of conventional type in which staple fibres of various textiles are mixed thoroughly and spun together to get the blended yarn or spinning two compatible polymers by mixing in the autoclave. In contrast, physical blending is carried out during weaving to produce union or blended fabrics; alternately it may also be done in synthetic fibre manufacturing stage by mixing and spinning two compatible molten or liquefied polymers coming out of two different spinnerets (Nunn, 1979).

When two or more fibres are blended, the resultant textile may be dyed with a uniform hue and shade in 1-bath if a specific class of dye show affinity for both the components, i.e. if component fibres are anionic they may react with cationic dyes, e.g. dyeing of acrylic–wool blend; dissimilarity in electrical nature in component fibres enables to produce various fancy shades as the resultant blend is not dyeable in 1-bath with only one class of dye. In fact, depending on type of fibres in blend, fancy dyeing effects can easily be produced. Attempts are to be made if the shade can be produced in 1-bath 1-step dyeing; such a move reduces handle of textile and wastage of energy, lesser water consumption, etc. Before moving to discuss dyeing of blends in detail, it is mandatory to know some basic information required to handle the dyeing process, viz. nature of component fibres and blends, types of shades and dyeing methods (Mittal, 1984). However, to do excellence in this part of chemical processing, the dyer must have good knowledge of fibre properties as well as basic technology of dyeing of individual fibres.

26.2   Classification of fibres

Textile fibres can be classified according to their dyeability as follows (Nunn, 1979):

A-fibres – Fibres which are dyeable with anionic dyes, e.g. cotton, viscose, cuprammonium, acid dyeable nylon, wool, silk, acid dyeable polypropylene, acid dyeable acrylic, modacrylic, anionic dyeable polyester.

B-fibres – Fibres which are dyeable with basic dyes, e.g. basic dyeable acrylic, wool, silk, modacrylic, nylon and cationic dyeable polyester.

D-fibres – Fibres which are dyeable with disperse dyes, e.g. polyester, nylon, acrylic secondary acetate and cellulose triacetate. Polypropylene is not dyeable with disperse dye.

In this classification, few fibres are common in different types, e.g. nylon, wool and silk occupies place in both A and B fibre types; nylon alone in all the types A, B and C as it belongs to man made class as well as possesses –COOH and –NH$_2$ reactive sites at terminal ends like those in wool and silk behaving as A and B fibre both. Conventional acrylic is included in types B and D. Disperse and basic dyes produce poor wash and light fast shades on nylon respectively. Nylon is invariably dyed with acid or metal-complex dye, is an anionic fibre in true sense and included in group A. Similarly, acrylic is not dyed with disperse dye due to production of only light shades irrespective of quantity of dye used; it is popularly dyed with basic dye and it is the true member of type B. Wool, silk are preferably type A fibres while cotton and other cellulosics belong to type A and so on.

26.3   Nomenclature of blends

26.3.1 Primary blend

Component fibres belong to same class:

(i)   A-blends: Blends in which constituent fibres belong to A group and are invariably dyed with anionic dye, viz. cotton–wool, cotton–silk, cotton–nylon, nylon–wool, wool–silk, nylon–anionic dyeable polyester, etc.
(ii) B-blends: Blends in which constituent fibres belong to B group and are invariably dyed with basic dye, viz. nylon–acrylic, wool–silk, nylon–wool, acrylic–cationic dyeable polyester, acrylic–wool, etc.
(iii) D-blends: Blends in which constituent fibres belong to group D and are dyed with disperse dye, viz. polyester–triacetate, polyester–acetate, polyester–nylon, acrylic–nylon, etc.

26.3.2 Binary blend

Two components of blend belong to two different fibre classes. The nomenclature of a binary blend will be based on sequencing the names in the blend, e.g.

D–A blends:	Secondary acetate – nylon or wool or cellulose
	Triacetate – nylon, wool or cellulose
	Polyester – wool, polyester/cotton, etc.
D–B blends:	Secondary acetate – acrylic or cationic dyeable polyester
	Triacetate – acrylic or cationic dyeable polyester
	Polyester – acrylic or cationic dyeable polyester
A–B blends:	Nylon – cationic dyeable polyester or acrylic
	Wool – cationic dyeable polyester or acrylic
	Acid dyeable acrylic – basic dyeable acrylic
	Acid dyeable acrylic – cationic dyeable polyester
	Cotton/acrylic, cotton – cationic dyeable polyester, etc.

26.3.3 Ternary blend

The blend must be composed of fibres from three different types:

DBA type–	Polyester–acrylic–nylon, polyester–acrylic–wool
	Polyester–acrylic–cotton, polyester–acrylic–viscose
	Polyester–cationic dyeable polyester–cotton, etc.

Blends of types DA, DDA, DDD, DAA, AB, AAB, AAA, ABB, B and BBB are practically primary or binary blends as only two types of fibres are involved in making the blend and hence are not ternary blends in true sense.

26.4   Types of shades

Presence of various fibres in yarn or fabric due to blending, chances of producing differential hues and shades, i.e. fancy effects become easier as because various fibres can be dyed with different dyes. A blend may be composed of two or three component fibres in practice; it is customary to accept a blend as having as least two components, if not otherwise mentioned. Shades on blends those can be produced are based on this concept. Blends may be dyed in different types of hues and shades depending on nature of component fibres. It is possible, at most, to form four different types of shades in dyeing blends (Nunn, 1979), viz.

26.4.1 Reserve shade

One of the components is dyed while the second component is kept reserved, i.e. left undyed. The resultant shade is a single hue with white dots spreaded throughout; intensity of white dots and the hue depends on share of component fibres in blend, e.g. dyeing of only cotton part in polyester–cotton blend. To produce a reserve shade, both the components must not have affinity for same class of dye.

26.4.2 Cross shade

Two components of the blend are dyed with two different dye classes in two different hues and shades, producing mixed effect. The component having lesser share will be visible as coloured dots on a different coloured ground. A cross shade can only be produced when both the components must not have affinity for same class of dye like that for a reserve shade. Dyeing of polyester with orange hue (say) and cotton in green hue (say) in polyester–cotton blend is an example. Hues are selected mostly with contrasting colour combinations, viz. orange-black, orange-green, navy blue-yellow, and royal blue-white, etc.

26.4.3 Shadow shade (shadow effect or tone in tone dyeing or tone to tone dyeing)

Both the components of the blend are dyed in such a manner that the shade formed on one component is a fraction of the shade on the other component. The net result is that a part of the shade looks like shadow of the shade on another component. In this case, both the components are dyed with same hue but at varying depth, e.g. dyeing of polyester–cotton blend with disperse-reactive dye system for say, 3% and 1% shades with same blue hue. Shadow shades are produced by default when blends having components possessing affinity for same dye are dyed for short time and shades at different depths are produced on these fibres due to their differential receptivity. However, prolonged dyeing develops a solid shade due to entering of more dye in both the components with increase in time of dyeing thus making practically no difference in depth on component fibres.

26.4.4 Solid shade or union dyeing effect

Both the components are dyed with same hue, depth and tone to develop a shade which may be visualized as a shade on single fibre based textile, e.g. in polyester–cotton blend, polyester and cotton may be dyed green hue separately with disperse and vat dyes respectively with say, 4% and 3% shades based on differential tinctorial power of various dyes, the coverage of the shade on both the components must be identical to produce a solid shade; any different of shade will result a shadow shade. Fibres having affinity for same class of dye will produce solid shade with ease.

26.5   Method of dyeing blends

There are three different dyeing methods, one among these are to be followed to dye a specific blend, viz. (i) 1-bath 1-step, (ii) 1-bath 2-step and (iii) 2-bath 2-step methods.

The 1-bath 1-step method, if followed, reduces cost of dyeing drastically due to nearly 50% reduction in use of water, heat, chemicals, time and manpower, etc., but components of the blend may not favour dyeing through this method, for example, nylon–wool, cotton–viscose may be dyed simultaneously from direct and acid dyebaths respectively; whereas polyester–cotton blend can not be dyed in this method (except very light shades). Polyester is dyed with disperse dye at high temperature in jet dyeing machine followed by dyeing of cotton part in the same machine or jigger or winch with cotton dye. Acrylic–cotton blend can not be dyed in 1-bath as both cotton and acrylic do not possess affinity for same dye, leaving scopes for 1-bath 2-step and 2-bath 2-step methods depending on the type of shade to be produced. In contrast, polyester–triacetate, nylon–wool, silk–wool, cotton–viscose, acrylic–cationic dyeable polyester, etc., blends can be easily dyed in 1-bath 1-step method due to affinity of same dye for both these fibres. Interestingly, when it comes to types of shades to be produced on these blends in this method, reserve and cross shade can not be produced at all or not so easily as any dye or dye class combination has affinity for both the component fibres. It can be concluded here that 1-bath 1-step method is the most suitable for reserve shades.

In 1-bath 2-step method, material handling is remarkably less, as the blend once loaded remain there till completion of dyeing for both the components; the same bath is used for dyeing both these fibres saving water and energy, etc. In this case, material is loaded in the machine; a dyebath is produced for one component. Once dyeing is over for it, the bath is not drained out rather dye for second component is added and favourable condition of dyeing is generated, like changing an acidic pH in earlier bath to an alkaline bath, etc., to dye second component. Polyester–cotton blends are popularly dyed in jet dyeing machine using disperse–reactive HE dye combination in this method. Other examples are acrylic–cotton, acrylic–pet blends.

The 2-bath 2-step method is popular in industries to dye blends made of two fibres of two different nature or three fibres out of which any two are dyeable with a class of dye while the third component is dyed with a second dye but overall dyeing cost is very high due to more handling of material, huge water consumption, doubled loading/unloading time, more manpower to carry the textile in front of two different machines at two locations, etc. The popularity of this method is mainly because dyeing of components in two separate baths eliminates complications arising from interaction of dyes and chemicals if 1-bath is used. Polyester–wool and polyester–cotton are two most popular blends or any other blend whose component fibres possess affinity for two different dyes are commonly dyed in this method.

The objective of this chapter is to manipulate these three sets of data, i.e. types of fibre and blends (primary, secondary and ternary), types of shades (reserve, cross, shadow and solid) and different methods of dyeing to predict viability of a method to develop the required shade/shades on a specific blend. The method that produces the shade at least cost, less time, less handling, superior fastness, etc., will be the most suitable one.

With existing natural and man made fibres, numerous blends can be produced most of which are only of academic interest and possess no commercial importance, neither practically produced at all on technical grounds. Only dyeing of commercially important blends will be dealt with in this chapter. In addition to, it is essential to know nature of individual fibres as well as their dyeing technology in detail for dyeing of blends effectively.

Keeping in view these discussions and data, possibility will be explored for producing various shades on various blends based on commercial interest.

26.6   Dyeing of primary blends

26.6.1 Dyeing of A-blends

Important blends and their end use

Commercially, nylon–cotton blends (10–50% nylon) are produced for dress wears and leisure shirts. Stretch fabrics with crimped nylon warp and wool or viscose staple weft is suitable for sports wears, specifically for ski wears and leisure clothing. Nylon warp stretch fabrics with cotton or nylon/cotton wefts are for swimwear and narrow fabrics. Cotton warp/nylon weft wovens are for uniforms and raincoats.

Nylon–wool velvet and other fabrics are important for upholstery sectors when nylon/wool (20–25/80–75) blends used in carpet industry and nylon/cotton (25–20/ 75–80) used in apparel purposes.

Weft-knitted toweling with nylon filament and cotton pile forms children's wear, beachwear and leisure shirts.

Feasibility of producing shades

All cellulosics, wool, silk, nylon are true A-fibres. Dyes commonly used for dyeing cellulosics are direct, sulphur, vat, reactive, solubilised vat and naphthol colours; these are anionic in nature in dyebath. Wool, silk and nylon are invariably dyed with acid and metal complex dyes which are anionic too. All dyes used on cotton are also applicable on wool, silk and nylon and are practically used depending on end use and impact on fibre strength. These last three fibres, viz. wool, etc., possess anionic and cationic groups on terminal ends of chain; form ionic bond between anionic part of dye and cationic part of fibre (ref 15.3) and wash fastness depends on molecular size of dye. In contrast, cellulosics can not be dyed with acid and metal complex dyes as acidic pH required for application of these dyes do not favour dyeing.

It is learnt that if the A-blend is made of only cellulosics, i.e. cotton–viscose, cotton–cuprammonium, etc., application of any dye suitable for cellulosics will dye both these components in 1-bath 1-step process but dye uptake as well as diffusion will be according to free volume and compatible regions available in component fibres. Dye uptake for short time dyeing will follow the order: viscose > cuprammonium > cotton and will develop shadow shade easily though prolonged dyeing will develop nearly a solid shade due to more dye uptake thus reducing the difference in final dye uptake among these components. Production of cross shade and reserve shade is not practically feasible. Similar will be the case of blends composed of wool–silk, wool–nylon, silk–nylon, etc., showing affinity for acid or metal-complex dye. Nylon can be dyed with disperse dye to produce reserve shades but is avoided as produced shades are poor wash fast.

Blending a cellulosic fibre with either of nylon, wool and silk makes the situation complicated. Cotton dyes if applied on blends like cotton–nylon, cotton–wool, etc., produce shadow shades with ease and solid shades in prolonged dyeing with difficulty. Interestingly, reserve shade can be produced on these blends if acid/metal complex dye is applied to dye only wool or silk or nylon part. Production of cross shades is possible on A-blends if a bigger molecular size direct dye is selected for cotton and an acid dye for nylon, wool, etc., which is impracticable under industrial set ups.

Nylon–wool

Solid shades: These can be developed with levelling, milling and metal-complex dyes as nylon content of blend is mostly 20–25%. Hetero-distribution of two components develop shadow shades only due to differential dye uptake, e.g. sports wears, swim wear and leisure clothing where warp is nylon filament and weft is staple wool as distribution of dye depends on dye structure, applied depth, pH, blend ratio and quality of fibres. Nylon reaches to its dye saturation limit quickly due to very less $-NH_2$ groups compared to wool, dye left in bath is exhausted on wool. 1:1 metal-complex dye shows lower substantivity on nylon and critical depth is relatively low, whereas 1:2 type dyes nylon more easily with poor coverage of shade. For heavy shades, nylon is dyed with 1:2 and wool with 1:1 in 2-bath method.

Shadow shades: Short time dyeing with metal complex dye produces this effect easily.

Reserve shades: Can not be produced. Disperse dye may be applied on nylon where wool will not be dyed. But this method produces a cross strain in wool, wash fastness of dyed nylon is poor and is never commercially practiced.

Cross shade: Not possible to develop commercially.

Nylon–cellulose (cotton or viscose)

Reserve dyeing: Possible only through dyeing nylon part with acid/metal complex dye in 1-bath method.

Solid shade: Possible through dyeing nylon with acid/metal complex dye and cotton part with higher molecular weight direct dye in a 1-bath 2-step or 2-bath 2-step method.

Another technique is to dye the blend with reactive dye. First, nylon part is dyed with reactive dye at pH~4–5 for 30 min at 80–90°C; electrolyte is added for exhaustion on cotton part followed by fixation with alkali on cotton. This is achieved in a single bath three stage process.

Disperse reactive/normal reactive dye combination may be used where nylon is dyed with disperse reactive in acidic pH and normal reactive is to dye cotton. Dyebath is set at 50°C with both the dyes added at one time; dispersing agent (1 g/l) for dispersion of disperse reactive dye and NaCl

(100 g/l) is applied for exhaustion of conventional reactive dye simultaneously on cellulose for 15 min at neutral pH; Na_2CO_3 (10–20 g/l) is added for fixation at 85–95°C for 1 h on cotton.

Shadow shade: This can be produced in short time dyeing with any of the techniques used for solid shade.

Cross shade: Can not be practically produced. However, dyeing of nylon with disperse dye and cotton with higher molecular weight direct dye may produce dull cross shade with poor fastness.

Cotton–viscose

This is a very important blend to compensate shortage of good quality cotton and lustre on cotton. Reserve and cross shades can not be produced. Shadow and solid shades are produced with any cotton dye in 1-bath 1-step method. The shadow effect is developed on short time dyeing whereas solid shade on prolonged dyeing.

26.6.2 Dyeing of B-blends

Very limited blends are produced in this class. Normal acrylic (basic dyeable) and cationic dyeable polyester are the only true members ; nylon, wool and silk though retain acid groups in structure and can be dyed with basic dyes, are not practically dyed because of poor light fastness but may be exercised if share in blend is too less. Modacrylic retains much anionic co-monomer than acrylic giving chances for differential dyeing rate for components in blend.

Acrylic–modacrylic

Shadow shade: These are produced on short time dyeing with basic dye.

Solid shade: 1-bath 1-step dyeing with basic dye for more time produces solid shades.

Cross and reserve shades can not be produced at all in any method.

Acrylic–cationic dyeable polyester

Acrylic is dyed with basic dye at 90–95°C while cationic dyeable polyester with same dye at 110–115°C. Acrylic turns yellowish beyond 110°C. Cationic dyeable polyester may be dyed with disperse dye too, but below 110°C (Renard, 1973).

Reserve shade: By dyeing cationic dyeable polyester with disperse dye below 120°C with thermal decomposition on acrylic.

Cross shade: Can never be developed due to affinity of basic or disperse dye for both these components.

Shadow effect: Dyeing with basic dye just above boil for short duration will produce this effect. Alternately, disperse dye may be applied to dye both just above boil.

Solid shade: May be produced in the same way as shadow shades are produced with selective disperse dyes for only light shades or with basic dye for all shades.

26.6.3 Dyeing of D-blends

Polyester, nylon, acrylic, acetate (secondary), triacetate and polypropylene are the members in this class and dyed with disperse dye; out of these, nylon is not dyed with disperse dye due to poor wash fastness while only light shades can be developed on acrylic with disperse dye and is seldom used. Polypropylene can not be dyed in existing exhaust methods due to typical compactness in structure, the opening of fibre structure even at very high temperature does not seem to be adequate to provide passage to disperse dye and so only dope dyeing is preferred.

Polyester is invariably mixed with acetate and triacetate to reduce overall cost; however, polyester, acetate and triacetate remarkably differ in their dyeabilities, e.g. with low energy disperse dyes (A and B), 50% exhaustion can be obtained at 40–50°C on secondary acetate, at 70°C on triacetate and at or above 100°C on polyester but with poor sublimation fastness. As far as compactness of fibres is concerned, it follows the order as: polyester > triacetate > acetate; due to this shadow effect is easily produced while solid shades may be produced with trouble. Cross shades can never be produced on D-blends while reserve shade may be produced depending *on case.*

Secondary acetate–triacetate

Shadow effect: Secondary acetate may be dyed at 50–60°C and triacetate at 75–85°C with low energy disperse dye to produce shadow shade.

Solid shade: Dyeing beyond 85°C adversely affects lustre of secondary acetate and solid shades can not be produced. Use of carriers may cause swelling of triacetate at lower temperature, but secondary acetate also swells up causing damage. Even low energy disperse dyes (A and B) can not produce solid shade.

Reserve shades may be produced through dyeing at low temperature with staining on triacetate.

Secondary acetate–polyester

Polyester is too compact and can not be dyed below 100°C while acetate is not dyed beyond 85°C to avoid damage.

Reserve shade: A and B type disperse dyes produce this shade at 60–70°C for 1 hour which do not dye polyester at such a lower temperature.

Shadow effect: Not possible without damaging acetate.

Solid shade: Not possible.

Triacetate–polyester

This is the most versatile D-blend commercially; fibres are most compatible in respect of dyeing. Class C and D disperse dyes are applied on both. Presetting lowers dyeability of triacetate and polyester both.

Reserve shade: Not possible. Triacetate may be dyed with B and C grade disperse dyes at or above boil but staining on polyester can not be avoided.

Shadow shades can be developed just above boil when triacetate will be fully dyed but little dye will go to polyester part.

Solid shades can be produced with greater ease due to compatibility in nature of dyeing. Dyeing can be done by using carrier at boil or in HTHP dyeing method for continuous dyeing or two stage dyeing in which polyester is dyed at 130°C then cooled the bath and add second dye and carry out dyeing at 100°C for acetate. This can be done with A and B-disperse dyes. Under practical situations, a little triacetate is blended with polyester and a single bath dyeing with grade C and D disperse dyes produces this shade.

26.7   Dyeing of binary blends

26.7.1   Dyeing of D–A blends

Important blends are polyester–cotton (terry–cotton) and polyester–wool (terry–wool), polyester–viscose (terry–viscose). In fact, most of woven shirting, suiting and other light dress materials are made of these blends. One component of these is hydrophobic (D-part) and non-polar fibre while the other is hydrophilic and polar (A-part). Due to this, no single dye has affinity for both the components enabling production of all shades.

Polyester–cellulose (cotton, viscose)

Polyester dyed with disperse dye and cellulose is dyed with vat, direct, reactive, sulphur, solubilised vat or insoluble azoics.

Reserve shade: Polyester is dyed with disperse dye and cellulose component left undyed or vice-versa. It is done in single stage-single bath method.

Cross shade: Two classes of dye formulations are selected, viz. one essentially disperse dye and another from vat, reactive, etc. 1-bath 2-step or 2-bath 2-step methods are popular. In 1-bath 2-step method, dyebath is set with disperse-reactive HE dyes or disperse–vat dyes and polyester is dyed first with disperse dyeing technology at 130±2°C followed by cooling down the bath at required temperature and dyeing the cellulose part with reactive or vat dye. Soaping and washing completes the process. However, if polyester part is dyed in jet dyeing machine in rope form, a dye having lesser affinity for cellulose is to be selected to produce levelled shade, such as reactive dye. As conventional reactive dyes (M and H brands) are hydrolysed at 130°C, either reactive dye is to be applied after cooling down the bath or reactive HE dyes, which are thermally stabilized, are to be applied. Vat dye will not be a potential choice for cotton part due to its higher affinity in reduced and solubilised state which may dye outer surface of rope with deep shade leaving the interior with light shade. Vat dye can form levelled shade on cellulose in open width.

In 2-bath 2-step method, polyester is dyed in jet dyeing machine with disperse dye, the rope is reloaded in a winch for dyeing of cotton with vat, sulphur, reactive, naphthol, etc., or opened up in a scutcher and loaded on jigger to dye in open width.

Shadow shade: Possible in 1-bath 2-step or 2-bath 2-step methods as stated earlier with calculated concentration of dye for each fibre. Selection of dye combination with same tone is very important. Another approach is to select a low molecular weight vat dye for polyester and cellulose both, which is applied on polyester in disperse dyeing method in 1-bath 2-step method followed by cooling down the bath and vat dyeing in same machine with left dye in bath. However, chances of unevenness exist.

Solid shade: Possible in 1-bath 2-step and 2-bath 2-step methods, where both the components are dyed in jet dyeing machine in 1-bath 2-step method or one component is dyed in one machine with a dye class; the blend is unloaded and reloaded in another machine for dyeing of second component.

Very unique method of producing only light solid shades is to dye the blend in open width in jigger with solubilised vat dyes to dye cotton part only with double depth. During finishing at high temperature, a part of the dye is sublimized-off, is condensed on polyester phase and diffuses inside lowering the final shade depth.

Polyester–cellulosic blends may be dyed in jigger in open width in light shades with A and B group disperse dyes to dye polyester at boil, the bath is drained and cellulosic part is dyed with other dyes. Carriers may be engaged during dyeing polyester part but in lieu of poor light fastness and more waste-water load.

In continuous dyeing, the blend is padded with a solution of solubilised vat dye (x%), Na_2CO_3 (3%) and $NaNO_2$ (2%), dried at 60–70°C and passed through a bath of H_2SO_4 (10%) to restore original dye at required temperature. The blend is washed, dried and passed through curing machine for thermo-fixation of dye on polyester at 180–210°C for 30–60 s. Only light shades can be produced in this method.

Selected vat dyes can be applied on polyester through thermosol treatment and light to medium shades can be produced with single class of dye. The pad liquor is prepared with dye, dispersing agent, wetting agent and migration inhibitor. The blend is padded, dried, thermofixed followed by loading in jigger to dye cellulose as per vat dyeing technology.

Polyester–wool

The main problem is to dye wool below 105–110°C to avoid its thermal decomposition, but polyester is dyeable at 130±2°C. Hence the blend can not be dyed with desired shade without some sort of degradation on woollen part. Due to this, wool and polyester fibres are invariably dyed in top form with desired shade and blended to produce fancy yarn and fabric.

Dye combination: Disperse–acid or metal–complex dye

Disperse dye heavily strains wool during dyeing of polyester which reduces the wash fastness of blend; the problem becomes severe if carriers are added. High temperature dyeing causes less straining as disperse dye goes to wool and then migrates to polyester.

Due to match in working pH, dyeing may be carried out simultaneously in 1-bath 1-step method by dyeing polyester at 130°C, the bath is cooled down to 80°C followed by dyeing of wool at boil. However wool is degraded up to some extent.

Polyester–nylon

These blends are produced through making molten polymer blend in autoclave on ground of better elasticity and moisture regain. Disperse dyeing of polyester causes dyeing of nylon too with poor wash fastness.

Reserve shade – Easily possible with only dyeing nylon with acid/metal complex dye.

Cross shade – Not possible to develop.

Shadow effect – Possible with disperse dye with poor wash fastness on nylon.

Solid shade – May be produced with prolonged dyeing with disperse dye again with poor wash fastness.

26.7.2 Dyeing of D–B blends

The only important members under this category are polyester–acrylic and polyester–cationic dyeable polyester.

Polyester–acrylic

Like wool, acrylic fibre also gets thermally degraded beyond 105–110°C developing yellowness. Due to this, it is better to dye polyester at boil with carrier and disperse dye followed by dyeing acrylic in the same bath with basic dye. Dispersing agents used to disperse the dye are mostly anionic in nature which may interact with basic dye causing precipitation. This in turn creates a two fold problem, i.e. interaction with basic dye will produce light shade on acrylic due to precipitation of a part of basic dye and at the same time a part of disperse dye will be precipitated due to inefficient dispersion through loss in effective concentration of dispersing agent. A non-ionic anti-precipitate may be used to handle this problem. Even cationic retarding agent can not be used due to this problem.

It is advisable to dye the blend in 2-bath 2-step method to dye components separately to avoid interaction among dye and chemicals.

Reserve shade: May be produced by dyeing acrylic with basic dye. Dyeing of polyester may stain acrylic and hence not viable.

Shadow shade: Possible through dyeing the blend with disperse dye at or nearer to boil with poor wash fastness. Dyeing at 130±2°C will improve fastness but in expense of degradation of acrylic and tone change in shade due to yellowing.

Cross shade: never possible

Solid shade: Viable for light shades only with disperse dye.

Polyester–cationic dyeable polyester

Cationic dyeable polyester is dyed at 110–115°C with basic dye and dyeing beyond 120°C degrades fibre while dyeing below boil requires addition of carrier in bath (ref 21.4.2). Disperse dyeing gives 10–20% higher dye uptake over that on normal polyester.

Reserve shade: Possible by dyeing only cationic component with basic dye.

Cross shade: Can not be produced as both are dyeable with disperse dye and dyeing of polyester will dye cationic part also.

Shadow shade: It is produced with disperse dye as dye uptake on cationic part is higher.

Solid shade: Difficult to produce. Only lower molecular weight disperse dyes which can be applied upto 120°C are used on both the components in 1-bath method (ref. 21.1).

26.7.3 Dyeing of A–B blends

Members are cellulose–acrylic, cellulose–cationic dyeable polyester, nylon–acrylic, nylon–cationic dyeable polyester, wool–acrylic, wool–cationic dyeable polyester, etc.

Cellulose–acrylic

These two fibres behave independently against suitable dyes for dyeing like that with polyester–cotton and so all types of shades can be practically developed. Vat, sulphur, etc., dyes applied on cellulosics are anionic while basic dye is cationic. Dyeing of both fibres should not be carried out in 1-bath 1-step and 1-bath 2-step methods otherwise basic dye–anionic dye complex will be precipitated. Vat and sulphur dyeing of cotton may be avoided to get rid of reducing action of $Na_2S_2O_4$ and Na_2S, which may decolourize basic dye. If these dyes are to be used, cotton part must be dyed first followed by thorough washing and dyeing of acrylic part.

Reserve shade: Dye cellulosic part with vat, sulphur, reactive or direct dye. Alternately, acrylic can be dyed with basic dye.

Shadow effect, cross shade and solid shades are produced in 2-bath 2-step method with respective dyes.

Cellulose–cationic dyeable polyester

All shades can be developed as in the case of cellulose–acrylic. However, cationic dyeable polyester is to be dyed first with low molecular weight disperse dye below 120°C or with basic dye at 110–115°C followed by dyeing of cotton. Cellulose is to be dyed first if vat and sulphur dyes are used.

Nylon–acrylic and wool–acrylic

In nylon–acrylic, both have affinity for basic as well as disperse dye but wool–acrylic is not used on technical grounds. Any attempt to dye acrylic with basic dye will dye nylon also but acid or metal complex dye applied on nylon will not affect acrylic. Acid dye and basic dye should not be added in 1-bath to avoid precipitation of dye complex. Dyeing of both components takes place in proportion to –COOH groups attached to these fibres.

Reserve shades are produced with acid or metal complex dye but staining of acrylic may occur.

Shadow effects can be produced easily with basic dye at boil.

Cross shades can never be produced while only solid light shades may be produced with difficulty.

Nylon–cationic dyeable polyester and wool–cationic dyeable polyester

Dyeing procedure is same as those for nylon–acrylic or wool–acrylic. Only difference is that, cationic dyeable polyester can be dyed with basic and disperse dye both but below120°C. Anionic sites (–COOH) in cationic dyeable polyester is more while that on nylon is limited.

Shadow effects can be easily produced with basic dye at 110–115°C.

Cross and reserve shades can not be developed as nylon and cationic dyeable polyester both possess affinity for basic as well as disperse dye.

Solid shades can be produced with either of disperse or basic dye. If basic dye is applied on both, only light shades can be produced as nylon has limited –COOH groups to form solid shade with poor light fastness. Disperse dyeing on both components can not produce solid shades so easily because of very high affinity of dye for cationic dyeable polyester; a difference in shade will always exist. Further, dyeings will show poor wash fastness.

26.8   Dyeing of ternary blends

Members are polyester–acrylic–cellulose, polyester–cationic dyeable polyester–cellulose, polyester–acrylic–nylon (or wool) and polyester–cationic dyeable polyester–nylon (or wool). Among these, polyester–cationic dyeable polyester–cellulose blend is very popular. Basic dye can produce very bright shades on cationic dyeable polyester which is not possible with disperse dye. 'Frosting' problem due to difference in abrasion resistance of components in polyester–cotton blend can be reduced by introducing cationic dyeable polyester in it with a ratio of 35%, 35% and 30% for anioinic polyester, polyester and cotton respectively and dyeing these components with deeper, lighter and intermediate shades respectively.

26.8.1 Polyester–cationic dyeable polyester–cellulose

Disperse and cotton dye combination may be used to develop solid shades in 2-bath 2-step method.

Reserve shade may be developed by dyeing cotton part; colour intensity may be increased by dyeing anionic polyester too with basic dye of same hue. Alternately, disperse dyeing may be carried out to dye both types of polyester leaving cotton undyed or dyeing of anionic polyester with basic dye only.

Cross shade may be produced with disperse–cotton dye combination of contrasting hues while shadow shades by same combination at varying depths. However, only selective disperse dyes are to be selected to dye anionic polyester below or at 120°C and affinity of disperse dye is 15–20% more on anionic polyester over conventional one.

26.8.2 Polyester–acrylic–cellulose

Dyeing is same as those for polyester–cationic dyeable polyester–cellulose, only difference is that acrylic accepts very little disperse dye.

26.8.3 Polyester–acrylic–nylon (or wool)

In place of cotton dye, a metal-complex dye is to be used for nylon or wool. Other aspects remain unchanged

26.8.4 Polyester–cationic dyeable polyester–nylon (or wool)

Nearly same dyeing procedure as that used for polyester–acrylic–nylon (or wool) with a difference of producing all types of shades and hues.

26.9   References

MITTAL R M (1984) *Processing of blends*, India, The Textile Association.
NUNN D M (1979) *The dyeing of synthetic polymer and acetate fibres*, Bradford, UK, Dyers Company Publication Trust.
RENARD C (1973) 'Dyeing properties of polyester fibres with modified affinity', *Melliand Textilber*, **54**, 12, 1328–1335.

27

Dyeing in super-critical carbon dioxide

Abstract: Water, the universal dyeing media poses various problems like, waster water load, inability in complete dissolution of dye and chemicals, post-dyeing drying, etc., which makes the technology of dyeing complicated and increases cost. Carbon dioxide in its super critical state is capable of solubilising dye and chemicals and leaves no waste water load at the end of dyeing. It is non-toxic and inexhaustible in atmosphere, no post-drying is required, non-flammable and most interesting is that it requires no auxiliary chemicals except dye cutting down cost of dyeing many-folds.

Keywords: Water, super critical state, carbon dioxide, temperature, pressure

27.1   Shortcomings of aqueous dyeing system

Water, the universal solvent in dyeing is associated with certain problems. Discharged bath pollutes sewage and disturbs ecological balance, absence of recycling affects economy, ability to dissolve chemical and dye is less, higher dyeing time and drying succeeds dyeing. Source of fresh natural water is too limited with gradual depletion of underground water level due to which industries are to install softening plants – a factor which adds to cost. In contrast, use of water as solvent simplifies dyeing process and do not interfere on functioning of chemicals in bath. In order to reduce shortcomings associated with use of water, it was thought long back to try with other solvents.

27.2   Super-critical fluid

A fluid can be taken to its super-critical state by increasing simultaneously its temperature (consequently the vapour pressure) and pressure; a closed system thus reaches critical values where no boundary between the liquid and gaseous state exists, termed as 'super-critical state' of the given fluid. A super-critical fluid may be better characterized by referring to the phase diagram as shown for CO_2 in Fig. 27.1 (Rostermundt, 1990). Further increase in pressure increases dielectric constant developing dissolving power of the system which is not possible at normal pressure and temperature as shown in Fig. 27.2 (Saus et al, 1993b).

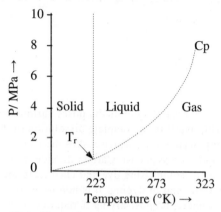

27.1 Phase diagram for carbon dioxide.

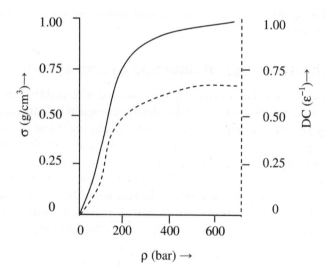

27.2 Dependence of density and dielectric constant of CO_2 on pressure.

The low viscosity of super-critical fluids and high diffusion properties of dissolved molecules are especially promising aspects of these dyeing processes. A super-critical dyeing fluid should easily dissolve solid dyestuff and penetrate even the smallest pores without the need of vigorous convection procedures (Rostermundt, 1990).

27.3   Benefits of dyeing with super-critical CO_2

Selection of super-critical CO_2 as a medium for dyeing is based on its numerous positive features, viz. (i) it is an inexhaustible resource, (ii) non-toxic at concentrations below 10%, (iii) very low concentration effective (10^{-5}–10^{-4} molar) (Knittel et al, 1993), (iv) it is produced on commercial scale and is readily available together with necessary logistics, (v) no disposal problems, (vi) easy to handle and incombustible at higher temperature,   (vii) the critical point of CO_2 is within the range which is readily manageable by technical means (Rostermundt, 1990), (viii) non-flammable, (ix) can be easily recycled in closed system, (x) no negative impact on strength of fibre, (xi) complete elimination of water and water pollution, (xii) no need of auxiliary chemicals, (xiii) gives high flexibility in dyeing time and promoting 'just in time' delivery, (xiv) no need for reductive cleaning, (xv) high degree of levelness and dyestuff exhaustion (Knittel et al, 1993), (xvi) polyester micro-filament fabrics, which are difficult to dye with levelled shades in conventional methods, achieve excellent levelness with super-critical CO_2 technique and show even better abrasion resistance than those dyed in conventional dyeing systems (Rostermundt, 1990), (xvii) foils and even thick polyester wires can be dyed evenly, (xviii) polyamides, polypropylene, elastomers with lighter shades than polyester can also be dyed effectively, (xix) dyeing of aramids like Kevlar and Nomex at 200°C yielding good degree of levelness, (xx) substitute for chlorinated hydrocarbon in dry cleaning processes, (xxi) polychromic dyeing with good results, (xxii) pressure and temperature can be adjusted to build-up desired shades in very short times, and (xxiii) since dye gets dissolved mono-molecularly in supercritical fluids, no filtering arrangement is required like that used in jet or HTHP beam dyeing machines for separation of dye aggregates.

High diffusion rates and low viscosity allow dye to penetrate into fibre shortening process time with enhanced performance. By reducing pressure at the end of process, dye and CO_2 can be recycled and reused. To use super-critical fluid in dyeing, some other properties also play crucial role, e.g. phase behaviour (Saus et al, 1993a). Phase state plane related to dyeing lie above Pc (7.38 MPa) and Tc (304.2 k) (Rostermundt, 1990), where Pc and Tc are critical pressure and critical temperature respectively. In contrast, CO_2 poses toxic effect at very high concentration (>10%), causes chocking attacks if contained in inspired air at high concentration with a threshold limit value is 5%. The initial investment is also very high.

27.4   Dyeing in super-critical CO_2

27.4.1 Dyeing equipment

A high pressure laboratory apparatus is used with a heated autoclave and pressure sealed stirrer. The system can handle up to 500 bar and 350°C (Saus et al, 1993a). Pressure is applied from CO_2 gas cylinder delivery with about 60 bar by means of membrane compressor. A second system for pressurizing consists of a hand spindle press used for dosing liquids into the system and for regulating working pressure inside a monitoring autoclave. This monitoring vessel which can be thermostated up to 350°C is equipped with sapphire windows and is used for spectroscopic measurements in the UV-VIS region (Saus et al, 1993b; Scheibli et al, 1993).

27.4.2 Dyeing process

Dyeing is performed in high-pressure stainless steel autoclave with volume of 80ml. Polyester sample to be dyed (around 10–25 cm) is wrapped around a perforated stainless steel tube and mounted inside the autoclave around the stirrer (Fig. 27.3) (Knittel et al, 1993). Dyestuff powder is placed at the bottom of the vessel and the apparatus is sealed, purged with gaseous CO_2 and pre-heated. Glycol is added as heat carrier. CO_2 is isothermally compressed to the chosen working pressure under constant stirring. Pressure is maintained for a dyeing period of 60 min and released afterwards. After dyeing, the dry sample is removed and rinsed with acetone to remove adhering residual dye (Saus et al, 1993b). Dye uptake is assessed spectroscopically after extraction of sample with hot C_6H_5Cl.

In order to obtain comparable exhaustion level for dyestuff, longer periods are required at lower working temperature and absolute values depend strictly on constitution of dye molecules; dyeing of polyester for 1min at 100°C results in dye uptake of 6 mol/g of fibre. Dyeing time, isobaric and isothermal period combined with time for expansion ranges from 5–20 min and is independent on dyestuff. Given this operation is carried out properly, reductive after-treatment of dyed material is omitted. In the range of 80–120°C, there is a steep rise in the isobaric exhaustion curves using pressure of 250 bar which decrease on further heating.

The pre-heating time is excluded and a stepwise expansion is preferred. This procedure gives enhanced dyestuff exhaustion and helps to avoid deposits of excessive dyestuff on surface of fabric. All polyester and polyamides showed excellent levelness while optimized constant parameters and dyeing cycles. Mostly a temperature range of 80–120°C and pressure of 250 bar is preferred with a variation in pressure at each

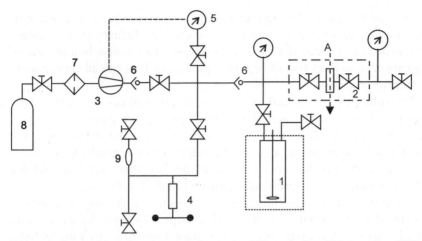

[(1) autoclave with pressure sealed stirrer, 300 cm³, P – 500 bar, T – 350°C, (2) pressurizable cell, 10 cm³, optical path length 100 mm, (3) membrane gas compressor, (4) manual spindle press, (5) pressure regulator, (6) back pressure valve, (7) filter, (8) CO₂ gas cylinder and liquid reservoir]

27.3 Layout of laboratory dyeing equipment with pressurizable observation cell.

step. At 120°C and pressure of 180 bar, successful dyeing can be achieved; beyond 180 bar dye uptake rapidly increases (Fig. 27.4) (Saus et al, 1993b).

A dyeing period of 10 min at 130°C or 40 min at 100°C are required to obtain exhaustion level of 98% resulting in dye uptake around 20 mol/g of fabric; about 3 moles of dye uptake can be reached within a period of 15 min. The state of dyestuff in super-critical solution can virtually be described as gaseous and is absorbed by fibre at a rate comparable to the high diffusion rates corresponding to a gas causing dye levelness and

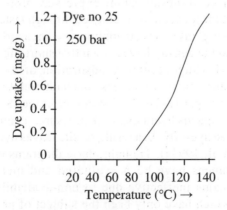

27.4 Isobaric dye uptake.

exhaustion of any residual dyestuff is easily recovered after expansion of the super-critical system. Dissolved dye quickly diffuses into boundary layers resulting in high degree of levelness and low convection in spite of high absorption rate. The adsorption equilibrium is achieved very quickly favouring levelness.

The solubility of dyestuff in a super-critical fluid can be continuously changed across a wide range during the stepwise reduction of density of super-critical CO_2. A short time over-saturated solution is formed which accelerates adsorption of dye by the fibre. Super-critical CO_2 is partly dissolved in the polymer since it has a softener like effect which accelerates diffusion process by increasing chain mobility of polymeric molecules, which in turn, reduces dyeing time and temperature. As soon as fluid has expanded to atmospheric pressure, it completely loses its capacity to dissolve dye. Any unfixed dyestuff will thus drop out in dry powder form and can be disposed-off. Dyeings leave the equipment in dry state and doesn't contain any solvent, because CO_2 being a gas, is completely eliminated automatically.

27.5   Mechanism of dyeing

Supercritical CO_2 performs two essential functions of aqueous liquor which includes transfer to dye stuff and transfer of heat to fibres (Poulakis et al, 1991). These two steps are further categorized into (i) dissolution of dyestuff, (ii) transfer of dye to the fibre, (iii) adsorption on the fibre surface, and (iv) diffusion into the fibre.

Solubilising power of CO_2 in critical state corresponds to weak solvents and the dissolving power of dyestuff depends on its particle size. Disperse dye is transferred to fibre out of a molecularly dispersed monomeric solution. As the viscosity is low, the fluid easily enters the pores and capillaries of fibre. Due to substantial lower pressure drop, high degree of evenness with minimum convection is achieved through faster mass transfer causing higher dyeing rate. Spectrophometric measurements during stepwise reduction in the density have shown that short life over-saturated solutions are formed which accelerate adsorption of dyestuff on fibre followed by diffusion by means of increase in chain mobility of polymeric molecules. CO_2 remains partly dissolved in polyester too (Saus et al, 1992). As soon as the fluid expands to atmospheric pressure, it completely looses its capacity to dissolve dyestuff. As a result, textile leaves dyeing equipment in dry state (Saus et al, 1993a). Though, phase diagrams of a number of hydrophobic substances have been evaluated and mentioned in the literature, these remains ineffective due to non-availability of suitable disperse dyestuffs which have only been the subject of preliminary tests and restricts commercialization of the technology (Poulakis et al, 1991).

Experiments monitoring the expansion steps by spectrometric control showed that a super-saturated phase of dyestuff is created which may lead to its deposition on fibre surface, from where temperature dependent diffusion into fibre is favoured (Poulakis et al, 1991).

It is an alternative method to high temperature high pressure dyeing of polyester without addition of carrier and within short time with high degree of levelness and good rubbing fastness rating of 4–5 (Saus et al, 1994).

27.6   Change in fibre structure

Super-critical CO_2 introduces carrier effect on PET causing swelling of the latter or plasticizing effect by lowering down T_g by 20–30°C; also results in formation of a network of stabilizing crystallites. No significant change of textile properties of super-critical fluid-treated polyester could be detected till now. No fibre damage also has been reported for nylon due to supercritical fluid dyeing using disperse dyes (Liao et al, 2000).

27.7   Non-textile applications and future prospects

Super-critical fluids characterized by exceptional physico-chemical properties are extensively used in large scale industrial applications, such as, extraction of caffeine from coffee beans and extraction of other natural substances for production of drugs, cosmetics, spices, etc.; also used in tobacco industry for extraction of nicotine and mechanical treatment of tobacco especially with super-critical CO_2; UV stabilized sand perfumes can be transferred to fibres; in pre-treatment for removal of fats from wool (Gebert et al, 1993). The applicability of super-critical CO_2 has been investigated for desizing too and in chromatographic applications. Other applications include extraction processes for removal of spinning oils. In other words, use of super-critical CO_2 in textile finishing plants is by no means limited to the dyeing of synthetic fibres. Insights gained during dyeing test can be principally employed for solution of several problems arising from finishing processes. Though large scale application has not been viable so far, smaller plants with short set-up can be designed to apply this technique.

Dyeing in non-aqueous process using super-critical CO_2 makes no difference in fibre strength and fastness of dyeing compared to traditional processes, at least in case of polyester. In case of polyester micro-fibres, super-critical fluid gives better working fastness than conventional treatments. This new process has a high probability of achieving wide applications for dyeing. However, at present, only the preliminary understanding and laboratory trials are in progress, further research is necessary to establish its commercial viability.

27.8   References

Gebert B, Knittel D and Schollmeyer E, (1993) 'Supercritical carbon dioxide as replacement for perchloroethylene', *Melliand Textilber.*, **74**, 2, 151–152.

Knittel D, Saus W and Schollmeyer E (1993) 'Application of supercritical carbon dioxide in finishing processes', *J. Textile Inst.*, **84**, 4, 534–552.

Liao S K, Ho Y C and Chang P S (2000) 'Dyeing of nylon 66 with a disperse-reactive dye using supercritical carbon dioxide as the transport medium', *J. Soc. Dyers Color.*, **116**, 12, 403–407.

Poulakis S, Spee M, Scheider G M, Knittel D, Buschmann H J and Schollmeyer E (1991) 'Dyeing of polyester in carbon dioxide above the critical conditions', *Chemiefasern / Textil Ind.*, **41/ 93**, 2, 142–147.

Rostermundt K H (1990) 'Dyeing and finishing of trevira microfibres', *Textil Praxis Int.*, **45**, 12, 1274–1276.

Saus W, Knittel D and Schollmeyer E (1992) 'Dyeing with supercritical carbon dioxide – an alternative to high temperature dyeing of polyester', *Textil Praxis Int.,* **47**, 11, 1052–1054.

Saus W, Knittel D and Schollmeyer E (1993a) 'Dyeing with supercritical carbon dioxide: physico-chemical fundamentals', *Textil Praxis Int.*, **48**, 1, 32–36.

Saus W, Knittel D and Schollmeyer E (1993b) 'Dyeing of textiles in supercritical carbon dioxide', *Textile Res. J.*, **63**, 3, 135–142.

Saus W, Knittel D, Hoger S and Schollmeyer E (1994) 'Dyeing from supercritical CO_2 – fastnesses of dyeing', *Melliand Textilber.*, **75**, 5, 388–394.

Scheibli P, Schlenker W and Strahm U (1993) 'Dyeing in supercritical carbon dioxide – an environmental quantum leap in textile processing', *Chemiefasern / Textil Ind.*, **43/95**, 5, 410–414.

28

Garment dyeing

Abstract: Garment dyeing refers to colouration of sewn garment; cutting and stitching is performed either in grey state or after required pre-treatment of fabric. It reduces wastage of dyed fabric, reduces lead time required with fast response to market demand, etc. Problems are also experienced in dyeing a garment having differential construction with collar, rib, cuff, tightly stitched button holes and main body which obstructs uniform diffusion of dye producing chances of unlevelled dyeing. Button, zip and other garment closures further intensifies these problems. Differential coloured effect can not be produced at all.

Keywords: Garment, collar, rotary dyeing, button, zip

28.1   Prospect of garment dyeing

A coloured garment is produced in either of the three basic ways, viz. (i) colouration and finishing of fabric followed by its conversion to garment (the ancient process, basically it is fabric dyeing), (ii) pre-treated fabric is transformed to garment followed by colouration and (iii) garment is prepared from grey fabric followed by its pre-treatment, colouration and finishing; the last two are commonly known as garment dyeing. Garment dyeing has proliferated itself in present scenario because of its superior multi-fold ability, viz. (i) to respond quicker fashion changes, (ii) reduced cost of production, (iii) better aesthetic properties, (iv) reduced inventory, (v) augmentation of easy monitoring techniques, (vi) replenishment of stocks with newer trend and fashion, (vii) reduced level of stock holding, (viii) complete computerized control, (ix) reduction in processing lead time by almost 33% as colouration is introduced late in the apparel process chain, (x) short pipeline necessary, (xi) reduced losses of dyed fabric waste through computer integrated manufacturing (Kaur, 2009) , (xii) no post-shrinkage, (xiii) lower energy, water and labour costs, (xiv) ability to produce smaller batches with ease, (xv) production of uniformly dyed garments with no variation in shade and (xvi) emerging additional feature of international bar coding (Houser, 1991; Chakraborty et al, 2005; Bone et al, 1988; Natusch and Comtesa, 1987; Teichner and Hollywood, 1987). Actual interest in garment dyeing has emerged to serve domestic market with exceptional speed of response as means of differentiating from traditional colouration of fabric, flexibility and other features conforming to the latest but short-lived fashion trends has fuelled the need for garment dyeing for production of quality garments to meet demands of customers more accurately. Certain limitations are associated with garment dyeing too, which is mainly the inability to produce complicated or differential and multi-coloured dyeing effect.

28.2   Structure of garment dyeing sector

It comprises mostly of conventional finishers, laundries, drycleaners and garment dyers. Garment dyeing activities may be divided into four types, viz. (i) fully fashioned garment dyeing by major commission dyers and finishers, (ii) cut and sew operations of garments covering woven and knitted fabrics dyed to required fastness standards, (iii) dyeing of pure cotton goods for 'boutique' trade with low colour fastness - only suitable for hand washing, and (iv) washing, desizing and bleaching denim goods followed by bio-polishing and over dyeing or highlighting effects. Cotton garments are often supplied either in greige or prepared for dyeing forms; the former requires thorough pre-treatment while for PFD (prepared for dyeing) garments, fabric of which was earlier pre-treated requires small amount of chemicals to enhance garment manufacturing processes, such as softeners, anticurls and lubricants; these must be removed to ensure level dyeing. Garment processing concerns fabric manufacturing in greige mills with apposite knowledge, cooperation, coordination and feedback between the greige mills, preparation and finishing departments, cut and sews operators and garment dyers. Otherwise, umpteen processing problems in garment size control, appearance and texture may evoke due to immoderate shrinkage, variation in yarn size, twist, warp cover in woven and courses / inch in knits. Stability of sewing threads and garment closures, viz. zips, buttons, etc., is also of utmost importance.

28.3   Methodology of garment dyeing

Since late 70s, there has been a tremendous change in garment dyeing process, replacing traditional fabric dyeing prior to garment manufacture to associate better appearance and rheology to the merchandise. This has enhanced the fashion look, appeal and comfort rendered by the merchandise and reduced lead time by leaps. In garment dyeing of woven fabrics, continuous preparation and drying of fabric is succeeded by its delivery to the garment manufacturer and then cutting and sewing of it to produce stock. The sale of the retail subsequent to dyeing and finishing of garment completes the cycle. Pre-treatment of knitted fabrics is carried out in pre-relaxed condition followed by relaxed drying and their delivery to garment manufacturer. Definitely this had an advantage of short lead time of about two weeks and the ability to respond to point of sale feedback for stock replenishment with dynamic response service has stimulated the growth of cotton garment dyeing in both woven and knitted fabrics as well as in multi-component designs, incorporating woven and knitted panels (Bone et al, 1988; Collishaw and Cox, 1986). For fully fashioned cotton garment, dyeing is slightly different from the case for multi-component cut and sew garment. For knitted garment blanks, dyeing is preceded by make up and stock preparation. Garment passage continues till its sale through intermediate drying and finishing. This considerably increases lead time to 5–6 weeks from 2–3 weeks in the prior route. Detailed study on dyeing of various types of garments may be seen elsewhere (Voss, 1993; Harper and Lambert, 1988, 1989; Gore and Settle, 1994; Brissie and McAulay, 1988; Riley and Jones, 1988; Dixon, 1989; Hargraves et al, 1991; Reinhardt and Blanchard, 1988).

28.4   Garment dyeing machineries

Cotton garment dyeing can be affected through either of (i) traditional side-paddle machines or (ii) rotary-drum machines.

28.4.1 Paddle dyeing machine

Fig. 28.1(a) shows a commercial paddle dyeing machine and the cross-sectional view in Fig. 28.1(b) (Hilden, 1988). There is slow circulation of both liquor and garments in a large vat by means of a rotating paddle. This machine offers relatively low productivity and is slow in filling, heating, cooling and draining, giving rise to longer lead time and requirement of higher liquor ratio to facilitate circulation and levelness (Hilden, 1988; Collishaw and Parkes, 1989). One limitation of this machine is lack of cooling facilities of dye bath which overcame in twin-paddle dyeing machines fitted with heat exchangers below the false bottom. Completely enclosed overhead paddle machines, recently developed, consist of horizontal cylindrical vessel with doors along the top sides and horizontal paddle extending the length of machine parallel to and along the axis of the vessel. Perforated and closed steam pipes are situated under a perforated false bottom. It features with a circulating pump which withdraws hot liquor from below the false bottom and injects it through the jet positioned along the side of machine, just below the liquor level to assist movement of goods, shown in Fig. 28.2 (Hilden, 1988).

28.1a Paddle dyeing machine. [Source: Parktek]

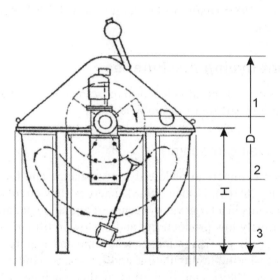

[(1) paddle, (2) passage of goods and (3) drain valve]

28.1b Cross-sectional view of a paddle dyeing machine.

28.4.2 Rotary dyeing machine

It consists of a horizontal perforated cylinder, radially partitioned into three or four compartments, rotating with periodic reversal inside a stainless steel box holding dye Liquor (Figs. 28.3a and 28.3b). Goods are loaded into the cage through sliding doors in the circumference. These machines

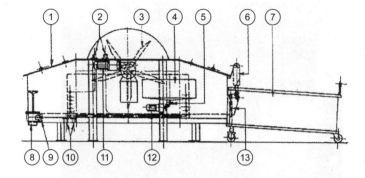

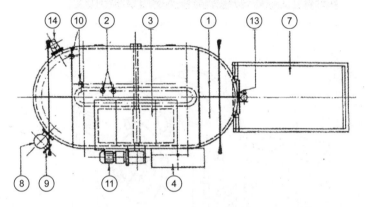

[(1) Inlet flap, (2) Cooling, (3) Paddle wheel, (4) Add tank, (5) Outlet, (6) Piston, (7) Material Truck, (8) Drain valve, (9) Water inlet, (10) Heating coil, (11) Gear, (12) Dye pump, (13) Fast unloading, (14) Over-flow]

28.2 Schematic diagram of an oval paddle dyeing machine (Source: Finolex). Liquor and goods are moved round an island.

are gentle in their action. Recent developments in these machines include addition of hydro extractors, re-inforced glass doors, variable cage speed and operation on a programmed automatic cycle in which liquor volume, time and temperature are fully controlled (Hilden, 1988).

28.4.3 Recent developments

During 70s, developments in garment dyeing equipments were too many, viz. (i) side paddle machines with variable speed paddles of variable length, (ii) 'smith drums' consisting of a rotating perforated drum normally with four compartments, (iii) modified drycleaning units for solvent dyeing, (iv) dyeing machines with toroidal action. Liquor was pumped up into a

28.3(a) Rotary dyeing machine. [Source: Trademe]

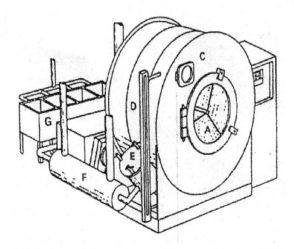

[(A) Perforated drum with or without partitioning, (B) Control unit,
(C) Sampling arrangement, (D) Liquor level indicator, (E) Lint filter,
(F) Heat exchanger, (G) Chemicals and dye liquor tanks]

28.3(b) Schematic diagram of a Rotary dyeing machine.

vessel containing garment carrier or plate and was deflected to the vessel circumference and then down into chamber with no mechanical assistance, movement being completely dependent upon the pumped action of water. Latest high temperature toroid dyeing machines work at liquor ratio as high as 40:1 to process 100% polyester garments at 140°C (Kramrisch, 1987), (v) dyeing in foam-in machinery specifically designed for 'sancowad low liquor process'. Other developments included a centrifuging and drying operation as an integral part of dyeing machine. However, advent of aqueous low liquor dyeing machinery with compromising liquor ratios of approximately 10:1 and flexibility to dye garments from all fibres led to obsolescence of this type of machinery, (vi) centrifuging machines for rotary dyeing with integral centrifuging and short processing times lends itself ideally to the quick turnaround objectives in today's market place. Rotary dyeing machine has evolved out with considerable modifications every time in a more sophisticated and state of art version with the highlighting features, viz. (i) optical basket configuration in which selective pockets or baskets are available whose configuration affects the degree and type of mechanical action applied to garment, (ii) an external circulation which ensures maintenance of uniform temperature throughout the machine essential for critical fibres, like acrylics. It involves circulation of dye liquor through a filter to aid liquor-goods interchange, thus compensating for reduced agitation when running the cylinder at very low speeds. This protects the garment surface while allowing good penetration into seams and thick areas. An additional filter is used to remove loose fibres and lints to prevent redeposition on garment, (iii) high speed centrifuge system ensuring that the moisture content is reduced to minimum prior to drying operation without need for additional handling and centrifuging in separate equipments in dyeing. Machines incorporated with centrifuge action have an added advantage of efficient rinsing in very short time, (iv) sophisticated heating and cooling with various means of accomplishing, e.g. direct steam injection into the dye liquor and addition of water directly to the machine suitable for simple dyeing processes; indirect heating and cooling by passing steam and water alternately through closed coils at the base of machine; indirect heating by means of a steam jacket around the body of the machine and use of an external heat exchanger for indirect heating and cooling. Heat exchangers work with steam, oil or hot water while cooling by direct addition of water or indirectly through heat exchangers, (v) flexible control systems urge the use of automatically regulated functions, process parameters and microprocessor control on centrifuging machines. Technology requires advent of newer paraphernalias including central control room with automatic addition of dyestuffs and chemicals from a central dispensing area, (vi) special sampling port which results in complete inspection of garments during process without reducing liquor level. Inspection of an entire garment from the batch lot at sampling stage ensures proper assessment of shade, seam penetration, degree of milling and physical appearance, (vii) easy loading / unloading is obtained through tilting mechanism in which the garment has to be loaded directly from overhead feed arrangements and unloaded without manual handling into a conveyor system, (viii) variable cylinder speed during dyeing cycle ranging normally from 5–35 rpm allowing greater agitation at critical stages in dyeing process and less agitation to protect garment condition and maximize quality standards, (ix) high temperature processing of garments ensuring garment dyeing at temperatures up to 135°C and (x) dye liquor recovery where spent dye liquor is transferred to a storage tank for reprocessing.

Today's garment dyeing has advanced into a unique art incorporating in it newer designs and features in which primitive paddle, rotary and other garment dyeing machines, etc., have been supplemented with automated and microprocessor controlled machines. Salient features in some of these machines are (i) low liquor ratio, (ii) microprocessors for effective control, (iii) heat exchangers, (iv) lint filters, (v) centrifugal hydro extraction, (vi) easy sampling, (vii) pocketed drums, (viii) tilting mechanism for easy loading and unloading, (ix) high temperature option, (x) large load sizes, (xi) cushioned suspension, (xii) variable drum speed, and (xiii) automatic balancing of drum.

28.5   Specific problems in garment dyeing

Unlike fabric which is uniform in thickness, compactness, etc., garments possess differential structure and composition. A garment itself has five differential areas, viz. collar, rib, cuff, tightly stitched button holes and main body. Collar is made up of a hard starchy compact layer tightly covered and stitched with fabric meant for garment. Ribs are folded and doubled layer area like edges on arm in a half shirt while cuffs are rather compact multi-layered area meant to cover wrist like that in full shirts. All these areas show differential diffusion during dyeing and inadequate handling may spoil the chances of a uniform shade. Keeping in view more time is required for distribution and diffusion of colour required for a levelled shade, a short time dyeing must not be practiced. In addition to that, garments retain closures tightly fitted on it, viz. zips, buttons, laces, etc., (Fig. 28.4). These are mostly made of metals or polymers, deteriorate in contact with dyebath chemicals, heat, etc., and show severe rubbing with the interior of the dyeing machine either getting damaged or severely affect quality of garment.

28.6   Selection of colour for cotton garment

Basic requirements in garment dyeing are (i) good level dyeing properties throughout the entire garment with differential thickness and compactness, (ii) higher migration and diffusion properties to assist seam penetration and (iii) suitability for short or automated process. Depending upon end use of garment, direct, fibre reactive, vat and sulphur dyes are generally used. Pigments have emerged to be used for exhaust dyeing too.

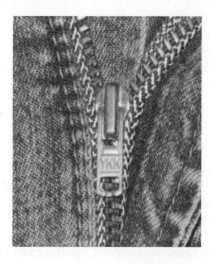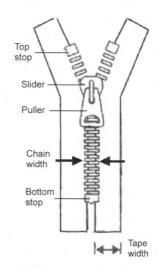

28.4 Various types of garment closures.

28.6.1 Direct dye

Direct dyes of A and B classes which have similar temperature of maximum exhaustion are selected to ensure even application and higher reproducibility. At a dyeing temperature of 95°C, electrolyte is added to enhance dye yield. Addition of calgon and a levelling agent is advisable for optimum results. Key application parameters for direct dyes involve dye bath exhaustion, rinsing and after-treatment (Kelley, 1987; Houser, 1987).

28.6.2 Reactive dye

Reactive dye is a better choice to produce levelled shade on garment due to its lesser affinity for cotton. Less reactive monochlorotriazinyl dyes are preferred with their high temperature application to optimize diffusion and migration, low reactivity to optimize level dyeing and ease of automation using linear addition profiles for both dye and alkali additions. Further details of garment dyeing with fibre reactive dyes have been discussed elsewhere (Bell, 1988).

28.6.3 Vat dye

Vat dyes are costlier but used to achieve all round fastness. Hot pigmentation technique is generally employed in which the dye is dispersed as an insoluble pigment under alkaline conditions at 60–80°C. After a levelling period, dye is reduced to leuco form and penetration into fibre is affected. After flood rinsing at pH 10, reduced dye is oxidized back to insoluble form to achieve high wash fastness. Garment handling can be improved by after-treatment with cationic softener.

28.6.4 Sulphur dye

Sulphur dyes cover a broad shade range with good to excellent wash and light fastness at relatively low cost as compared to other dyeing systems. Black and medium to heavy shades of brown, blue and green are most commonly used; the range does include some fairly bright colours as well. Nowadays, liquid sulphur dyes, called sodyesul liquids, are used which offer excellent performance and economy in exhaust dyeing. These are much advantageous than fibre reactive dyes as these have lower costs, shorter dye cycle, better cotton coverage and much less salt with comparable fastness properties. 'Sandoz antioxidant B', a sugar based reducing agent, results comparable reduction as that is obtained with sulphides. This allows an all-in procedure without the controls needed for sodyesul liquids. Presence of iron, calcium and magnesium upto 20 ppm may alter dye yield and shade; sequestering agents can overcome problem of weak dyeings, poor crocking and streaks caused by the metal contamination (Dixon, 1988).

28.6.5 Pigments

Pigments are used in printing and pad dyeing; being non-substantive in nature, binders fix pigment on fibre surface. It was believed that the exhaust dyeing with pigments might not be practically viable due to low exhaustion and poor levelness. Development of suitable auxiliaries has lead to the induction of necessary affinity of pigments by pre-treating textile for exhaust dyeing with pigments. Exhaust dyeing can be carried out in the following sequence: pre-washing, pre-treatment, exhaustion, stone washing (optional), fixation and thermal treatment (Kass, 1958; Chong et al, 1992; Walker and Bynum, 1991). Thorough removal of impurities from garment is necessary to ensure best results. A critical part is pre-treatment, where cotton is pre-treated with cationic compound to develop affinity for pigment; adequate uptake and uniform adsorption of cationic agent is essential for levelled dyeing. During dyeing, an ionic attraction is developed between pre-treated fibre and aggregated pigment particles. Exhaust dyeing is carried out in a pigment dispersion followed by fixation with binder to improve fastness properties. Stone washing, if required, is carried out before binder fixation. During curing, binder polymerizes to form cross-links to develop required fastness (Chong et al, 1992).

28.7   Dyeing of woollen garment

Woollen knits and woven garments are dyed in rotary drum dyeing machines as well as side and overhead paddle machines. Machine washable fabrics are used to make garments which are to be dyed. Polyamide accessories used are also dyed with the same tone as that of woollen garment for complete matching. Common problems encountered during woollen garment dyeing are (i) seam penetration, especially on pre-chlorinated garments where chlorination within the seams may be uneven, (ii) fibre damage during prolonged dyeing, (iii) reaching to the required level of colour fastness, and (iv) cockling effect on the garment. Metal complex, milling acid dyes, reactive dyes and chrome dyes, if properly used, can minimize these problems. Specific dyeing auxiliaries used in woollen garment dyeing are (i) acid-donor system, (ii) low temperature dyeing, (iii) after-treatment, and (iv) anti-felting agents. Felting tendency of wool poses some problem on garment dyeing as it leads to surface felting. Anti-felting protective agents are used in 'over-dyeing' to minimize surface distortion. Use of anti-felting agents, like Meropan NF (Chemische Fabrik Tübingen) and Nofelt WA (Tanatex) in dye bath temporarily protects unchlorinated garments during dyeing and are removed during rinsing (International Wool Secretariat, 1986). Anti-felting agents are also used successfully in dyeing chlorinated lambs wool garments. In latest developments, 'Dylan GRB' and 'Hercosett' shrink-resistant finishes have been introduced to meet the demand for machine washable woollen knitwears. These are based on application of resins to the surface of wool fibre after the latter has been modified by chlorination. Some reactive dyes have been introduced for wool to get adequate fastness and are restricted to reds, yellows, browns and intermediate colours only.

28.8   Dyeing of polyester garment

Polyester fabric stability, aesthetic appeal and resistance to wear demand increase in use of textured polyester yarns for outer garments. Polyester garments are dyed as usual with disperse dyes of improved fastness properties keeping the maximum dyeing temperature at minimum to preserve texture.

28.9   Quality control

Boost in export of readymade garments requires maintenance of time schedule for shipments and production of high quality garments in order to improve piece realization. Production and marketing of high quality garments thus depend upon their conformation to rigid standards of export, compliance with the latest fashion changes, appropriate quality control and sale's demand. 'Price vs quality' is a major aspect in this concern. Price of finished garment chiefly depends upon three factors, viz. history of the fabric turned into garment, technical inputs and use of appropriate processing machineries. Major areas for quality control of garment are as follows (Bone et al, 1988; Kamat and Menezes, 1993a, 1993b).

28.9.1 Anti-corrosion techniques

Casual wears such as jeans are incorporated with metallic components, like zips, buttons, rivets, and labels made up of brass, iron, copper, aluminium or zinc which imposes difficulties during dyeing. Peroxide bleaching is avoided for such prepared fabrics. Certain problems are encountered, such as dye resist mark or physical breakdown of the fabric, particularly in the area of direct contact with the metal component. Ferrous based materials have the risk of developing rust stains; use of sequestering agents prevent shade alteration due to chromophore sensitivity. Sandocorin 8160 protects the metal from corrosion. Best option is to use nickel plated brass as metal component; inhibitors based on phosphate ester, triazole gives protection against metal corrosion, although care should be taken to avoid retardation in build-up of shade.

28.9.2 Seams, elasticized areas, waist bands and scuffs

These areas must be fairly loose and seams should not be prepared taut otherwise this may lead to poor penetration of dyestuff in over-swollen fibres. Possible remedy to this problem is application of high temperature which results in better diffusion, penetration, running of cloth and facilitating liquid flow. This guarantees fruition of migration potential and is essential for multi-layered seams or elasticized waists. Normally, a temperature of 95°C is maintained before cooling for fixation.

28.9.3 Shrink behaviour

Excessive or uneven shrinkage of garment where knitted and woven fabrics are mixed, leads to seam puckering; it is essential to pre-relax knitted fabric and pre-shrink woven part through various shrink proof treatments for high quality garments.

28.9.4 Chafe marks and creases

These are defects developed particularly on delicate garments due to mechanical stress in drum dyeing machine, which deteriorate quality of final product. As a remedial measure, garments prone to chafe marks or pilling effect should be turned inside out and dyed in presence of non-foaming lubricant. Additional preventive measures may also be adopted by avoiding overloading of dyeing drum.

28.9.5 Accessories

Prudent choice of zips and metals accessories, like buttons and studs is inevitable to prevent corrosion or breakage. While iron make are avoided on rusting grounds, nickel and its alloys are preferred unless the garment is bleached or reactive dyed with high concentration of electrolyte or alkali. Choice for cellulosic buttons may lead to its undue breakage while polyester buttons are non-dyeable and are used as neutral colours in coordination with umpteen shades. Nylon buttons can be coloured in subsequent dyeing process with extra effort. Ribs of T-shirts are occasionally made of natural rubber or polyurethane (lycra) for better fitting while lycra mixed stretch garments have come up in the market and have acquired popularity over rigid one due to its incomparable fit and comfort, freedom of movement, crease recovery, stretchability, enduring shape retention, consumer brand affinity and quality assurance, etc. These fibres are associated with some inherent problems. Natural rubber is adversely affected in presence of certain metals like copper, while lycra is degraded by strong oxidizing agents like chlorine. A solution to these traumas is achieved by coating the metal accessories with corrosion protective agent. Sandcorin 8160 liquid, an anionic organic corrosion inhibitor, is an obvious choice as corrosion retardant. It prevents non-ferrous and white metals from oxidation and tarnishing by hindering the action of sensitive dyes onto the metal ions.

28.9.6 Sewing threads

Sewing thread must possess desirable properties like strength and fineness to produce neat seam which might enhance longevity of garment. Unmercerized cotton thread used in sewing of cotton garments produces a solid dyed inconspicuous seam. Polyester threads purposely enhance the effects in the seams due to their dye uptake resistance and mercerized cotton threads increase the dye uptake, resulting in darker appearance in shade. For cotton thread sewn garments meant for dyeing, it is not possible to switch-off directly from conventional spun polyester or core-spun threads to 100% cotton threads and retain the same seam characteristics. Equivalent seam strength can be obtained using heavier cotton threads. Similar degrees of elasticity in seam can be obtained by increasing stitch density but care must be taken to ensure damage of garment due to it. Seam pucker resulting from thread shrinkage during garment dyeing can be avoided by slacker stitches on lock stitch and chain stitch operations.

Selection of fibre to produce sewing thread depends on type of fabric to be sewed. Generally, a common selection is made for the thread used for fabric preparation and sewing both. However, this is applicable mainly for cotton, as sewing threads made of other fibres do not possess affinity for cotton dyes. This unarguably hikes the cost of garment manufactured as well as the production speed.

28.9.7 Foreign substances

Foremost objective of efficient garment dyeing is its value addition. This objective caters need for adaptation of preventive measures to hinder value loss of the article. Protection of garments from stains due to oils, greases and other lubricating agents is a must. Other impurities, like sizes and resins containing additives such as elastomers and oil repelling agents could also denigrate the appearance of garment.

28.9.8 Interlinings

Purpose of interlinings is to stabilize and enhance the appeal of outer fabric. With post-dyed garments, certain special properties are to be inflicted into the interlinings to ensure better performance. Special properties inherited by the interlinings includes equivalent dye uptake by it as that of outer fabric. The adhesive used for bonding the outer fabric to the interlining must remain intact during and after dyeing and so is the handle. Aftermath of dye sorption should not be adverse on colouration.

28.9.9 Care labelling

Augmentation of fastness properties of dyed garments is essential for customers demand. Major attributes to be taken into consideration are dimensional stability, wash fastness and rubbing fastness. To meet these parameters effectively, care labelling is introduced which alerts the customer to adopt precautions to be taken while cleaning the garment. These care labels are embellished with specific instructions with respect to method of washing, bleaching, drying, ironing and suitability for dry cleaning.

Quality level in garment dyeing is also augmented through justified selection of fabric and dyestuff to survive the rigors of processing, proper sizing, effective training of staffs to ensure professional garment handling, good programming of dyeing equipments and innovations to increase production.

28.10   Novel effects in garment dyeing

Until the advent of stone-washing, garment dyeing remained a relatively small but stable business. Stones soaked in oxidizing agents are forced to rub garments resulting in worn out look. Colour receptivity of garments can also be modified using cross-linking agents in finishing (Harper and Lambert, 1992). Stone-washed cotton denim gives distressed fancy effect and a stylish look, which has changed the demand for garment dyeing in today's market. 'Ice washing' is another technique for creating unusual colouring effects, where dyed garments are stone tumbled with chlorine or sodium permanganate – impregnated rock (pumice stone) for a precise period and then treated in sodium bisulphite bath. Different dyes, e.g. direct and reactive dyes, are affected differently by chlorine and sodium permanganate, even some dyes are completely discharged by these chemicals. Various techniques of garment washing, optimization of parameters, enzymatic washing or bio-polishing and various other effects obtained including those for denim have been reported in detail (Lennox, 1989; De, 1998; Shah, 1995; Chong and Ma, 1996; Tyndall, 1992; Lantoo et al, 1996; Rucker et al, 1992; Harper and Lambert, 1993; Kochavi et al, 1990).

Special exhaust type fabric softeners for after-treatment of dyed garment have been developed. These softeners are normally added to the last cycle to improve handle of finished garment and restore some physical properties which might have been reduced during dyeing (Cheek, 1987). It is quite possible to cause a drastic change in look of dyed garments with foam finishing, post-cure pre-treatment, wrinkle resistant finishes, flock coating or high frequency fixation (Harper and Lambert, 1991; Metzler, 1995; Karnes, 1993; Brown, 1994; Wagner, 1980; Jola, 1980).

28.11   Recent challenges in garment dyeing sectors

Recent resurgence in garment dyeing is probably because of fidelity of the article to recently evolved quality control processes to adjudicate the loopholes. Certainly, export of developing country's readymade garment is becoming a more challenging job with the passage of time. This is reflected due to increase in trade barriers, growing consumer awareness, rigid quality standards, multi-fibre agreement and more recently abdication of 'quota' system. ISO 9000 and 14000 quality assurance certification and red listed chemicals are two new concepts required for enhancement of serviceability and quality of product. Red listed chemicals are hazardous pollutants which are accumulated on fibre during various processing phases.

28.12   Advantages of garment dyeing

There are several inherent advantages in garment dyeing, viz. (i) handling of smaller lots economically, (ii) unlimited selection of colour, thus dyeing to order almost instantaneously, (iii) a specific product can be dyed in any number of shades, thereby increasing productivity, (iv) distressed look with wrinkles, (v) fuller handle due to shrinkage during dyeing, (vi) quick turnaround at the retailer's counter, and (vii) ability to respond quickly to the changes in colour taste of the fastidious and fickle customers (Jola, 1980).

Cost justification of finished garment is an essential aspect of garment dyeing sector to add to its advantage. Major cost components in garment dyeing are (i) dye and chemical costs, varying considerably upon shades, fibre and dyeing technology, (ii) utilities primarily based on cost of natural gas to heat water and determine dye–liquor ratio and (iii) depreciation or labour (Reed, 1987; Besnoy, 1989).

28.13   Problems of garment dyers

Most of the garments received by the garment dyers are cut and sewn from previously prepared cloth which aggravates certain problems, viz.

(i) garment is frequently cut from differently pre-treated pieces of cloths, (ii) garment has woven and knit pieces sewn together, (iii) garment has been sewn with greige, prepared, bleached with optical whitener treated pieces together and (iv) a finish like softener, resin, etc., has been imparted to garment (Murphy, 1987).

Preparation must be thorough, consistent and reproducible. This is extremely important since the cut and sew operation will need to be oversized keeping in mind subsequent shrinkage. Variation in preparation will result in inconsistency of final garment size after dyeing and drying.

The most common mistakes made in garment dyeing are (i) prolonged drying of goods resulting in wastage of energy and time, over drying and subsequent pilling, (ii) excessive heating, creating wrinkles and at very worst resulting in melting of goods, (iii) inadequate cooling down, (iv) insufficient loading to dryer, (v) not running dryer at constant production rate, (vi) unnecessarily long load or unload times, (vii) unclean filters, (viii) non-maintenance of seal of dryer, and (ix) inaccurate production and cost figures (Diedling, 1989).

28.14 References

BELL S J (1988) 'Garment dyeing with fibre reactive dyes', *Am. Dyestuff Rep.*, **77**, 5, 36–51.

BESNOY R (1989) 'What's next in garment dyeing', *Am. Dyestuff Rep.*, **78**, 5, 30, 47.

BONE J A, COLLISHAW P S AND KELLY T D (1988) 'Garment dyeing', *Rev. Prog. Color Related Topics,* 18, 37–46.

BRISSIE R L AND MCAULAY T (1988) 'Advances in radiofrequency garment drying', *Am. Dyestuff Rep.*, **77**, 7, 28, 49.

BROWN R O (1994) 'Enhancing the performance of wrinkle resistant cotton garments', *Am. Dyestuff Rep.*, **83**, 9, 106–109, 131.

CHAKRABORTY J N, PAL R AND MEGHA P R (2005) 'Garment Dyeing' *Indian J. Fibre Textile Res.*, **30**, 4, 468–476.

CHEEK D (1987) 'Guidelines for better garment dyeing', *Am. Dyestuff Rep.*, **76**, 8, 74–75.

CHONG C L, LI S Q AND YEUNG K W (1992) 'Pigment dyeing study of cotton garments by exhaustion', *Am. Dyestuff Rep.*, **81**, 5, 17–26, 63.

CHONG C L AND MA C W (1996) 'Ozone fading of indigo-dyed denim', *Am. Dyestuff Rep.*, **85**, 11, 15–20.

COLLISHAW P S AND COX K P (1986) 'Progress in the use of reactive dyes for the dyeing of cotton fashionwear in garment form', *J. Soc. Dyers Color.,* **102**, 10, 298–301.

COLLISHAW P S AND PARKES D T (1989) 'Cost effective processing developments in the batchwise dyeing of cotton', *J. Soc. Dyers Color.*, **105**, 5/6, 201–206.

DE P (1998) 'Denim washing and finishing: a review', *Man Made Textiles India*, **41**, 3, 129–131.

DIEDLING M (1989) 'Avoiding the ten most common mistakes in garment drying', *Am. Dyestuff Rep.*, **78**, 9, 86–87.

DIXON M (1988) 'Role of sulphur dyes in garment dyeing', *Am. Dyestuff Rep.*, **77**, 5, 52–55.

DIXON M W (1989) 'Improving the 'neat look' in garment dyeing applications', *Am. Dyestuff Rep.*, **78**, 5, 24–28.

GORE D C AND SETTLE J H (1994) 'Improving processing applications in garment dyeing and finishing', *Am. Dyestuff Rep.*, **83**, 5, 24–28.

HARGRAVES R, EISSELE E AND PISARCZYK K (1991) 'Innovation in pellet technology for garment dyeing', *Am. Dyestuff Rep.*, **80**, 5, 28–32.

HARPER R J AND LAMBERT A H (1988) 'Print dyeing: an opportunity for garment dyers', *Am. Dyestuff Rep.*, **77**, 2, 13–18.

HARPER R J AND LAMBERT A H (1989) 'Single-side crosslinking: an approach for garment-dyeable cotton fabrics', *Am. Dyestuff Rep.*, **78**, 5, 15–20, 47.

HARPER R J AND LAMBERT A H (1991) 'Foam finishing for the 'white look' in garment dyeing', *Am. Dyestuff Rep.*, **80**, 5, 43–47.

HARPER R J AND LAMBERT A H (1992) 'Stone finishing and dyeing of cotton garments', *Textile Chem. Color.*, **24**, 2, 13–18.

HARPER R J AND LAMBERT A H (1993) 'Reverse denims via resist finishing', *Am. Dyestuff Rep.*, **82**, 5, 17–23.

HILDEN J (1988) 'Garment dyeing: opportunities, risks and technical developments in equipment', *Int. Textile Bulletin*, **34**, 4, 30–45.

HOUSER N (1987) 'Optimizing garment dyeing procedures', *Am. Dyestuff Rep.*, **76**, 10, 25–28, 49.

HOUSER N E (1991) 'Garment dyeing: is it here to stay, *Am. Dyestuff Rep.*, **80**, 5, 18–23, 50.

INTERNATIONAL WOOL SECRETARIAT (1986) 'Wool on the wards', *Textile Horizons*, **6**, 6, 32.

JOLA M (1980) 'Determination and control of the bleaching agent (peroxide) concentration by on line thermometry', *Melliand Textilber.*, **61**, 11, 931–936.

KAMAT S Y AND MENEZES E W (1993a) 'Quality control in garment dyeing', *Colourage*, **40**, 6, 24–26.

KAMAT S Y AND MENEZES E W (1993b) 'Garment dyeing- I', *Colourage*, **40**, 11, 41–44.

KARNES S G (1993) 'Advances in post-cure permanent press impact on fashion', *Am. Dyestuff Rep.*, **82**, 9, 74–75, 95.

KASS W (1958) 'Possibilities and Limitations of the Application of Pigments to Textiles', *J. Soc. Dyers Color.*, **74**, 1, 14–21.

KAUR A AND CHAKRABORTY J N (2009) 'Computer Integrated Manufacturing: An indispensable technology for Apparel Industry', *Asian Textile J.*, March, 58–61.

KELLEY J J (1987) 'Innovative and analytical approaches in dyeing cotton garments', *Am. Dyestuff Rep.*, **76**, 11, 44–47.

KOCHAVI D, VIDEBACK T AND CEDRONI D (1990) 'Optimizing processing conditions in enzymatic stonewashing', *Am. Dyestuff Rep.*, **79**, 9, 24–28.

KRAMRISCH B (1987) 'Garment dyeing: ready to wear-fashion from the dyehouse', *Int. Dyer*, **172**, 10, 9–15, 23.

LANTOO R, OINONEN A M AND SHOMINEN P (1996) 'Backstaining in denim wash with different celluloses', *Am. Dyestuff Rep.*, **85**, 8, 64–65, 72.

LENNOX K P (1989) 'New mercerizing unit adds indigo dyeing flexibility', *Textile World*, **139**, 2, 61.

METZLER R B (1995) 'Improving the look of no press garments', *Am. Dyestuff Rep.*, **84**, 8, 85–89.

MURPHY J M (1987) 'Improving preparation techniques for garment dyeing', *Am. Dyestuff Rep.*, **76**, 11, 41–43, 48, 50.

NATUSCH K AND COMTESA V P (1987) 'Garment dyeing shortcuts effect savings', *Am. Dyestuff Rep.*, **76**, 5, 31–34.

REED W (1987) 'Cost justification in garment dyeing', *Am. Dyestuff Rep.*, **76**, 11, 35–38, 50.

REINHARDT R M AND BLANCHARD E J (1988) 'Potential for dyeing easy-care cotton garments', *Am. Dyestuff Rep.*, **77**, 1, 29–34, 53.

RILEY S W AND JONES C (1988) 'Setting up laundry facilities for garment dyeing', *Am. Dyestuff Rep.*, **77**, 10, 23–24, 52.

RUCKER J W, FREEMAN H S AND HSU W N (1992) 'Evaluation of factors contributing to the light-induced yellowing of whitewashed denim-II: Effects of various treatments on metal content and photo yellowing', *Textile Chem. Color.*, **24**, 10, 21–25.

SHAH D L (1995) 'Modern quick wash denim garment', *Man made Textiles India*, **38**, 11, 434–435.

TEICHNER A AND HOLLYWOOD N (1987) 'Garment dyeing in the United States today', *Am. Dyestuff Rep.*, **76**, 7, 38–41.

TYNDALL R M (1992) 'Improving the softness and surface appearance of cotton fabrics and garments by treatment with cellulose enzymes', *Textile Chem. Color.*, **24**, 6, 23–26.

VOSS E (1993) 'Experience with automatic pH control for dyeing piece goods and made-up garments: Very high degrees of exhaustion for polyamide and wool in bath dyeing', *Melliand Textilber.*, **74**, 6, 564–566.

WAGNER K (1980) 'Automatic flock coating of T-shirts', *Melliand Textilber.*, **61**, 7, 609–611.

WALKER B AND BYNUM D (1991) 'The weathered pigment – a 90s trend for garments and hosiery', *Am. Dyestuff Rep.*, **80**, 9, 56.

29

Assessment of fastness of dyeings

Abstract: Fastness of coloured textiles is of utmost importance for their efficient end use. Various fastness properties, viz. light, wash, rubbing, thermal, etc., are based on mainly dye, fibre and their combined systems beside numerous other parameters too. Poor wash fastness causes staining of adjacent garment during domestic washing with simultaneous fading in depth of textile under wash; poor light fatness results quick fading on prolonged exposure to daylight. To perform better in swimming pools, dye used in dyeing costumes must have good chlorine fastness while sweating should not cause fading of colour on exposed areas of garments, etc. It is essential to assess specific fastness grades of coloured textiles to ascertain its performance for specific end use. This chapter describes parameters influencing various fastness aspects and tests to ascertain fastness of coloured textiles.

Keywords: Fastness, grade, grey scale, fading, staining

29.1   Basic considerations

All apparel and clothing are invariably exposed to light, heat, intense rubbing, chlorinated water, etc., or treated with various chemicals and detergents for washing which may initiate slow fading of colour for selective dyes. It is to be granted for sure that colour fastness is a feature governed by coloured textiles but not of a colourant in isolation; fastness of a dye without mentioning the textile which retains it, is meaningless. For example, direct dyes on cotton show very poor wash fastness due to its water solubility as well as attachment with cotton with physical forces, while the same direct dye on nylon or wool, etc., will show better wash fastness because of binding of direct dye with chemical forces. Hence, as such, direct dye alone possesses no fastness properties. This is true for other dye–fibre systems too, which we often miss to interpret. Even wet fastness and wash fastness are very different as wetting simply implies dipping in water or other agents and does not take care of removal of impurities, while wash fastness is tested in presence of surfactants.

326

Photofading of textile dyes is the most complicated phenomenon and need more research to establish the facts. Light fastness, a crucial property of dyed textiles, is governed by (i) wavelength of incident radiation, (ii) compactness of dye and fibre structures, (iii) dye–fibre system (Oakes, 2001; Allen, 1987; Bentley et al, 1974), (iv) degree of dye aggregation (Giles et al, 1977), (v) effective humidity (combination of air, surface temperature and relative humidity governing the moisture content on the surface of fibre), (vi) dye induced catalytic action, (vii) temperature, (viii) availability of oxygen (photo-oxidation) or UV light (photo-reduction), (ix) amount of colourant on fibre, i.e. deep or light shades (Nunn, 1979), (x) presence of impurities, viz. carriers, dispersing agents, dye-fixing agent, metals and various dilutants, (xi) time of exposure, and (xii) exposed surface area under light. The shorter the incident wavelength, the higher the energy release on coloured fabric surface and obviously higher will be rate of fading. Anthraquinoid structures or simply vat dyes show excellent light fastness on cotton due to compactness of dye structure, which is not possible with other classes of dyes. More compact fibre obstructs pores not to allow passage of oxygen or moisture inside to suppress fading. Basic dye produces light fast shades on acrylic but fails to provide that on wool, silk, etc., is an example of dye–fibre attachment. Dyes of bigger sizes takes generally more time to allow initiation of fading while effective humidity facilitates diffusion of oxygen to the exited dye or act as a highly dielectric reaction medium. In dye induced photofading, vat yellow promotes fading of blue when in mixture; the later remain stable if exposed alone. The higher the temperature the higher will be rate of fading; prolonged exposure to visible light causes photo-oxidation on availability of oxygen. Extent of fading over short time can not be visually assessed in a deep shade, which will be easier on a light shade due to more percent of dye destruction in the latter on basis of total dye in fibre. Presence of impurities, which are essentially to be used in dyeing, like carrier, dispersing agent, etc., or metals in fibre of dye structure promotes fading. It is a known fact that carriers like o-phenylphenol reduce light fastness of dispersed dyed polyester. Complete time of exposure is a vital parameter too; on short exposure dyed textile gets enough time to release energy to get back to its ground status from exited one. A coarser fibre fades slowly due to lesser surface area to volume ratio implying lesser dye on surface; the rate of fading will be obviously more for flat filament and micro-fibres.

Sublimation temperature of dye must be higher if the dyed textile has to successfully pass through high temperature treatment in curing, finishing, domestic pressing or to know migration behaviour; otherwise a part of colour sublimates-off making the shade lighter. Capability of dye to withstand high temperature is directly proportional to its molecular weight and the intermolecular binding force.

Washing performance of dyed textiles depends on numerous factors, viz. dye chemistry, size and solubility of dye, nature of dye–fibre attachment, location of dye in fibre structure and detergent formulation used in washing, etc. Reactive dyes are water soluble as well as hydrolysable in bath. Monochlorotriazine and vinyl sulphone reactive dyes are monofunctional, are either attached with fibre or hydrolysed resulting good wash fastness provided thorough wash has been imparted; a reason why these dyes are highly preferred in printing. Dichlorotriazine reactive dyes form partially hydrolysed dye showing poor wash fastness. Direct dyes are water soluble and so after-treated with external chemicals to remain trapped in situ cotton forming a bigger insoluble complex. Opening of polyester structure during dyeing should be adequate to ensure passage of dye at the interior. Dyeing machines are heated up injecting steam which is supplied by boiler through connecting pipes. Leakages or condensation in supply pipe during typical winter causes drop in effective steam pressure as well as inadequate opening of fibre structure restricting location of most of dye at surface only. Extent of aggregation of insoluble dyes in cotton enhances wash fastness. Better formation of aggregation by anthraquinoid vat dyes promises excellent wash fastness compared to relatively smaller aggregates of sulphur dyes, which show only very good fastness. In both these cases, fibre is cotton and mode of attachment is by means of physical forces.

Dyed or printed fabrics, if shaped to swimming costumes, dye must be resistant to action of chlorine which is frequently injected in swimming pools to rectify infection. Most of the sulphur and reactive dyes are liable to be decolourised against chlorine putting a restriction on these dyes in dyeing costumes.

Human Perspiration is an emulsion consisting of trace elements, ammonia, urea, amino acid, glucose, chlorides and the major constituent is lactic acid with a pH around 6.14–6.57 (Mcswiney, 1934). It may be the synergistic combined effect of metal and heat on dye or the emulsion, which causes fading of colour at selective points.

People carrying out welding works face reducing fumes which may cause destruction of colour on their uniforms if the dye belongs to azo classes, i.e. direct and insoluble azoic colours; azo chromophore is broken down to convert the whole dye molecule in smaller colourless chemicals.

Rubbing or crock fastness may not be adequate because of superficial dyes or weak dye–fibre attachment at the surface layer or formation of coloured lake at the fibre–air interface. Wet rubbing fastness remains mostly inferior to dry rubbing, may be due to solubilisation of a part of dye and its migration to the surface of coloured specimen.

In a word, coloured textiles are liable to face various adverse situations through which these have to pass satisfactorily during their use. Assessment of fastness in terms of light, wash, perspiration, chlorinated water and sublimation, etc., are essential depending on the end use of a specific textile (Smith, 1994).

29.2   Standards of fastness grades

Light fastness is graded from 1 to 8, interpretation of these grades in ascending order are very poor, poor, moderate, fairly good, good, very good, excellent and outstanding (Pugh and Guthrie, 2001). Weightage of each grade is almost double or little more over the preceding one except 7 and 8, i.e. grade 2 is more than twice as fast as grade 1, grade 3 is more than twice faster over grade 2 and so on. However grade 7 is four times faster over grade 6 and grade 8 is stable infinitely (Trotman, 1994). These grades are evaluated against control blue woollen dyed samples of known light fastness along with specimen, whose fastness is to be assessed. Specification of eight blue woollen standards (controls) used to assess light fastness grades specimens are obtained through dyeing of well scoured and bleached woollen fabric with blue dyes of known light fastness grades, viz. for control samples of grade 1 to 8, dyes used are Acilan Brilliant Blue FFR (C I Acid Blue 104), Acilan Brilliant Blue FFB (C I Acid Blue 109), Coomassie Brilliant Blue R (C I Acid Blue 83), Supramine Blue EG (C I Acid Blue 121), Solway Blue RN (C I Acid Blue 47), Alizarine light Blue 4GL (C I Acid Blue 23), Soledon Blue 4BC powder (C I Soluble Vat Blue 5) and Indigosol Blue AGG (C I Soluble Vat Blue 8) respectively.

In contrast, other fastness properties, e.g. wash, perspiration and sublimation, etc., are related to staining of adjacent white sample and simultaneous fading of coloured specimen under test. Two different grey scales are used to compare fading of specimen as well as staining of two adjacent white samples of different fibres placed and stitched on either side of the specimen. The grades are coded from 1 to 5, viz. poor, moderate, good, very good and excellent respectively, e.g. if a dyed sample during wash fastness test shows no change in colour, there will be no staining of white sample too – the wash fastness will be expressed as 'change in colour -5 and degree of staining -5' (no fading and no staining, i.e. excellent wash fastness or wash fastness rating 5) and so on. Look of a grey scale for a rate of fading as well staining 2 is depicted in Fig. 29.1 separately. In fact, the grey scales have nine possible grades of fading or staining for better precision, viz. 5, 4–5, 4, 3–4, 3, 2–3, 2, 1–2 and 1. Precision of result depends on accuracy and expertise of evaluator and is also influenced by viewer's vision standard as all coloured specimen are to be compared with grey coloured scales.

      (a)   (b)

29.1 Grey scales to assess fading and staining.
(a) Grey scale to assess staining and (b) Grey scale to assess fading
[Reproduced with permission of the Society of Dyers and Colourists]

29.3   Light fastness

29.3.1 Characteristics of various sources of light

Light fastness of coloured sample is assessed by exposing it in an most appropriate artificial source of light whose energy distribution pattern simulates with that of sunlight, e.g. carbon arc, xenon arc, fluorescent sun lamp (FS 40) or mercury tungsten fluorescent (MBTF) together with eight blue dyed woollen controls of known light fastness and is graded by matching with fading pattern of a specific known control (Pugh and Guthrie, 2001; Friele, 1970; Sato and Teraj, 1990; Innegraeve and Langelin, 2002). The test becomes too lengthy if performed under daylight whose intensity is inconsistent. Although various manufacturers of light fastness tester attach either of these artificial light sources, but from technical point of view, it is the xenon arc which radiates energy with better reliability and a closer pattern to that by terrestrial sunlight as compared in Fig. 29.2 (Atlas, 2009; Park and Smith, 1974). To get more accurate information of light fastness grades, heat developed on coloured samples is to be removed by a cooling attachment. Consistency in humidity is another important factor, which shows its impact on fastness grades as because the test is performed over a considerable length of time.

Details of various fastness assessments are available in standard handbooks prescribed by many international and national scientific bodies. Few popular test procedures are explained below for some of the important fastness criteria and may be conducted without any such remarkable difference in precision.

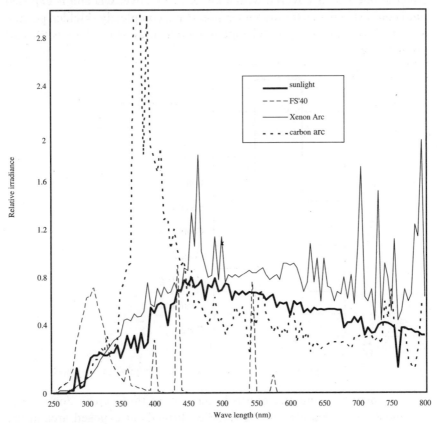

29.2 Spectral energy distribution pattern by various light sources. [Courtesy of Atlas Material Testing Technology LLC]

29.3.2 Assessment of light fastness

Standard method

Eight known controls and the specimen, 1 cm × 4.5 cm each are placed side by side on a hinged opaque cover, the central one-third of each is covered with a black cover AB as in Fig.29.3 (Trotman, 1994) and exposed to xenon arc (heat filter may be placed between sets of samples and the xenon arc to expel UV radiation). Any change in colour in the specimen is inspected in times by lifting the black cover till the contrast between covered and exposed positions of the specimen is equivalent to grade 4 on the 'change in colour' gray scale. This is roughly the light fastness rating of the specimen. If control 7 also fades to grade 4 on the same scale, the test may be terminated at this stage. If control 7 does not fade to grade 4, one half of all controls and the specimen along with half of central black cover at one side is covered with another black cover CD and is exposed until contrast between throughout exposed and completely shielded portion are equal to grade 3 on the 'change in colour' gray scale. This is the end point of the test.

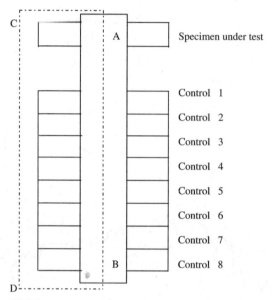

29.3 Arrangement of control and specimen for light fastness test.

The specimen and the controls have now three clear zones, e.g. completely shielded area in the middle, throughout exposed area at the right side of AB and partially exposed area at the left; the zones are compared with the control and if the two degrees of fading on specimen do not correspond with the control, the fastness should be mean of the two.

Routine checking

In day to day test requirements, a simple and comprehensive test procedure may be adopted. Eight controls of 3 cm × 1 cm size are placed on black paper side by side. Specimens of same size, which may be large in numbers and require simultaneous testing, are also placed at one side of these controls on the same paper (Fig. 29.4). Half of the set is shielded with black paper. The combined system is exposed to xenon arc. At suitable intervals of time which depends on experience of concerned person, the black cover is lifted and any sort of fading between exposed and shielded parts of samples is compared. When control 1 starts fading, the fading of specimen is checked; if no fading is there till control 1 had completely faded, the light fastness of specimen is considered as beyond 1. When control 2 starts fading, the same on specimen is checked, if not, awaited till control 3 fades and so on. To confirm extent of fading at any time, the black cover on lower half is to be temporarily removed, contrast between exposed and shielded portion is of controls as well as specimen are checked and fixed again. The light fastness grade of a specimen becomes equal to that of a control, when both of a specific control and one or more specimen start fading simultaneously.

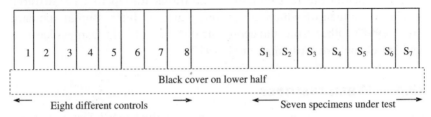

29.4 Arrangement of samples for routine test of light fastness.

Most of the popular models of light fastness tester come alongwith a water cooling system to avoid unnecessary heating up or sublimation of dye due to prolonged exposure to heat that is released by light source. Attention must be paid to see the specimen under test is not heated up upto any such extent.

Assessment without woollen standards

Light fastness test can also be conducted without using blue woollen controls. To do that, first requirement is to control temperature inside the tester against which the samples are exposed when the instrument is 'on', as action of light is synergised in presence of heat and reduces time required for fading of a specific sample or control (ATIRA, 1996). If the temperature inside a light fastness tester is maintained at $42\pm1°C$ (the sunlight temperature), time required for fading of eight known 1–8 blue controls are 5–6, 10–12, 22–24, 75–80, 155–160, 250–270, 380–400 and beyond 400 h respectively. If the temperature is $50\pm1°C$, these values get changed to 3–4, 5–6, 16–18, 50–55, 110–120, 200–210, 325–350 and beyond 350 h respectively. If the temperature inside the light fastness tester in use differs from these temperatures, the same may be calibrated using eight blue woollen controls. The timings are for a contrast on grey scale 'change in colour' of 3 grade, when the test is to be terminated. However, the source of light must remain identical to keep match in fading with time.

Specimens, whose light fastness grades are to be assessed, are placed on black sheets as shown earlier with one-half covered. After exposure for desired time, the samples are uncovered, checked and if no fading is found, again covered and exposed. Enormous samples can be tested simultaneously in this method without blue woollen controls and limited life of the lamp can be well utilized.

Recently developed software operated light fastness tester, 'The Megasol Apparatus' by James H Heal and co reduces complications associated with use of conventional testers. The Megasol apparatus performs test through automatic setting and monitoring of test conditions; the instrument sets up and controls the test chamber conditions for up to fifty different standards, sixteen of which are pre-programmed international standards (Heal, 1998). Other rapid and economical testing of light fastness has been detailed elsewhere (Adelman et al, 1977).

29.4   Wash fastness

There are five recommended test methods to evaluate wash fastness grades. One-half of dyed specimen (10cm × 4 cm) is covered on both sides with two pieces of undyed white fabric (5 cm × 4 cm) made of two different fibres and the three pieces are stitched round the edges, leaving 5 cm × 4 cm exposed as shown in Fig. 29.5 (Trotman, 1994).

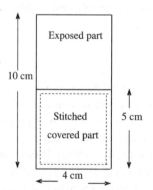

29.5 Arrangement of specimen for wash fastness test.

For first three test methods, while one white fabric is identical in nature with that of the dyed, the type of the second depends on the dyed sample under test. If the dyed sample is made of cotton, silk, linen, wool, viscose, cellulose acetate, nylon, polyester or acrylic, then the second white sample will be made of wool, wool, wool, cotton, wool, viscose, (wool or viscose), (wool or cotton) or (wool or cotton) respectively. Five sets of composite samples are to be prepared and treated in five tests separately to assess fastness grades 1–5 (Trotman, 1994). First composite sample is tested for staining or fading grade 1, on getting negative result, the second composite sample is treated for grade 2 and so on. Alternately, all the five composite samples are treated in five different baths under varying conditions. After conducting a specific test, the stitches are removed and the three fabric pieces are assessed separately. The exposed portion of dyed sample is matched with 'change in colour' grey scale to assess extent of fading of colour, while the white fabric pieces are matched with 'change in staining' grey scale to get degree of staining. The results are expressed as 'degree of fading' and 'degree of staining'. Generally these two degrees become identical.

ISO method 1: The composite sample is treated in soap solution (5 g/l) at liquor ratio 1:50 at 40 ± 2°C for 30 min in a wash fastness tester followed by washing and drying.

ISO method 2: Same as test method 1, only difference are temperature is 50 ± 2°C and time 45 min.

ISO method 3: Same as test method 1 but Na_2CO_3 (2 g/l) is added in soap solution (5 g/l); temperature 60 ± 2°C and time 30 min.

To conduct ISO methods 4 and 5, the specification of white samples will be different but one of it will be essentially made of same fibre as that of specimen while the second one depends on the dyed one. If the dyed specimen is made of cotton, linen, viscose, cellulose acetate, nylon, polyester or acrylic, the second white sample will be of viscose, viscose, cotton, viscose, (viscose or cotton), (viscose or cotton) or (viscose or cotton) respectively.

ISO method 4: The composite sample is treated in soap (5 g/l) and Na_2CO_3 (2 g/l) at liquor ratio 1:50 at 90 ± 2°C for 30 min.

ISO method 5: Same as in method 4 but time is 4 h.

In all these five tests, liquor ratio plays a vital role. The effect of liquor ratio on cross-staining and change in colour, when correlated between the results from laboratory test and those obtained in a domestic washing machine, it was found that the correlation obtained for cross-staining but not shade change is improved when the liquor ratio is reduced below 50:1. A revised setting of 20:1 is recommended for an international ring test to confirm repeatability and reproducibility (Park, 1976; Phillips et al, 2003).

Out of all the five tests described, the ISO-3 is the most important method and followed invariably as this has a match with domestic washes conducted with alkaline soap or detergent at or below 60°C, especially in typical summer. All international wash fastness reports are based on ISO-3 test, if otherwise not mentioned.

29.5   Perspiration fastness

Coloured fabric may get locally faded out or decolourised when comes in contact with perspiration. For assessment of perspiration fastness, dyed fabric specimen of 10 cm × 4 cm, is placed between two different specific white unfinished fabrics of 5 cm × 4 cm at one half as done in wash fastness testing and stitched at the periphery followed by treatment with two different artificial perspiration solutions. While one piece of white fabric is of same fibre as that in coloured specimen under test, the second white fabric is a different one. If the coloured specimen is made of cotton, wool, silk, linen, viscose, acetate, nylon, polyester or acrylic fibres, then the second white specimen will be of wool, cotton, cotton, wool, wool, viscose, wool or viscose, wool or cotton and wool or cotton respectively.

Two composite specimen are thoroughly wetted out in two different artificial perspiration solutions at a liquor ratio 1:50 for 30 min at room temperature and are placed between two glass plates, pressed with a force or weight of 4.5 kg for 4 h at 37 ± 2°C (body temperature). Specimens are removed, dried at or below 60°C and change in colour as well as degree of staining on white are compared with respective gray scales (Trotman, 1994). Alternately, specimens saturated in the above recipe separately for 30 minutes at 30°C are kept for 14 h under 4.5 kg weights in perspirometer.

The two different artificial perspiration solutions are prepared as shown below (Trotman, 1994):

	Solution A	Solution B
1-Histidine mono hydrochloric monohydrate	0.5 g	0.5 g
NaCl	5.0 g	5.0 g
Disodium hydrogen orthophosphate	2.5 g	2.2 g
Distilled water	1000 ml	1000 ml
0.1N Sodium hydroxide	adjust pH to 8	–
0.1N Acetic acid	–	adjust pH to 5.5

29.6   Fastness to chlorinated water

This test is used to assess resistance of dye against chlorine which remains present in swimming bath at very lower concentrations as disinfectant.

NaOCl is used as source of chlorine which is first titrated to assess g/l available Cl_2. A standard 1 litre of NaOCl solution is prepared by mixing NaOCl (150g/l available Cl_2), NaCl (120–170 g/l), NaOH (20 g/l), Na_2CO_3

(20 g/l). A final test solution of 20 mg/l active chlorine buffered at pH 8.5 is prepared from this stock solution. The 10 cm × 4 cm specimen is immersed in test solution at liquor ratio 1:100 at room temperature for 4 h. It is than rinsed thoroughly with water, dried at room temperature and any change in colour is assessed using gray scale (Trotman, 1994).

29.7   Rubbing fastness

Testing is done by a crock meter having two main parts: (i) a rubbing finger consisting of a cylinder of 1.6 cm diameter on which the coloured specimen under test is mounted tightly with the help of a clamp and (ii) a white unfinished cotton fabric of 5 cm × 5 cm size tightly mounted on another head at the base and just below the cylinder. When started, the cylinder can move to and fro on straight lines and can track 10 cm path on the white sample at the base with a downward force of 9.0N causing rubbing. The instrument is run only for 10 s to complete 10 complete to and fro movement of the cylinder (10 cycles). Rubbing fastness is assessed by comparing the stained white sample mounted on base head with 'grey scale for assessment of staining' ranging from 1 to 5.

Rubbing fastness is assessed in two ways: (i) dry rubbing and (ii) wet rubbing. In dry rubbing, the white fabric used is a dry one during testing, while wet rubbing fastness is tested by using a wet white cotton fabric of 100% expression with water.

29.8   Sublimation fastness

The coloured specimen of 10 cm × 10 cm is stitched between two specific white unfinished fabrics of same size. While one white fabric is of same fibre like that of coloured one, the second white is generally made of polyester. The composite is converted into roll form and placed in a test tube of size 15 mm diameter, the mouth of tube is tightly closed with cork and is heated up separately at 120 ± 2°C, 150 ± 2°C, 180 ± 2°C, and 210 ± 2°C for 120s, 30s, 30s and 30s respectively. Different sets of specimen are prepared for the purpose. Degree of staining of white adjacent fabrics is reported by comparing with grey scale as the sublimation fastness at that temperature and time.

Sublimation fastness of pile fabrics is not assessed through this test procedure.

In another test, a long strip of dyed specimen is placed on a sublimation tester; the latter has a few thermostatically controlled heating points (generally 7) at different temperatures from 130 to 250°C to heat up the specimen at different points for required time and to assess degree of fading at different temperatures.

29.9   Fastness to dry-cleaning

The composite sample is prepared as described in wash fastness test and is treated it in C_2HCl_3 (trichloroethylene) at liquor ratio 1:100 at room temperate for 30 min with vigorous agitation; squeezed to remove excess solvent, thoroughly washed and dried. The result is expressed as 'change in colour' and 'degree of staining' with the help of grey scales. The method stated is for dyed woollen and synthetic textiles.

29.10   References

ADELMAN M, KITSON C J, FORRESTER S D, GILES C H and HASLAM R (1977) 'Comparison of light fastness assessments with the Microscal (mercury vapour illuminant) light fastness tester and xenon arc illumination', *J. Soc. Dyers Color*, **93**, 6, 224–226.

ALLEN N S (1987) 'Photofading mechanisms of dyes in solution and polymer media', *Rev. Prog. Color Related Topics*, **17**, 61–71.

ATIRA (1996) Light Fastness Tester manual, India, Ahmedabad Textile Industries Research Association.

ATLAS (2009) *Light Fastness Tester manual*, USA, Atlas Material Testing Division.

BENTLEY P, MCKELLAR J F and PHILLIPS G O (1974) 'Photochemistry of dyes, fibres and dye/fibre systems', *Rev. Prog. Color Related Topics*, **5**, 33–48.

FRIELE L F C (1970) 'Changes in shade of the blue dyeings of the ISO light fastness standards and their dependence on the amount of incident light radiation', *Textilveredlung*, **5**, 12, 899–911.

GILES C H, WALSH D J and SINCLAIR R S (1977) 'Relation between light fastness of colorants and their particle size', *J. Soc. Dyers Color*, **93**, September, 348–352.

HEAL J H AND CO, (1998) 'The Megasol apparatus', *Textile Ind. Dyegest Southern Africa*, **17**, 1, 10.

INNEGRAEVE H and LANGELIN L (2002) 'Light fastness predictive analysis', *Ind. Textile*, **1347**, 62–64.

MCSWINEY B A (1934) 'The composition of human perspiration', Proceedings of the Royal Society of Medicine, London, March 16, 839–848.

PARK J and SMITH D J (1974) 'Comparison of sources for light fastness testing', *J. Soc. Dyers Color.*, **90**, 12, 431–435.

PARK J (1976) 'Rapid and economical testing of light fastness', *Int. Dyer*, **155**, 5, 5th March, 220.

PHILLIPS D, PERCIVAL R, SCOTNEY J, BEVAN G and LLOYD J (2003) 'Effect of liquor ratio on the shade change and cross-staining observed in the ISO 105-C08 test', *Coloration Tech.*, **119**, 3, 177–181.

PUGH S L and GUTHRIE J T (2001) 'The development of light fastness testing and light fastness standards', *Rev. Prog. Color Related Topics*, **31**, 42–56.

OAKES J (2001) 'Photofading of textile dyes', *Rev. Prog. Color Related Topics*, **31**, 21–28.

SATO T and TERAJ K (1990) 'Studies on the instrumental method for assessing the light fastness of coloured materials-II: The colorimetric evaluation of fading of ISO type and AATCC type blue standards, Journal of the Japan Research Association for Textile End uses, 31, 8 (1990) 387–394.

SMITH P J (1994) 'Colour fastness testing methods and equipment', *Rev. Prog. Color Related Topics*, **24**, 31–40.

STEVENS C B (1979) *The dyeing of synthetic polymer and acetate fibres*, Ed. Nunn D M, Bradford, UK, Dyers Company Publications Trust.

TROTMAN E R (1994) *Dyeing and Chemical Technology of Textile Fibres*, 6th Edition (1st Indian Edition), Delhi, B I Publications Pvt Ltd.

30

Identification of dye class on cellulosics

Abstract: Identification of dye is important keeping in view reproduction of dyed sample doing well in the market or based on customers demand. The piece of dyed specimen supplied to the sales section is forwarded to dyeing department to reproduce it which requires to be analyzed to ascertain dye class applied on it earlier. A series of experiments are to be performed to reach to the conclusion. This chapter describes tests those are to be conducted on dyed specimen.

Keywords: Identification, test, observation, inference, confirmation

30.1   Reasons to identify dye

Dye houses remain often engaged in analyzing coloured sample supplied by customer for its exact reproduction. Analysis of dye for its identification is essential to achieve that and also to explore the possibility to change it with a cheaper one, keeping properties of the newly dyed consignment almost parallel to that of the supplied, if possible. Methods used to identify the dye on all types of textiles are not same. Initially primary tests are conducted to ascertain the dye; on getting response, confirmatory tests are conducted to confirm it. Identification tests are specified in standard handbooks by authorized bodies in various countries and a test number is assigned to each test to mark its authenticity.

Various textile fibres are dyeable with different classes of dyes (see Table 2.1), out of which only a limited number of dyes are selected in practice to achieve desired properties on dyeings, cost too plays important role. Based on this very fact, cellulosics are invariably dyed with direct, sulphur, reactive, vat, naphthol, solubilised vat colours, though in exceptional cases, mineral, phthalocyanine and oxidation colours are also used. Protein fibres, viz. wool and silk are dyeable with acid, basic, metal complex, direct, reactive and vat dyes; practically acid dyes are generally not applied on poor wash fastness ground while basic dyes are liable to produce only light to medium shades of poor light fastness. Vat dyes are applied from highly alkaline bath at specific temperature with chances of degradation of these fibres. Direct dyes produce only dull shades, thus paving way for mainly acid, metal complex and reactive dyes to dye these fibres. Nylon possesses resemblance with wool and silk and is obviously dyed with same dyes as those are applied on wool and silk, but never dyed with disperse dye as only poor light fast shades are produced. Acrylic is dyed with only basic dye based on use of anionic co-monomers during polymerization and is never dyed with disperse dye due to limitation of only light shade formation. Polyester is exclusively dyed with disperse dye, which is specifically designed to dye it.

It is understood from the discussion that identification of dye on dyed cellulosics is a tuff job, as a long list of dyes are to be tried. In contrast, wool, silk, nylon are dyed with very limited number of dyes making the identification process rather simple. Acrylic and polyester are dyed with only basic and disperse dyes respectively making the process a known one.

Troubles are frequently encountered during identification of a dye. A dyed textile may be directly put under test or the dye may be extracted and tests can be run on it. This extraction technique may help to identify the fibre too and hence the dye in certain cases. Specifically to improve fastness of dyeings of cellulosics, certain after-treatments are imparted which makes the job too complicated. Direct dyed textiles are frequently topped with basic dye, naphthol dyed yellow hues are topped with reactive blue dye to develop green hue, phthalocyanine blue shade is topped with naphthol yellow hue or with reactive yellow dyes, mineral khaki is topped with vat olive dyes to improve fastness and brilliancy, aniline black is topped with vat black – these are some of the frequent practices run in shop-floor level either to get a better fast shade, reduce cost of dyeing or to get best use of stock of dyes. In a number of cases, faulty dyeings are redyed with black, navy blue or other deep shades. Expertise is required to identify dye in such cases.

In this chapter, only identification of dye on cellulosics will be discussed as a wide range of dyes are used on this fibre and in some cases, two dyes are separately applied in a two stage dyeing process. It is advised to definish the supplied dyed sample to ascertain dye on it by treating it in HCl (1%, w/v) at boil in two successive baths followed by neutralization and thorough washing as dyed cotton fabrics are invariably finished with resins (DMDHEU) to improve anti-crease property or after-treated with resin or other simple chemicals to improve fastness. The tests mentioned here are viable for day to day routine analysis; detailed test methods are to be had of these handbooks.

Only a small part of the control is to be tested to identify dye on it, rest is to be preserved to match with the reproduced sample.

30.2   Identification of dyes

30.2.1 Primary tests

Tests may be started with the following one as it gives information about a few dye classes in a single test, whether the dye belongs to azo class or not.

Primary test	Observation	Inference
(1) Specimen is boiled in a solution of $Na_2S_2O_4$ and NaOH for 1 2 min.	(A) Distinct change in colour, original colour is restored on exposure to air	(A) Vat dye or Aniline black
	(B) Colour is permanently discharged	(B) Azoics, diazotised and developed, after-treated azo dyes
	(C) Specimen is discoloured or changes to yellow but original colour is restored on exposure to air	(C) Sulphur, indigoid vat, indigo, hydron blue and few basic colours
	(D) Little or no change	(D) Logwood black and most of mordant colours

Based on inferences, subsequent tests are carried out to identify whether the dye is vat and aniline black (A-1 and A-2) or azoics or after-treated direct dyeing (B-1, B-2, B-3 and B-4) and for sulphur or basic dyes (C-1 and C-2).

Primary test	Observation	Inference
A-1. Repeat test 1	Sample shows change in colour. Original color is restored in H_2O_2 or on exposure to air	Vat dye
A-2. Dye is extracted by treating the specimen in conc H_2SO_4. The extract is diluted with water. A filter paper is soaked in this extract followed by passing through NaOH solution	Filter paper turns red violet	Aniline black

If tests against inference 1-A indicate presence of those dyes, confirmative tests are to be carried out. Otherwise, tests against inference at 1-B are to be conducted as follows:

Primary test	Observation	Inference
B-1. A little of specimen is burnt to ash, a little of Na_2CO_3 +$NaNO_3$) (1:1) is added and fused	Fused mass becomes orange–yellow when hot and greenish yellow on cooling.	Direct dye after-treated with Cr-salts

B-2. A little of specimen is burnt to ash and dissolved conc.HNO_3, little water is added, boiled, cooled and NH_4OH is added to it	Blue color	Direct dye after-treated with Cu-salts
B-4. Specimen is boiled in pyridine	(i) Stripping of colour (ii) Specimen either does not bleed or bleeds slightly	(i) Azoic dyes (ii) Diazotised and developed

On getting no response for the said dyes against test-1 inferences at A and B, the following tests are to be conducted against inferences at 1.(C)

Primary test	Observation	Inference
C-1. Specimen is boiled in little Na_2CO_3 and Na_2SO_4 solution. To the collected extract a pinch of NaCl is added. White bleached cotton is boiled in it.	White cotton is dyed. Original specimen and dyed white cotton are oxidized. White cotton is dyed to original hue while the specimen restores colour	Sulphur dye
C-2. Specimen is warmed in dil CH_3COOH. Add little water and boil. Remove sample, add 25 mg	Stripping of colour. Mordanted cotton or wool is boiled in the extract and boiled. Mordanted cotton or wool is dyed	Basic dye

If response is not obtained in previous tests, the following tests are to be conducted.

Primary test	Observation	Inference
(2) Specimen is boiled in 5% NaOH	(A) Decolouration occurs	(A) Basic, chrome yellow, chrome orange or Prussian blue. Chrome dyes turn black with ammonium sulphide, while colourless alkaline extract of prussian blue restores colour when acidified.
	(B) Violet extract, turns yellow on acidification	(B) Alizarin red
	(C) Colour bleeds	(C) Direct or acid dye
	(D) Too little or no change	(D) Sulphur, azo, vat or aniline black
(3) Specimen is boiled in 5% HCl	(A) Hue changes to black or blue. Original hue restores on treatment with NH_3	(A) Congo red
	(B) Yellow colour is developed which turns violet with alkali	(B) Turkey red, vat, acid, sulphur and azo dyes.

	(C) Little or no change	(C) Most of direct dyes, vat, acid, sulphur and azo dyes.
(4) Specimen is warmed with $CHCl_3$	(A) Distinct coloured solution (B) No change	(A) Indigo, Indigoid vat, azo, mordant and few basic dyes (B) Anthraquinoid vat, dcc, hydron blue, logwood black
(5) Specimen is treated bleaching powder or sodium hypochlorite solution	(A) Colour destroyed (B) Colour of sample changes to pale yellow upto brown (C) Slow destruction of colour (D) Colour of sample changes to brown or olive (E) Little or no change	(A) Direct dye, diazotized and developed, sulphur, basic, indigo, mordant dyes (B) Sulphur black topped with aniline black (C) Hydron blue (D) Aniline black (E) Vat, mineral, insoluble azoics and direct colour
(6) Specimen is placed on watch glass and a few drops of conc H_2SO_4 is added when the fibre gets dissolved. Dilute the solution with water	(A) Light green solution (B) Yellow solution (C) Pink colour is developed at the junction	(A) Aniline black (B) Alizarin, Turkey Red (C) Insoluble azo dyes (naphthol-AS)
(7) Specimen is boiled in separately in 50%DMF, conc DMF and conc gl. CH_3COOH + rectified spirit (1:1 v/v), wash and squeeze	No stripping or partial stripping of colour	Reactive, phthalocyanine, oxidation black, may be basic dyes
(8) Specimen is boiled in 1% NH_4OH	(A) Stripping of colour. Tests at (i) and (ii) are to be conducted as follows: (i) Bleached cotton is dyed with stripped out colour alongwith little NaCl at boil. White sample is dyed (ii) Few drops of CH_3COOH are added to stripped out colour. Bleached cotton and scoured wool are dyed at boil. Wool is only dyed. (B) Very little or no stripping	May be Acid, Basic or Direct dye May be direct dye Basic dye Absence of acid and basic dyes
(9) Specimen is boiled in dilute HCl. Specimen is taken out and the test is repeated.	Treated specimen is tested for presence of direct dye, if responds	Direct dyeing after-treated with resin
(10) Samples not showing response against above treatments.	–	Pigments

In addition to these primary tests, a few more tests may be carried out if required before conducting confirmatory tests.

30.2.2 Confirmatory tests

Direct dyes (BIS, 1988)

Test	Observation
(1) Specimen is boiled in dilute NaOH solution alongwith a small piece of mercerized cotton.	Mercerized cotton is dyed and can not be stripped off with 1% NH_4OH. (Direct dyes possess sulphonic acid groups and are soluble in dilute NaOH solution causing stripping out of dye)
(2) Specimen is shaked in ethylene diamine, the extract is diluted, white cotton is dyed at 80°C in it, NaCl is added, dyeing is continued for some time and cooled.	White cotton is dyed and not stripped with NH_4OH.

Sulphur dyes

Test	Observation
(1) Specimen is boiled with stannous chloride or zinc dust in a test tube and the tip of the tube is covered with a piece of filter paper soaked in lead acetate (most authentic test to detect sulphur)	Deep brown or black stain on filter paper. All sulphurised vat dyes too show same result. H_2S formed due to reduction of free sulphur in dye with stannous chloride or zinc gets converted to brown black PbS. Any sulphur free reducing agent may be used in place of Zn or Sn_2Cl_2
(2) Specimen is treated in NaOCl	Colour discharged

Vat dyes (Trotman, 1994; Kaikeung and Tsim, 1994)

Test	Observation
(1) Specimen is boiled in a solution of sodium hydrosulphite and NaOH	Colour of specimen is changed but is restored on oxidation
(2) Specimen is boiled in ethylene diamine and glucose	Colour of specimen is changed but is restored on oxidation
(3) Specimen is treated in dimethyl sulphoxide (DMSO) for 5–10 min	Indigoid vat dyes are completely stripped out after 1–2 hrs while almost all anthraquinoid vat dyes are stripped out when heated at 50–60°C for 5–10 min

Insoluble azoic colours (Giles et al, 1962)

Test	Observation
(1) The specimen is boiled in ethylene diamine and the extract is collected. (A) 1st part of extract is boiled with $Na_2S_2O_4$ (B) 2nd part is diluted and boiled.	Colour discharged The liquid becomes turbid.
(2) Specimen is boiled in $Na_2S_2O_4$ and NaOH. (the decolouration is due to breaking down	Colour is permanently discharged and is not restored on oxidation of N=N linkages in dye)

Phthalocyanine blue

Test	Observation
(1) Specimen is boiled in methyl pyrolidone, cooled and little NaOH and $Na_2S_2O_4$ are added and boiled.	Colour is changed to violet and in 20% CH_3COOH changes to blue again
(2) Phthalocyanine green Test as that for blue	Colour of specimen reduces to dark violet and remains same in 20% CH_3COOH
(3) Specimen is boiled in NaOH and $Na_2S_2O_4$	No change
(4) Treat specimen in conc. H_2SO_4	Dye is solubilised in acid and is precipitated when diluted with water

Oxidation colours

Test	Observation
(1) Specimen is treated in conc H_2SO_4 and diluted.	Deep Green colour
(2) Sample is imparted NaOCl treatment	Black turns to brown

Basic colours

Test	Observation
(1) Warm specimen with little CH_3COOH. If colour is extracted, dilute the bath with equal amount of water. Enter a piece of scoured and mordanted cotton in bath with slow heating.	Mordanted cotton is dyed with same hue and shade
(2) Specimen is boiled in 5% NaOH for few seconds, add 4–5% NH_4Cl and reboil. Collect extract and conduct following tests. (i) Wool is added in 1st part of extract and cooled	Wool is dyed

Test	Observation
(ii) Add 10% CH_3COOH in 2nd part and tannin reagent	Coloured precipitation
(3) Add 10% CH_3COOH in 3rd part followed by 1% $FeCl_3$.	Black precipitation

Reactive dyes (Giles et al, 1962; Jordinson and Lockwood, 1962, 1968)

Test	Observation
(1) The specimen is boiled in non-ionic surface active agent for 15 min	
(2) The specimen is boiled in glacial CH_3COOH and ethyl alcohol (1:2, v/v) mixture for 4 min	No stripping of colour in all these four tests
(3) The specimen is boiled in diluted DMF (1:1, v/v) for 4 min	
(4) The specimen is boiled in DMF for 4 min	

The tests mentioned above are based on properties of dyes, which is shown against action of a specific chemical on it. To identify dye on a dyed specimen, it is not required to conduct all the tests mentioned here, only a few of the required tests will suffice to primarily establish the dye later backed up by confirmatory tests. Some tests result multiple observations and due attention is to be paid in those cases to ascertain the dye. Under industrial circumstances, a dye with a specific hue, e.g. red, yellow, etc., may be available in other cotton dye series but with a tonal difference. Experienced dyer working in dye houses remains familiar with the situation and even can predict the dye on dyed sample without conduct of any such tests. Some tests are highly hazardous, e.g. use of conc H_2SO_4, $CHCl_3$, etc., and utmost care is to be taken in such cases. Another important factor is that, only a small dyed sample is supplied by client which is not adequate to conduct such a large number of tests on it, only selective tests are to be carried out. This requires knowledge of dye and related technology of dyeing.

Identification of dye class is most important; that of individual dye does not hold any such importance as dyes, if in combination too, belong to same class to facilitate application with a specific technology and within a class, at most 2–3 dyes with same hue exist with varying tones.

To ascertain specific dye on cellulosics, readers are referred elsewhere (Green, 1946) or chromatographic tests may be conducted for the purpose (Trotman, 1994).

30.3   References

BIS (1988), *Handbook of Textile Testing: Part 4 – Identification and testing of dyestuffs and their colour fastness on textile materials*, 1st Revision, New Delhi, Bureau of Indian Standards, A/1-3.

GILES C H, BASHIR AHMAD M, DANDEKAR S D and MCKAY R B (1962) 'Publications sponsored by the Society's identification of dyes committee-II : Identification of the application class of a colorant on a fibre', *J. Soc. Dyers Color.*, **78**, 3, 125–127.

GREEN ARTHUR G (1946) *The analysis of dyestuffs and their identification in dyed and coloured materials, lake–pigments, dyestuffs*, London, Charles Griffn & Company Ltd.

JORDINSON F and LOCKWOOD R (1962) 'Publications sponsored by the Society's identification of dyes committee-I: Identification of reactive dyes on cellulosic fibres', *J. Soc. Dyers Color.*, **78**, 3, 122–125.

JORDINSON F and LOCKWOOD R (1968) 'Identification of newer types of reactive dyes on cellulosic fibres', *J. Soc. Dyers Color.*, **84**, 4, 205–211.

KAIKEUNG M and TSIM S T (1994) 'A new method to determine the indigo concentration', *Int. Textile Bulletin*, **40**, 3, 58–59.

TROTMAN E R (1994) *Dyeing and Chemical Technology of Textile Fibres*, 6th Edition (1st Indian Edition), Delhi, B I Publications Pvt Ltd.

31
Dyeing machineries

Abstract: Dyeing of any textile is carried out in machines for mass production. Based on forms of textile, i.e. fibre, yarn or fabric, various machineries have been developed. Even within the same form, different types of machines are available to dye the material, viz. fibre (loose, top), yarn (package, hank, etc.) and fabric (open width or rope form). Also selection of machine depends on type of process to be carried out, viz. batch, semi-continuous or continuous. All these machines do not necessarily work under the same principle. World of dyeing machine and make are endless. This chapter explains some first hand information on some basic dyeing machines.

Keywords: Machine, dyeing, working principle, form of textile

31.1 Working liquor ratio in machines

Textiles are often dyed in various forms, viz. fibre, yarn or fabric and in different shapes, e.g. loose fibre, fibre tops, hank, various yarn packages, fabric in rope form or open width, using numerous machineries to produce a levelled shade. The volume of liquor to be used in each machine also follows some guidelines. Keeping in mind the fact that almost all textiles are wetted before dyeing, if a jigger has full capacity of 300 l, then during loading of cotton fabric on the machine (say 100 kg), the fabric is passed through water to wet it in order to make a uniform roll on one roller of jigger and during this process the fabric absorbs water up to 100% expression, i.e. 100 kg. Once loading is over, a bath is prepared for dyeing which might have a volume as high as 300 l implying presence of 400 l of water including the water retained by fabric and finally a liquor ratio of 1:4. During change in direction of movement of the fabric, the machine puts a jerk which may throw-off a part of dye liquor if maintained up to full brim and so a lower liquor ratio is always preferred. On reducing weight of fabric to 50 kgs (may be as per requirement of sales section), the liquor ratio increases and may rise as high as 1:8. A jet dyeing machine has generally a capacity of 800 l with maximum loading capacity of 100–

120 kgs of polyester or blended fabric implying a liquor ratio around 1:8. Various manufacturers market different models for dyeing of textiles in a specific form and a detailed study of all such machines have been documented elsewhere (Bhagwat, 1999). In this chapter, a few most commonly used types will only be discussed here. It is to be noted that most of industries have introduced PTR (pre-treatment range) which omits installation of basic J-box, kier, etc., for pre-treatment and CDR (continuous dyeing range) which outhouses most of the discussed machineries in this chapter. These processing ranges are based on repeated padding and passing through rollers in open width; suitable for processing longer length of fabric only. The basic machines discussed here will remain in existence due to their invaluable contribution in dyeing.

Various dyeing machines work on three different principles, viz. (i) textile remain stationary while dye liquor gets circulated throughout it, (ii) dye liquor or bath remain stationary while textile moves through dye bath repeatedly and (iii) where both dye liquor and textile remain in constant touch with each other through circulation and movement. In the first case, i.e. dye liquor is circulated throughout the packed textile in fibre, yarn or fabric form, a two way circulation of liquor is preferred, viz. in-out and out-in. For in-out case, the liquor reaches in the central perforated tube pumped by the motor and is circulated through packed material from inside to reach to the outside for specific period and after that the direction of circulation is changed to out-in, i.e. liquor reaches to outer layer of packed textile and passes through it to reach at the core of packing and is collected through central pipe. A specific dyeing machine is manufactured based on either of these concepts ad are discussed below. The description deals with simple construction of these machines without any engineering specification to enable to know their actual working features.

31.2   Loose fibre dyeing machine

Working principle

Dye liquor is circulated throughout packed stationary material

Construction

A perforated stainless steel cylindrical container surrounds a central perforated hollow cylindrical tube and the container fits on a conical seating at the base of the tube (Shenai, 1993). The central tube receives the dye liquor and circulates it in an in-out mode for dyeing. A top cover with a lid is raised or lowered down to load and unload the machine (Figs. 31.1 and 31.2). The whole system is placed in a circular enclosed dye vessel. A heat exchanger heats up or cools down the liquor while passing through tubes inside it. The flow of liquor is affected by means of a two way valve for in-out and out-in action and is circulated by a centrifugal pump; direction of circulation can be reversed by the two way valve. A dye tank is placed at one side of the machine where dye solution is prepared and pumped in dyebath. The dye tank may also be used for any addition, if required.

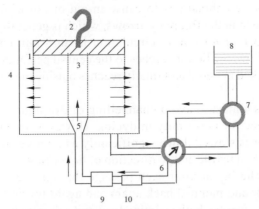

(1) Circular perforated metallic container, (2) Top cover with lid, (3) Perforated central tube, (4) Circular dye vat, (5) Conical seating, (6) Two-way valve, (7) Centrifugal pump (8) Dye tank/reservoir, (9) Heat exchanger and (10) Filter

31.1 Cross-sectional view of a loose fibre dyeing machine.

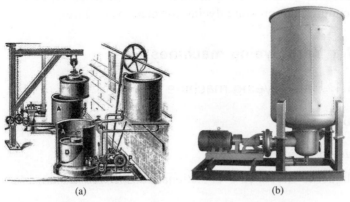

31.2 Loose fibre dyeing machines: (a) Obermair, (b) Quanrun. [Courtesy: Obermair and Quanrun]

Working

The lid is lifted to pack loose cotton, wool or polyester fibre uniformly in the perforated container followed by closing of lid and right positioning of the container on the conical seating. Dye liquor, prepared in the dye tank is pumped inside the machine by the centrifugal pump. The direction of pumping the liquor is changed automatically in a required cycle by the pump – two way valve combinations to cause in-out or out-in flow for levelled dyeing as shown in the figure by arrows, which is generally 5 min for in–out flow and 3 min for out–in circulation in an 8-min complete cycle. During in–out flow, the liquor reaches to the central perforated tube, passes through the packed material and finally reaches outside the container causing dyeing.

Uniform packing is a must to avoid channelling of dye liquor through point of least resistance for uniformity in dyeing. However, in doing so, the speed of the liquor is slowed down significantly as it moves from the inside to the outer layer of textile and so direction of liquor flow is reversed for levelled dyeing. The liquor after each passage through the textile is collected by the motor and pumped back again and again for build-up of shade. At the end of dyeing, the bath is drained out and a thorough washing is imparted followed by unloading. During passage of liquor through connected pipe, it passes through a heat exchanger, where either steam or cold water is passed through neighbouring narrow pipes to heat up dye liquor during dyeing or to cool down at the end. The filter in the path collects impurities from liquor to ensure level dyeing and requires cleaning before successive cycle. A dyeing cycle takes nearly 2–2.5 h to complete. The modern version of the machine is of enclosed type to cause high temperature high pressure dyeing for man made fibres.

31.3   Yarn dyeing machines

31.3.1 Hank dyeing machine

Hussong machine

Working principle

Dye liquor is circulated over stationary suspended hanks.

Construction

The machine consists of a rectangular vessel divided in two parts by a perforated plate; the steel frame carries poles for loading of hanks and is attached to the lid (Fig. 31.3). The dyeing compartment comprises of a false bottom which is sequentially perforated to allow consistency in liquor flow and to arrest any impurities like broken threads, etc. The smaller compartment, called addition tank is used to receive dye solution or any subsequent addition of dye or chemicals is affected through this tank; it in houses an impeller to spread the dye liquor slowly to the main dyeing tank. In order to obtain maximum uniformity of liquor flow, the holes in the false bottom are so arranged that there are more holes at the end (remote from the impeller) than adjacent to it. Steam and water supply are fitted at the lower part of the machine. The machine is not suitable for wool hank dyeing as wool felts badly in water. An improvement of the machine is Pegg pulsator machine which avoids the necessity of flow reversal and the flow is in the downward direction causing good uniformity due to tightness of hanks and good penetration of dye at the point of contact between pole and the hank being assisted by introducing a unidirectional pulsating flow.

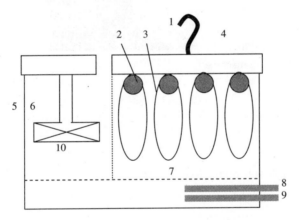

(1) Opening lid, (2) Poles/rollers, (3) Hank, (4) Stainless steel frame, (5) Dye vat, (6) Impeller, (7) False bottom, (8) Water supply, (9) Steam supply and (10) Addition tank

31.3 Cross-sectional view of hank dyeing machine.

Working

This machine is used for dyeing of cotton, silk and rayon hanks (Fig. 31.4). The steel frame carrying the poles is taken out by lifting the lid; hanks are loaded from poles and the frame is tightly placed into the large compartment. The impeller in the addition tank causes turbulence in the dye liquor when added and the latter diffuses through the perforations into the main dyeing tank. The flow of liquor is reversed from time to time. In practice, the circulation from the false bottom to upward is used for most of the times, because it presses the yarn into a compact mass against the pole frame and the liquor is compelled to pass uniformly through the mass.

31.4 Hank dyeing machine.

Klauder-Weldon machine

Working principle

Yarn moves through stationary dye liquor.

Construction

It consists of two wooden wheels placed parallel to each other and joined with an axle like that of in a cart; the axle is driven by a motor which in turn cause rotation of two wheels simultaneously at identical speed, shown in Fig. 31.5 (Shenai, 1993). Poles situated on the periphery of wheel and at the centre are used for loading of hanks. The two wheels are half–immersed in dye solution and the poles are mechanically attached to get independent rotation. On completion of dyeing, the cage at the top is opened and the wheels are taken out for unloading and reloading of hanks.

Working

Hanks to be dyed are mounted on poles at the periphery and at the centre, which in turn rests on cage. Well prepared dye solution is added to the

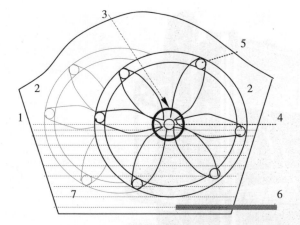

(1) Wooden dye vat, (2) Two parallel wooden wheels, (3) Drum-like cage,
(4) Axle, (5) Wooden poles or dye sticks, (6) Heating coil, (7) Dye solution

31.5 Cross-sectional view of Klauder-Weldon machine.

bath. During rotation of wheels, hanks pass regularly through the dye liquor
which is heated up by steam through open and close pipe heating systems.
The unique feature of the machine lies in its higher production rate as
well as little loss of heat being a closed one; relatively lesser floor area is
required to install (Fig. 31.6). Problem is experienced in dyeing with vat
and sulphur dyes which are prone to immature oxidation during dyeing
causing unlevelled shade formation or oxidation marks as the half of wheels
remain exposed in air during dyeing in open models.

31.3.2 Package dyeing machine

This machine is to dye yarn in cheese, cone, beam, mule and top form.

Working principle

Dye liquor is circulated throughout stationary yarn packages.

Construction

A package dyeing machine possesses a few perforated stainless steel
spindles rest on a perforated base in the dyeing vessel (autoclave) on which
yarn packages are fitted sequentially and tightened (Fig. 31.7). A rotary
pump circulates dye liquor throughout the yarn packages. A sampling unit
at the top of autoclave helps to inspect the shade developed and if any
further addition is required; pressure gauge helps to maintain pressure.

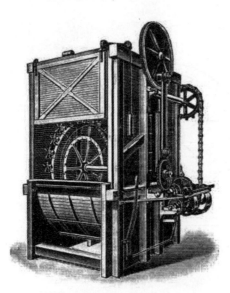

31.6 Klauder-Weldon machine. [courtesy: Gutenberg]

An addition tank accepts the dye liquor for dyeing as well as any further addition if to be made, which is pumped in and sent through the tubes to the packages by the rotary pump. The expansion tank accommodates any

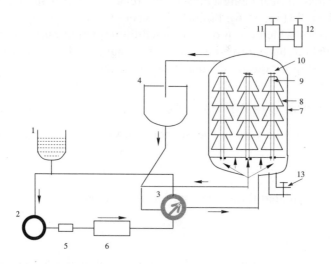

(1) Dye addition tank, (2) Rotary pump, (3) Reversal arrangement, (4) Expansion tank, (5) Filter, (6) Heat Exchanger, (7) Dyeing vessel, (8) Package, (9) Perforated frame/carrier, (10) Screw, (11) Pressure gauge, (12) Sampling unit and (13) Discharge valve. Other parts are safety valve, thermometer, liquor level gauge, time and thermostatic control.

31.7 Cross-sectional view of a package dyeing machine.

excess liquor due to expansion of the same in autoclave while a heat exchanger regulates heating or cooling of liquor and a filter cleans the liquor (Fig. 31.8).

31.8 Package dyeing machine. [courtesy: Wooyangmc]

Working

These machines are extensively used for dyeing yarn in various forms of package, i.e. cheese, cone, beam, etc. (Tsui, 2000). Yarn packages are placed on perforated steel spindles and tightened by screws. Pre-dissolved dye liquor when added to the addition tank, is pumped in by the rotary pump and sent to the yarn packages for dyeing. It is essential that dye must remain in well dispersed state. Dyeing takes place continuously with automatic control of reversal flow i.e. from out-in and in-out of package. At the end, sampling unit is opened to take out dyed sample for inspection. Any excess liquor coming out from the autoclave due to expansion at high temperature and pressure flows to expansion tank which is again pumped in by the motor (Burley et al, 1985).

The machine is popular due to little handling of yarn with practically no movement and friction causing reduced number of broken fibres, filaments and waste. However, problem of filtering out of dye particles exists and can not be rectified so easily due to compactness of packages. To ensure level dyeing, selected dye must have good solubility with lesser strike rate (Sharma, 1977).

Some precautionary measures are necessary for better results: (i) packages must be built up with a rapid cross wind and not with a parallel one as the latter is extremely densed and difficult to penetrate by the dye;

(ii) the packages must not have wound too hard to cause difficulty in dye penetration or too loose causing channelling of liquor forming uneven shade. Even, the pump must give a good flow of dye liquor. Automation has been introduced in modern package dyeing machines engaging a robot for various works (Mehlhos, 1986).

31.4   Fabric dyeing machines

31.4.1 Jigger

Working principle

Movement of fabric in open width through dye liquor.

Construction

Jiggers are very popular because of multi-purpose use of this machine starting from dyeing to desizing, scouring, bleaching, etc. The V-shaped stainless steel made dye vessel is fitted with two rubber coated metallic batch rollers of around 6? diameter, viz. take up and let-off rollers at the either end at the top as well as three guide rollers; all these rollers run across the whole width of the machine (Fig. 31.9). Water and steam supplies are also fitted to the dye bath at the bottom and well below the guide rollers to avoid fabric damage (Shenai, 1993; Trotman, 1994). The take up and let-off rollers are fitted with brakes to control tension on fabric for smooth transfer from one to other roller.

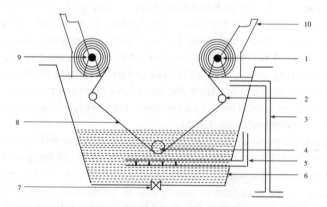

(1) Take-up roller with brake, (2) Guide roller, (3) Water supply, (4) Centre roller, (5) Steam supply, (6) Dye vessel, (7) Exhaust pipe, (8) Fabric, (9) Let-off roller with brake and (10) Final roll up bracket. In addition to these, two brakes are fitted at one extension of take up and let off rollers.

31.9 Cross-sectional view of a Jigger.

After each cycle, i.e. one complete passage of fabric through dye solution, frequently referred as 'one turn' or 'one end', which takes around 15 min, the fabric is wounded up on the second roll and again moves back to the first one being controlled by hand-lever or auto-reverse mechanism (intermittent reversal system without disturbing the fabric) in the machine. On requirement, after a pre-determined number of turns, dye or other chemicals may be added or the bath may be drained for after-treatment of the dyed textile. The exhaust valve helps to discharge the bath for fresh recharge. The surface speed of let off and take up rollers must be same at different rpm to facilitate smooth loading and releasing of fabric. Roll up brackets are used to wound up the fabric at the end of dyeing or to place padded fabric to load in jigger (Fig. 31.10).

31.10 Jigger. [courtesy: Tradeindia]

Working

During dyeing, a few yards on both sides remain out of reach to the dye liquor and so the leading as well as tailing ends of the fabric are stitched with two extra pieces of fabric (known as chindi, may be grey or pre-dyed one depending on hue and shade to be produced), followed by passing of the leading end over first batch roller to the second one via guide rollers to load the fabric wetting with water. The tailing end is placed tightly on the opposite roller. Loading is done at uniform tension and in open width without any fold or crease marks. Dye solution is prepared in the jigger trough at specific temperature; brakes fitted on the rollers are adjusted (the tension on take up roller must be on little higher side while that on let-off roller on lower one for smooth transfer of fabric at same surface speed) and the fabric is allowed to move through the dye liquor for required but even number of turns. Though no rope or crease marks are formed as the fabric remains under tension in open width and a better levelled shade is developed due to exposure of each point of fabric against the dye liquor, problems are associated with this machine too.

As diameter of two batch rolls are not same at any moment, there is tension on the fabric causing damage – only coarse woven fabrics are suitable for dyeing in this machine and knitted / fine woven fabrics are not preferred; stopping of machine during sampling or due to power cut at the initial stages of dyeing causes the lower end to be dyed with deeper shade; if batching is not with proper tension, a dark selvedge is produced. The last problem often encountered is variation in distribution of heat throughout bath causing variation in temperature at different points of fabric. Jigger is very popular in dyeing industries, due to which enormous developments were made on the design as well as its working (Anon, 2000; Hilden, 1996; Ceolotto, 1989, 1990; Kusters, 1992; Norton, 1985); even models are in market based on computer programming (Fig. 31.11). Even design was modified for effective removal of oil stains (Zacharia et al, 1986), models for large batch dyeing (Anon, 1980) and dyeing of very sophisticated silk fabrics (Henriksen, 1989). Liquor to be used in jigger to obtain levelled shades (Chakraborty, 1991), technique to avoid oxidation stains (Gokhale and Patel, 1988) and new vacuum technology was also discussed (Anon, 1997). To dye bigger batches, jumbo jiggers have been in action which can process at least 4–5 times higher capacity over that of a conventional jigger (Fig. 31.12).

31.11 Computerised Jigger. [courtesy: Chinasaga]

31.12 Jumbo Jigger. [courtesy: Maebe]

31.4.2 Automatic tensionless jigger

Principle

Movement of the fabric through stationary dye liquor.

Construction

These jiggers are made of stainless steel and in houses light weight rollers running on ball bearing (Fig. 31.13). Both the take up and let-off rollers are simultaneously driven through different control mechanisms with adjustable speed so that the tension on the cloth can be accurately controlled; a differential driving mechanism controls surface speed of the two rollers such that the cloth passes through dye bath at uniform speed regardless of the varying diameter of the roll of cloth. The jigger may set to reverse and stop automatically after passing through required number of turns. The cloth expanders are fitted to keep the fabric free from crease marks and at full width so that it is dyed more uniformly. Steam supply is fitted at the top and bottom parts of the dye vessel for effective control of temperature in bath, even to work at high temperature (Fig. 31.14). Squeezing rollers are fitted over the let-off and take up rollers to squeeze out excess liquor from fabric during dyeing as these machines practically works on low-liquor principle. This helps to maintain a constant liquor ratio in bath.

Working

The outer cover of the machine is opened and the fabric is loaded on take up roller in open width in crease free state. Rest is as that occurs during dyeing with conventional jiggers. In some modifications, a 'completely

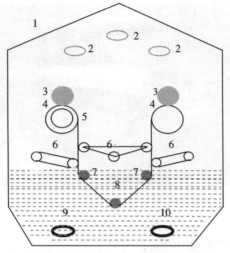

(1) Dye bath, (2) Closed steam pipe, (3) Squeezing rollers,
(4) Batch rollers, (5) Fabric, (6) Cloth expanders, (7) Guide rollers,
(8) Central guide roller, (9) Heating arrangement and
(10) Split pipe. Other parts are time, temperature and
automatic control.

31.13 Automatic tensionless jigger.

31.14 High temperature high pressure jigger. [courtesy: Daegu]

immersed jigger' has been specifically designed for dyeing of cotton with
sulphur and vat dyes to avoid premature oxidation during dyeing; also a
'completely enclosed jigger' prevents cooling during the dyeing and
distributes heat uniformly.

31.4.3 Winch

Principle

Movement of fabric rope through stationary dye liquor.

Construction

A winch consists of a trough containing the dye liquor across and over which is fitted a circular or elliptical roller known as driving reel (Shenai, 1993). Fabric is stitched in the shape of endless ropes of equal length; a few such ropes are passed over the driving reel and as the later rotates, it carries forward the ropes. Up to twelve or more pieces can be dyed side by side simultaneously being kept apart by driving rods (known as fingers) projected from the driving reel. A compartment known as salting box or addition tank is separated by a perforated baffle plate from main dyebath (Fig. 31.15). The outer look of a commercial winch is shown in Fig. 31.16. The addition tank is inaccessible to the cloth and houses water and steam pipes as well as provides facility for addition of dye and chemicals (Schuierer, 1981; Nezu, 1981). A guide roller at the back of the driving reel picks up the ropes bunched at the bottom of bath pushed earlier by the reel. The elliptical driving reel folds the rope in wide layers to make these free from random

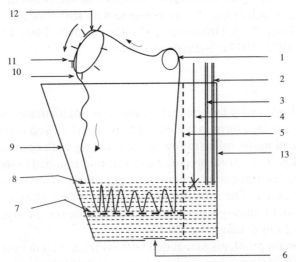

(1) Guide roller, (2) Steam supply, (3) Water supply, (4) Impeller, (5) Buffle plate, (6) Discharge valve, (7) False bottom, (8) Dye solution, (9) Dye vessel, (10) Fabric, (11) Finger, (12) Elliptical roller or Driving reel and (13) Addition tank

31.15 Cross-sectional view of a winch.

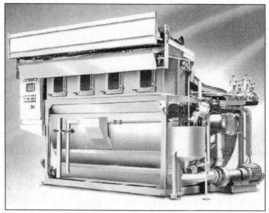

31.16 Winch. [Courtesy: Daegu]

creases as these falls into bath. A false bottom at the lower part of the machine protects the ropes from coming in contact with hot surface directly as well as traps giant impurities like broken ends and fibres. Short liquor winches are very important keeping in view requirement of lesser steam and water with a consequence of lesser waste-water generation (Anon, 1979). The driving reel is made elliptical in shape as it favours plaiting of fabric on the bottom plate to avoid entanglement which could not have been possible with a circular reel. A few other modifications also were introduced due to which modern winch machines can compete with soft-flow and jet dyeing machines (Minnet, 1976; Gruninger, 1975; Carpet Dyer, 1980; Thies, 1985; Anon, 1969, 1972, 1979a; Netherwood, 1975).

Working

Fabrics which are easily deformable – especially sophisticated woven and knits are dyed in rope form in winch. The fabric to be dyed is passed over driving reel and guide roller followed by stitching both ends to give it a shape of endless rope. Well-prepared dye solution is added to the addition tank and heated up if required. Liquor ratio is suitably maintained followed by dyeing of fabric. The impeller in the addition tank allows controlled passing of liquor through perforated baffle plate in the dyebath and maintains uniform dyeing (Fig. 31.17).

Some common problems associated with winch are (i) in case of power cut or during initial stages of loading, a large part of fabric is accumulated at the bottom causing shade variation, (ii) longer rope or uneven plaiting may cause entanglement at the bottom, (iii) unnecessary wastage of heat, (iv) temperature of dye solution at various points vary due to radiation of heat and (v) pre-treatment of fabric must be adequate otherwise poor absorbency causes to dye different layers of rope in differential shades.

31.17 Winch. (in operation)

31.4.4 HTHP Winch

Working Principle

Circulation of dye solution and movement of material; dyeing is done at or above atmospheric pressure.

Construction

This machine has a few common parts like those possessed by a conventional winch, e.g. driving reel, guide roller; the only difference is that there is a circular enclosed chamber to house other parts enabling it to work at and above atmospheric pressure as well as temperature extending use of this machine in dyeing of man made fibre fabrics too. The machine is widely used for dyeing of knitted and textured fabrics (Fig. 31.18).

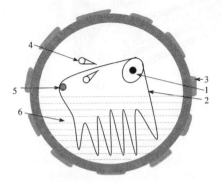

(1) Elliptical roller, (2) Fabric, (3) Dye vessel, (4) Rail guides or fingers, (5) Guide roller and (6) Dye solution

31.18 Cross-sectional view of a HTHP winch.

Working

Large pieces of fabrics are stitched together at the ends to form endless ropes and are drawn continuously through dye liquor by passing over the rotating elliptical roller (Trotman, 1994). Internal sprays by mean of jet help in uniformity of dyeing without formation of crease marks. However, movement of fabric is not highly efficient and bunches up frequently at the bottom causing uneven packing and variation in shade.

31.5   High temperature high pressure dyeing machines

31.5.1 Jet dyeing machine

Principle

Fabric in rope form is carried forward by circulation of dye liquor jet.
    Status of fabric: Fully tension less

Construction

A jet dyeing machine consists of numerous parts, varies from one model to another in look and is complicated in construction. Various types of partially and fully flooded jet dyeing machines are popular worldwide and have been well documented (White, 1998; Paterson, 1973). Look of a typical jet dyeing machine is shown in Fig. 31.19.

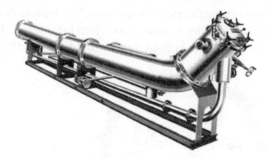

31.19 Jet dyeing machine. [courtesy: Hisaka]

A loading port at the foremost part accompanies the lid; the latter is opened up to load pre-measured length of fabric in each cycle and loading is completed by stitching both the ends. The machine has a large dyeing tank made of stainless steel with a supporting tube to carry the fabric forward in a continuous manner. The dyeing tank is also connected to several other connecting tubes to circulate dye liquor through heat exchanger, filter and finally to reach to the jet continuously. A heat exchanger at the path of circulation heats up or cools down the liquor with supply of steam or cold water through narrow pipes inside it. A filter eliminates impurities in the liquor coming from fabric and dye solution in the form of broken threads or insoluble chemicals. A driving reel placed in front of the loading port is a highly polished small roller used to facilitate smooth movement of fabric without touching the main dyeing vessel which otherwise causes degradation as well as damage of fabric due to high temperature and friction, respectively. A small view port helps to visualize the smoothness of the process, i.e. if the fabric has been entangled or to locate a stitch mark from where a sample is to be removed for shade matching. The schematic diagram of a jet dyeing machine showing various parts is shown in Fig. 31.20 (Star, 1985; Anon, 1979b).

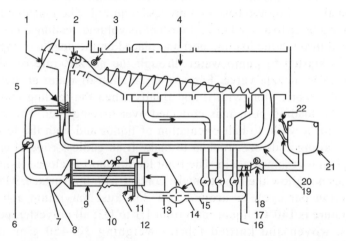

(1) Front door, (2) Driving reel, (3) Window, (4) Dyeing tank, (5) Nozzle, (6) Nozzle-valve, (7) Fabric, (8) Steam/cold water, (9) Heat exchanger, (10) Pressure gauge, (11) Drain, (12) Cooling water, (13) Filter, (14) Vacuum breaker, (15) Main pump, (16) Feed valve, (17) Non-return valve, (18) Booster pump, (19) Main tube, (20) Feedback valve, (21) Service tank, (22) Water filling valve

31.20 Various parts of a Jet dyeing machine.
[courtesy: Star Industrial & Textile Enterprises Pvt Ltd]

The unique part of a jet dyeing machine is the nozzle, which is a combination of a few conical identical metallic containers open at both the ends of two different diameters (like a frustum). Dye liquor coming from service tank or the dye vessel is pumped through the upper open side of nozzle with larger diameter to discharge it through the narrow open end causing tremendous increase in speed of flow, forcing the fabric to move along with circulation of liquor. Absence of tension on fabric ensures quality without distortion. Rate of flow of liquor through nozzle is controlled by adjusting the nozzle valve which in turn helps to control movement of fabric. Synchronized adaptive control technology with dual-nozzle O-type jet-flow dyeing machines is superior in this context (Cheng, 2003). In newer versions, translational motion technology (Melocchi and Francesco, 2003), design to control dye adsorption (Comelo et al, 2003) and rapid dyeing technology (Hinge, 1981) have been introduced too.

Working

Depending on percent shade to be produced, various programmed micro-processors with respective dyeing cycle are fitted in the machine at the start, which controls most of the operations automatically. There is scope of manual handling too based on process required. The machine is filled up with water up to around three-fourth of its volume; leading end of fabric is pushed through the loading port and placed on the nozzle trumpet. The motor is started to pump water through the nozzle, rate of which is controlled by nozzle valve. Forward circulation of water at high speed carries the fabric alongwith it and finally when the leading end comes before the loading port, it is collected over driving reel. The motor is stopped for a while to halt circulation of liquor and both the leading as well as tailing ends are stitched to make it an endless rope after which circulation is resumed again. The front loading port is closed tightly. The reel does not allow the rope to make contact with hot surface. Maximum production per cycle is around 200–250 kgs; maximum achievable temperature is 140°C, liquor ratio from 1:1 to 1:3; all polyester, polyester blends, woven and knitted fabrics weighing 30–450 g/m^2 may be successfully dyed in jet dyeing machine.

Closing of the discharge valve is ensured after which, the dye liquor, prepared in addition tank is pumped inside the autoclave (main dye vessel) by main pump through heat exchanger and filter, the liquor jets are accelerated by acceleration device (nozzle valve) and send the dye solution to the dye vessel. The liquor is heated up to 90°C quickly followed by heating at 1°C/min to 120°C and finally to 130°C for 1–2 h for dyeing of polyester or as per the dyeing cycle desired. A part of liquor comes down from autoclave through side tubes to the pump, which in turn pushes the liquor through the heat exchanger, filter and through the nozzle valve to the nozzle. At the end, the machine is cooled down by passing the exhausted liquor through cold water tubes in heat exchanger. Once cooled down to 80°C, the bath is drained out and soaped or reduction cleared in a fresh bath.

The machine is highly suitable for polyester and its blends, viscose rayon, wool and cotton woven, knitted and textured fabrics. A few popular makes are 'Scholl' (Germany), 'Gaston county' (USA), 'Hisaka' (Japan), 'Platt and long close' (UK), 'Staffi' (India), etc.

A jet dyeing machine offers some special advantages, e.g. the fabric is transported entirely by dye liquor jets without mechanical device causing no alteration in handle and freedom from distortion as no tension prevails on the fabric rope and no friction is there between the rope and body or any part of machine. The rope also remains completely immersed in the liquor. Other advantages are (i) high fabric transport speed (~300 m/min) by adjusting nozzle valve to cause level dyeing, (ii) large quantity of liquor flow, (iii) high heating and cooling capacity (2–3°C and 5°C, respectively) shortening time of dyeing cycle, working at high pressure (<4.8 kg cm^{-2}) and (iv) chances of foaming is less due to absence of air in the machine. Main drawback of this machine lies in forming typical crease marks which are developed during high temperature dyeing making difficult to remove in finishing. If carriers are used for levelling of shades, which is a common practice, the carrier gets evaporated-off from the bath, gets condensed on the top of the machine interior and drops down on fabric to form spots. One typical problem in some models is positioning of dye vessel and carrier pipe; main dye vessel must be on the top side as shown in fog. 31.20, otherwise in case of sudden power cut, the liquor from carrier pipe (if placed at the top) comes down to dye vessel at the bottom and the fabric under dyeing get adhered to hot metal surface. Other aspects of jet dyeing were paid due attention in several literatures (Holme, 1998; Henriksen, 1988; Von der Eltz, 1987; Naik and Gandhi, 1987; Shah and Danave, 1984; Anon, 1979c; Roaches, 1993). Latest jet dyeing machines are even capable of dyeing heavier fabrics like nylon carpets (Kerr, 1989).

31.5.2 Multi-nozzle soft flow dyeing machine

This machine works under the same principle as that for a jet dyeing machine. A multi-nozzle soft flow dyeing machine offers shorter dyeing cycle with higher productivity at lower liquor ratio (Fig. 31.21). It is a modified version of jet dyeing machine in which multi-nozzle dyeing technology has been introduced and works under both the features of soft flow as well as jet dyeing machine. The main advantage of this development is to increase the machine versatility to dye all types of fabrics. Mainly two types of nozzles work for liquor flow; the soft nozzles are of bigger diameter and are controlled by soft-flow valve while the main nozzle opening is adjusted by its own nozzle valve that in turn controls flow and pressure of liquor based on fabric quality. When required, the pressure and flow rate of liquor is reduced by more opening of all valves causing a soft or gentle flow of liquor through all the nozzles (Fig. 31.22).

31.21 Multi-nozzle soft-flow dyeing machine. [courtesy: Devrekha]

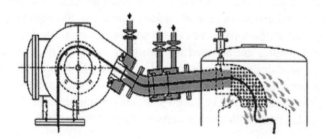

(1) Counter-flow reverse nozzle, (2) Fabric, (3) Fabric drive reel
(4) Main nozzle, (5) Nozzles

31.22 Cross-sectional view of nozzle action area on fabric.
[courtesy: Devrekha]

To run it as jet dyeing machine, the soft-nozzle control valves are closed; the main nozzle
valve is swung into action to decrease the nozzle gap that increases the pressure and reduces rate of water flow with high speed. For dyeing polyester, the machine is run as jet dyeing machine and to dye cotton and knitted fabrics, it is run as soft flow machine.

Chances of pilling on textured and lightly twisted staple yarn made fabrics is greatly reduced due to liquor flow at reduced pressure. Because of a shorter path of fabric, 10–300 kg of fabric can be dyed per cycle.

31.5.3 HTHP beam dyeing machine

Principle

Circulation of dye liquor through stationary beam of fabric in open width.

Construction

The machine is used to dye fabric in open width. The fabric to be dyed is batched on perforated stainless steel hollow perforated polished beams with the help of a batching machine with uniform tension and placed in the beam dyeing machine for dyeing. The perforated batch rollers must be wider than that of fabric to be dyed to avoid by-passing of dye solution (Hilden, 1988).

There is a rail inside the autoclave upon which the batched rollers are placed. The opening port with lid helps to open or close the autoclave; sampling unit and pressure gauges are on the lid for inspection of dyed samples and steam pressure inside machine, respectively (Figs. 31.23 and 31.24). The opening at the leading part of the beam if tightly fitted with conical fitting when placed in machine to facilitate in-out flow of dye solution, while the opening ends of both beams are tightly joined at the middle. A heat exchanger and filter are at the path of dye solution through supply pipes. A dye solution preparation tank and an addition or over-flow expansion tank are connected to a centrifugal pump; the latter changes the direction of pumping the liquor through in-out and out-in sequence for 5 and 3 min respectively to ensure level dyeing. The actual construction of a HTHP beam dyeing machine is illustrated in Fig. 31.25 which is practically too complicated in construction based on automation in working.

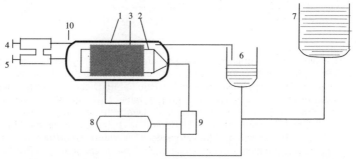

(1) Autoclave, (2) Perforated stainless steel beam, (3) Fabric, (4) Sampling unit, (5) Pressure gauge, (6) Over-flow Expansion/add dye vessel, (7) Preparation tank, (8) Heat Exchanger and (9) Centrifugal Pump and (10) Opening port with lid

31.23 Simplified schematic diagram of HTHP Beam dyeing machine.

31.24 HTHP Beam dyeing machine. [courtesy: Texfab]

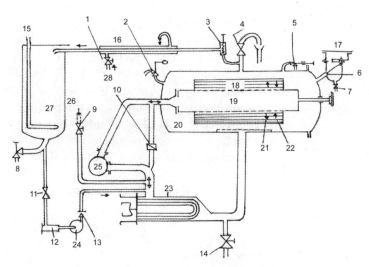

(1) Cooling water admission valve, (2) Safety valve, (3) Pressure regulation valve, (4) Scouring over-flow valve, (5) Cover safety valve, (6) Cut-off valve for sampling sluice, (7) Discharge valve for sampling sluice, (8) Discharge valve reserve tank, (9) Water admission valve, (10) Throttle valve, (11) Cut-off valve reserve tank, (12) Filter, (13) Non-return valve, (14) Dyebath discharge valve, (15) Water admission for reserve tank, (16) Cooler, (17) Sampling device, (18) Batch of material, (19) Dye beam, (20) Autoclave, (21) Liquor circulation: Inside–outside, (22) Liquor circulation: outside–inside, (23) Heat exchanger, (24) Pressure pump, (25) Circulation pump, (26) Water admission, (27) Reservoir and (28) Cooling water

31.25 Cross-sectional view of a HTHP Beam dyeing machine.

Working

Cloth is batched on the perforated hollow beams at constant tension in open width, is pushed horizontally in the autoclave after which the lid is tightened. A white piece of sample is also placed in the sampling unit. Dye liquor is prepared in dye tank and is pumped inside the machine by the centrifugal pump, through the heat exchanger, filter and finally through the beam; in-out as well as out-in cycles periodically (generally 5 min and 3 min, respectively). The dye liquor is heated up during dyeing passing steam or cooled down in post-dyeing phase with cold water through the heat exchanger. During dyeing, the temperature of dye liquor is raised to $130\pm2°C$ for polyester, dyeing is carried out for required time depending on depth of shade and on completion of the cycle, the bath is cooled down to 80°C, discharged followed by reduction clearing for deeper shades and soaping. Latest beam dyeing machines have beams of larger diameter affording a bigger area of perforated wall inside the beam through which the dye liquor can be fed to the fabric. A good result of this development is that larger and heavier load can be dyed with better level shades. However, beam dyeing machines relatively consume more water and energy both and increases cost of dyeings (Kretschmer, 1981; Naik and Gandhi, 1986).

As the fabric is being dyed in open width, rope and crease mark are not developed; batching under constant tension does not permit dyeing of knitted fabrics. Suitings are specially dyed in this machine to avoid formation of crease marks. However, dyeing of polyester in a HTHP beam dyeing machine is a relatively complicated process as it depends on some pre-requirements and precautions, which are very important to know and so are detailed below (Joshi and Shah, 1979).

(i) *Overlapping of fabric*. Fabric is to be batched on beams a little beyond perforation marks on both sides to prevent by-passing of dye liquor; batching too beyond the perforations will leave selvedges undyed. Even selvedges, being generally thicker causes uneven batching and consequently variation in shade as dye liquor under pressure always finds a path of least resistance. Extent of overlapping depends on quality and thickness of fabric as well as thickness of batch. A plain woven polyester fabric requires an overlap of 3–5 cm beyond perforations for quality shades.

(ii) *Perforation marks*. During dyeing, the in-out circulation of dye liquor causes deposition of impurities in dye and agglomers on the first few inner layers producing perforation marks equal to the size with those on beams; may be due to batching under variable tensions too. The problem can be handled by wrapping the beams first with 8–10 layers of jute or some other thick open lapping fabric followed by fabric under dyeing; this also prevents slippage of fabric during batching.

(iii) *Tailing effect*. This arises when only in-out dyeing cycle is taken into action and the out-in circulation does not work. The inner layers are dyed with deeper shades while the outer layers get lesser dye and a lighter shade resulting in a consistent and progressive lightness in shade visualized throughout length of fabric. Both the flow reversals (out-in and in-out) must be executed up to 120°C, till surface deposition of dye gets completed and after this only in-out flow is sufficient till completion of dyeing to bypass the problem.

(iv) *Unlevel dyeing*. Formation of levelled shade depends on (a) rate of rise in temperature, (b) ability to reach to optimum temperature, (c) right flow reversals, (d) compatibility of dyes in mixture and (e) selection as well as concentration of auxiliary chemicals. Disperse dyes show affinity for polyester beyond its T_g, (~85–90°C), when dye starts to enter into solution phase slowly keeping a balance with dye adsorbed on fibre surface; 1°C/min rise in temperature upto 120°C may be the right choice for even deposition and distribution of dye. In tropical areas, especially during chilling winter, pressure of steam depends on the distance to be covered by it from boiler and quality of insulation on the supply pipes. Ironically, a huge fall in pressure is observed which neither allow raising temperature uniformly nor up to 130±2°C causing insufficient diffusion due to inadequate opening of fibre producing lighter shades with poor fastness. In addition to this, compatibility is a vital factor in selection of dye in mixture to result maximum exhaustion with right shade, hue and tone. Dispersing agents, if in excess, enhances solubility of disperse dye causing poor exhaustion and hence a lighter shade.

(v) *Moiré effect*. A wavy coloured effect appears on fabric which is primarily due to either improper heat setting before dyeing or severe out-in flow of liquor. Improper heat setting causes shrinkage of fabric during dyeing; setting at 190–210°C for 40–45 s with 7–8% overfeed or alkaline treatment of fabric before dyeing at room temperature reduces this problem. If the out-in flow of liquor is only taken into action, the liquor can not penetrate the beam efficiently rather spreads over a few outer layers causing the problem. Inconsistency in tension applied on fabric during preparation of beam is another reason. The fault once appeared after dyeing may be corrected by NaOH treatment or redyeing the fabric with 3–4 g/l carrier along with 10% of dye or dye combinations used earlier at 135°C for 30 min.

(vi) *Recrystallisation of dye*. Dispersing agents promote consistent dispersion of dye for uniform absorption on polyester. At the end of dyeing, especially in dyeing deep shades, a small part of dye is left in bath and recrystallises during cooling from 130–80°C due to reduction in extent of solubility or inadequate dispersing agent in bath causing poor rubbing fastness. The exhausted bath may be discharged at as higher temperature as possible or excess dispersing agent may be added at the end of dyeing and before cooling to escape the problem or by imparting reduction clearing to the dyed fabric with $Na_2S_2O_4$ and NaOH (3 g/l each) at 50–60°C for 20–30 min; sequestering agent (1–2 g/l) may be used if water contain minerals otherwise dye may get precipitated. The pH of fabric must be either faintly acidic or neutral; presence of alkali precipitates dye.

(vii) *Agglomeration of dye.* Interaction among chemicals impairs dispersion stability of dyes, reduces electrostatic repellency and increases collision by particles causing agglomeration of dye. Dye manufacturers introduce required dispersing agent in dye before marketing, necessitating only dye and acetic acid for quality dyeing. Agglomeration appears if carriers, sequestering and levelling agents are added in bath and can be handled by adding these auxiliaries, if essential, only when exhaustion ceases and majority of dye molecules occupy fibre surface.

One development of beam dyeing machine is to use beams of larger diameter which facilitates dyeing of larger and heavier batch with levelled shades.

31.6   Continuous dyeing of textiles

Textiles in yarn or fabric form are continuously dyed for a longer length which is achieved by padding the yarn or fabric with required expression in a dispersion of dye solution consisting of other chemicals free from interaction among themselves. Continuous dyeing of yarn is invariably followed in dyeing denim warp sheet in open, or rope form (see 9.2). Fabric padding follows drying, steaming or curing and other after-treatments; a drier (hot flue, float drier, stenter or infra red dryer), steamer and other machines are attached in tandem for the purpose. Selection of padding mangle depends on thickness of fabric, dye system and extent of penetration of the liquor inside the fabric; more the number of dips and nips, the better the penetration, more dye will be required for a given shade.

Drying of padded fabrics is essential for dyes possessing no affinity for fibre using Steaming may be at normal pressure with saturated or superheated steam for solubilisation and transportation of dye at the interior.

31.6.1 Padding mangle

Continuous dyeing of textiles is carried out by padding fabric in a liquor constantly supplied to a small trough from a reservoir; the fabric is dipped in the liquor with the help of guide rollers, squeezed (nipping) by padding rollers to achieve desired expression and after-treated in some other machine for fixation of dye (Fig. 31.26). The grey and black coloured paths of fabric are options to use the mangle as 1-dip 1-nip and 2-dip 2-nip, respectively (Fig. 31.27).

Nomenclature of a padding mangle is based on how many times the fabric is dipped (treated in dye solution) and nipped (squeezed); more the dipping and nipping, better is the penetration of dye (Fig. 31.28). The

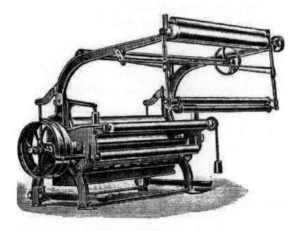

31.26 Commercial padding mangle. [courtesy: Gutenburg]

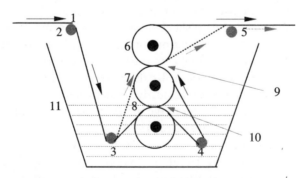

(1) Fabric, (2–5) Guide rollers, (6–8) Padding rollers, (9 & 10) Nips
(11) Padding trough and (12) Dye liquor

31.27 Cross-sectional view of a padding mangle.

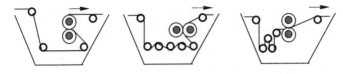

1-dip and 1-nip padding mangle

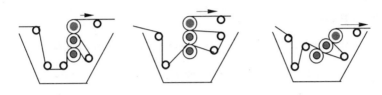

2-dip and 2-nip padding mangle

31.28 Various types of padding mangles.

latter is important when heavy fabrics are dyed through padding, especially with a dye having practically negligible affinity for fibre or fibre having poor absorbency.

Uniform distribution of dye is achieved through efficient dispersion of dye in the padding liquor; otherwise unlevelled shade is bound to develop as padding is followed by drying and fixation with no scope for further distribution through migration of dye. Pressure in the nip must be uniform across the width and consistent throughout the length of fabric as the 'pick up' or 'percent expression' is based on this; pick up of liquor is also governed by type of cloth, temperature, viscosity of the solution and the time of nipping.

Padding mangles consist of a small v-shaped stainless steel trough having excess length to accommodate wider fabrics. The fabric to be padded is passed over guide rollers immersed in liquor to facilitate picking up liquor while some guide rollers above the liquor help the fabric to enter or come out of mangle.

Padding rollers are covered with hard rubber or polymer and are placed over one another to squeeze out excess liquor and to control percent expression on fabric; metal rollers without compressible cover may crush fabric. The pressure on padding rollers is controlled by pneumatic system attached to an air compressor. A dye preparation tank is attached to the trough for continuous supply of dye liquor. Alternately, dye solution prepared outside is manually added to trough for relatively shorter length of fabric.

In semi-continuous 'pad roll' technique, padded fabric is batched, covered with a polyethylene sheet and the batch is rotated slowly for desired time to avoid channelling of liquor followed by fixation. Heavy fabrics are padded, dried (optional), dyed and developed in jigger with excess padding liquor followed by washing.

Latest developments in continuous dyeing include continuous dyeing ranges (CDR) which consists of several padding, dyeing and treatment chambers depending on a specific dye–fibre system.

31.7   References

ANON (1969) 'High-temperature winch beck with fluid expander', *Int. Textile Bulletin*, **3**, 214.

ANON (1972) 'Production increase in winch dyeing', *Textil Praxis Int.*, **27**, 5, 303–304.

ANON (1979) 'Short liquor ratio winch: system, application and costs comparison', *Int. Dyer*, **162**, 9, 441–443.

ANON (1979a) 'New generation of winch dyeing machines', *Melliand Textilber*, **60**, 3, 256–258.

ANON (1979b) 'Stafi jet dyeing machine', *Indian Textile J.*, **89**, 6, March, 57–58.

ANON (1979c) 'New high-temperature jet dyeing machine', *Melliand Textilber.*, **60**, 9, 763.

ANON, (1980) 'Energy conservation stressed for new jumbo dye jigger', *Textile Month*, May, 66.

ANON (1997) 'Jigger dyeing with vacuum technology', *Tinctoria*, **94**, 1, 70–73.

ANON (2000) 'Modern dyeing jigger for small lots', *Melliand Textilber.*, **81**, 9, E 183-E 184 + 761.

BHAGWAT R S (1999) *Handbook of Textile Processing Machinery*, Mumbai, Colour publications private ltd.

BURLEY R, WAI P C AND McGUIRE G (1985) 'Numerical simulation of an axial flow package dyeing machine', *Applied Math. Modelling*, **9**, February, 33–39.

CARPET DYER INTERNATIONAL (1980) 'New developments in winch dyeing', *Int. Dyer*, **164**, 3, 22nd August, 1–2, 8.

CEOLOTTO I (1989) 'Jigger: development of batch treatment', *Tinctoria*, **86**, 9, 96–99.

CEOLOTTO I (1990) 'Evolution of the modern dye jigger', *Textile Month*, March, 39.

CHAKRABORTY J N (1991) 'Exact volume of water to be used in jigger dyeing', *Textile Dyer Print.*, **24**, 23, 13th November, 31–33.

CHENG Y C (2003) 'Synchronized adaptive control technology with dual-nozzle O-type jet-flow dyeing machines', *AATCC Rev.*, **3**, 7, 8–12.

COMELO M F, NOBBS J H AND CARBONELL J (2003) 'Control of dye adsorption and color reproducibility on a jet dyeing machine', *AATCC Rev.*, **3**, 9, 18–22.

GOKHALE S V AND PATEL D L (1988) 'Oxidation stains in jigger dyeing: can they be avoided', *J. Textile Asso. (India)*, **49**, 4, July, 133–135.

GRUNINGER G (1975) 'Past and future of winch dyeing machines in piece dyeing-I, II', *Textilbetrieb*, **93**, 8, 40–46; .**93**, 9/10, 74–79.

HENRIKSEN A V (1988) 'Practical experiences with a jet dyeing machine', *Melliand Textilber.*, **69**, 4, April, 281.

HENRIKSEN A V (1989) 'New jigger for silk dyeing', *Textile Horizons*, **9**, 4, April, 21.

HILDEN J (1988) 'The beam dyeing machine–an old dyeing machine principle keeps up to date', *Int. Textile Bulletin*, **34**, 2, 53–59.

HILDEN J (1996) 'The jigger: a highly topical dyeing machine enjoying a renaissance', *Int. Textile Bulletin*, **42**, 2, 37–40.

HINGE S T (1981) 'Evolution of rapid jet dyeing machine-process and design, *Colourage suppl.*, **28**, 14, 2nd July, 22–27.

HOLME I (1998) 'Piece dyeing advantages with softflow machines', *Wool Record*, **157**, 3648, 55.

JOSHI B C AND SHAH J K (1979) 'Correction of faulty dyeings of polyester fabrics in high temperature beam dyeing machine, *Colourage*, **26**, 13, 21st June, 27–30.

KERR P L (1989) 'Jet dyeing machine for carpets', *Melliand Textilber.*, **70**, 11, 877.

KRETSCHMER A (1981) 'Crucial points of water and energy saving in beam machine dyeing', *Textil Praxis Int.*, **36**, 1, 30–36.

KUSTERS MASCHINENFABRIK GMBH AND CO K EDUARD (1992) 'Universal jigger for pretreatment dyeing and special treatments', *Melliand Textilber.*, **73**, 10, 840.

MEHLHOS L (1986) 'Automation of a package dyeing machine with robots', *Deutscher Farber Kalender*-1987, 91: (ISBN-3-87150-244-8) 50–55.

MELOCCHI L AND FRANCESCO M D (2003) 'Translational motion technology in jet dyeing machines', *AATCC Rev.*, **3**, 9, 36–39.

MINNET S (1976) 'Dyeing of cotton blend fabrics: Jet or winch, *Textilveredlung*, **11**, 2, 57–60.

NAIK D S AND GANDHI R S (1986) 'Economics of beam dyeing machine', *Colourage*, **33**, 20, 2nd October, 15–18.

NAIK D S AND GANDHI R S (1987) 'Economical jet dyeing machine, *Colourage*, **34**, 12, August, 24–25.

NETHERWOOD R (1975) 'Improved winch dyeing system', *Wool Record Survey: Dyeing and Finishing*, 7th March, 26–27.

NEZU O (1981) 'Energy saving by heat insulation method on winch dyeing machine, *J. Textile Machinery Soc. Japan*, **34**, 6, 284–286.

NORTON INTERNATIONAL SIR JAMES FARME (1985) 'Microprocessor-controlled dye jigger developed by UK machine builder', *Int. Dyer*, **170**, 2, 14.

PATERSON M (1973) 'Jet dyeing', *Rev. Prog. Color Related Topics*, **4**, 80–90.

ROACHES ENGINEERING LTD (1993) 'Jet dyeing machine', *High Perform. Textiles*, December, 9–10.

SCHUIERER M (1981) 'Rationalized winch beck dyeing', *Melliand Textilber.*, **62**, 4, 338–347.

SHAH S A AND DANAVE S M (1984) 'Foam control in jet dyeing', *Man made Textiles India*, **27**, 12, 615–616.

SHARMA D R (1977) 'Two new approaches for reducing the consumption of sodium hydrosulphite in the vat dyeing of cloth in continuous dyeing machine and vat dyeing of yarn in package dyeing machine', *Colourage*, **24**, 11, 26th May, 31–35.

SHENAI V A (1993) *Technology of Dyeing*, Mumbai, Sevak publications.

STAR (1985) '*Jet dyeing manual*', Mumbai, Star Industrial & Textile Enterprises Pvt Ltd.

THIES GMBH AND CO (1985) 'Replacement for the winch developed', *Textile Month*, February, 9.

TROTMAN E R (1994) *Dyeing and Chemical Technology of Textile Fibres*, 6th Edition (1st Indian Edition), Delhi, B I Publications Pvt Ltd.

TSUI W (2000) 'Progress in Fong's package dyeing machine', *Colourage*, **47**, 4, 78–83.

Von der Eltz H (1987) 'From short-liquor jet dyeing machines to the aerodynamic system', *Textiltechnik*, **37**, 12, 683–687.

White M (1998) 'Developments in jet dyeing', *Rev. Prog. Color Related Topics*, **28**, 80–94.

Zacharia J, Ray R D and Nirgude S K (1986) 'Modification of jigger for effective removal of oil stains', *BTRA Scan.*, **17**, 2, June, 15–16.

32

Waste-water problem in textile industry

Abstract: In exhaust dyeing, which is in fact the most popular method, dyebath is not exhausted completely; in most cases, only substantial exhaustion is achieved. This unexhausted dye alongwith auxiliary chemicals pollute natural water bodies, when discharged. Most of the dyes and chemicals are highly toxic, even a few are non-degradable. Few chemicals pollute air in shop-floor too. The nature of waste water generated from drainage of baths is typical in nature and varies from one dye–fibre system to another.

Keywords: Waste water, degradability, toxic, oxygen demand, harmful factor

32.1 Introduction

The rheology of waste water discharged by textile mills is highly complex in nature, sometimes difficult for a specialist to grasp and understand it. Out of 70% total water phase across the world, only a negligible part of it is in pure form (Park and Shore, 1984). Waste-water load is mainly generated by chemical, lather, colour and paint Industries, home launderings, acid rain, agriculture (fertilizer, hormones. pesticides), hardness, marine accidents and pathogens. Out of these, chemical processing industries, viz. chemical manufacturing, textile pre-treatment, dyeing, printing and leather processing units share the major load; which is generated due to use of harmful and toxic chemicals, excess of which are drained out after specific use. Textile chemical processing sectors are the leading consumer of water. Only processing of polyester and cotton necessitates 100–200 kg and 250–350 kgs of water respectively per kg of fabric and contribute 4–80% and 4–37% to total waste-water load respectively (Copper, 1978; USEPA, 1974).

Natural purification of a water body is influenced by micro-organisms, which being present in water bed digest impurities in presence of sunlight and dissolved oxygen to form CO_2 and water, called process of biodegradation, extent of which is denoted by BOD (biological oxygen demand) and is expressed in terms of 'milligrams of oxygen required for each l of waste-water' (mg O_2/l). Biodegradation is a slow natural process and indeed takes longer time for complete degradation. Some chemicals are so compact and toxic that these are not even biodegradable. Rate of degradation is influenced primarily by two natural factors, viz. sunlight and dissolved oxygen in water, while structure of chemical under degradation is of utmost importance. Structure of colour and chemicals should be preferably linear, without branch, organic, aliphatic and may have hydrophilic groups. Concentration of dissolved oxygen plays a crucial role as it controls rate of conversion of toxic products to CO_2 and water. Increase in temperature (T) decreases level of dissolved oxygen (DO) in water, i.e. $T \propto DO^{-1}$. In summer, rate of biodegradation will be suppressed due to lesser availability of oxygen and in winter, though dissolved oxygen will be more but rate of degradation will be slower due to fall in working temperature of micro-organisms, e.g. DO in water at 0°C is 14.6 mg/l but at 35°C is only 7 mg/l at atmospheric pressure. Presence of reducing chemicals in discharged water absorbs a part of dissolved oxygen retarding purification process. In adverse situations, micro-organisms get collapsed keeping the biodegradation at bay. Micro-organisms or pathogens are double activated over each 10°C rise in temperature. As a result, presence of oxygen, if in excess or is supplied externally, helps to enhance biodegradation. Increase in temperature too disturbs aquatics life and even kill microbes.

32.2   Implication of chemical processes on waste-water load

Starch removed in desizing has higher organic carbon content with consequent higher biological as well as chemical oxygen demand and not easily degradable. Scouring contributes residual starch, oils, grease, fats, dirts / dusts, minerals, impurities and excess unused alkali in sewage. Mercerisation discharges excess NaOH, which if not recovered, disturbs normal growth and nutrition of plants and trees when alkaline water is drawn from river for agriculture.

In dyeing, one of the main sources of water pollution is dyes and pigments of different classes. Drained out dyebaths consist of unused colour having compact structure and hazardous chemicals and are not so easily degradable. Exhaust dyeing creates more trouble as commencing of the succeeding step requires a vacant trough within shortest possible time. Exception is reactive dye which is self-hydrolysed in water body. Though chlorine, peroxide, ozone treatment and various other methods are effective in removing colour from waste water, sequence of operation are complicated, uneconomical and time consuming.

Typical problems are associated with use of $Na_2S_2O_4$ and NaOH in vat dyeing, Na_2S in sulphur dyeing, NaOH and NaCl in reactive dyeing; use of $K_2Cr_2O_7$, $Na_2Cr_2O_7$, $CuSO_4$ in direct and sulphur dyeing, copper complex in phthalogen blue dyeing, resins used to after-treat direct dyeing, etc.; these chemicals are liable to generate more load in sewage.

In dyeing of polyester, acetic acid is used which is not easily degradable. Use of dispersing agents, retarding agents, carriers, etc., increases this load tremendously. Use of carriers have been restricted due to high toxicity, irritation on skin and reduced light fastness of dyeings but still in use to rectify shades.

In printing, higher waste-water load producing chemicals are rongolite, citric acid, starch, anthraquinone paste, zinc oxide, binder, catalyst, etc. Kerosene in pigment printing causes air pollution. Though printing contributes lesser load in compared to that in dyeing, still it shares a reasonable percentage of the total load.

In finishing, resins used for anti-crease finish, like urea-formaldehyde, dimethylol ethylene urea and dimethylol dihydroxy ethylene urea, etc., release formaldehyde during application which block respiratory tracts and cause uneasiness in breathing. Flame-retardant chemicals like 'tetrakis' (hydroxymethyl phosphonium chloride) and 'tris' cause cancer. Acrylate and methacrylate copolymers containing free carboxyl groups are invariably used as soil release agents on polyester and its blends, are resistant to biological attack. Mothproofing agents, viz. dichloro diphenyl trichloroethane (DDT), eulan and mitin FF are also non-biodegradable.

Soap used in scouring and cleaning of dyeings results higher waste-water load. Anionic soaps are superior to non-ionics in performance; though linear chain dodecyl benzene sulphonates are relatively easy to degrade, the branched chain products are difficult to digest by bacteria. In dye houses, workers normally do not use eye glasses and nose covers. Flying out of dyes and chemicals causes irritation in eyes and blocking of respiratory tracts.

32.3   Classification of pollutants in textiles

Pollutants in textile effluent are mainly of five different types based on their ease in bio-degradation (Marquardt, 1974; Schefer, 1982, 1978, 1979; Sroka and Reineke, 1973a, 1973b; Durig et al, 1978).

(i)   Relatively harmless inorganic contaminants, viz. alkalis, mineral acids, neutral salts (chlorides, sulphates, phosphates and silicates) and oxidizing agents like peroxides, chlorine and chlorine dioxide (H_2O_2, Cl_2, ClO_2).

(ii)   Moderate to high BOD but are readily degradable, viz. starch sizes, vegetable oils, fats and waxes, biodegradable surfactants (linear-alkyl anionics, short chain EO adducts), organic acids (HCOOH, CH_3COOH, COOH.COOH) and reducing agents (Na_2S, $Na_2S_2O_4$, sulphites).

(iii) Dyes and polymers difficult to biodegrade (too high BOD value with too slower degradation), viz. dyes and fluorescent brighteners, fibres and polymeric impurities, polyacrylate sizes, synthetic polymer finishes and silicones.

(iv) Moderate BOD and difficult to biodegrade, viz. wool grease, PVA sizes, starch ethers and esters, mineral oil (spin finish), surfactants (branched anionics, long chain EO adducts), anionic and non-ionic softeners.

(v) Negligible BOD but unsuitable for conventional biological treatment, viz. HCHO and N-methylol reactants, chlorinated solvents and carriers, cationic retarding and softening agents, biocides (pentachlorophenol, organometal complexes, insecticides), sequestering agents (EDTA, NTA) and heavy metal salts (Cu, Cr, Cd, Sb).

Though individual dyes or chemicals generate waste-water load at varying extents, their combination changes the scenario. Waste-water load generated from various exhaust dyeing processes are summarized in Table 32.1 (Fiebig and Konig, 1977b). Pollution is a sign of inefficiency in industrial production; it is the money that is going up the chimney, down the sewer and out of plant in waste trucks.

Table 32.1 Waste-water load from various dyeings at a glance

Dyeing process	liquor ratio	BOD_5 (mg O_2/l)	COD (mg O_2/l)	BOD_5 / COD	g COD / kg dyeings
Dyeing of cotton with vat dye					
Winch vat	40:1	–	402	–	112.6
Winch vat	20:1	144	523	1.0:3.6	83.7
Jigger	5:1	298	1076	1.0:3.6	59.2
Dyeing of cotton with sulphur dye					
Winch vat	40:1	198	534	1.0:2.7	149.5
Winch vat	20:1	153	447	1.0:2.9	80.5
Dyeing of cotton with reactive dye					
Winch vat	40:1	64	358	1.0:5.5	114.6
Winch vat	20:1	123	369	1.0: 3.0	51.7
Jigger	5:1	143	596	1.0:4.17	29.8
Dyeing of cotton with direct dye					
Winch vat	40:1	39	118	1.0:3.0	23.6
Winch vat	20:1	–	162	–	16.2
Dyeing of wool with acid dye					
Winch vat	40:1	–	206	–	24.7
Winch vat	20:1	135	335	1:2.5	26.8

Dyeing of nylon with acid dye					
Winch vat	40:1	–	275	–	33.0
Winch vat	20:1	177	496	1:2.8	29.8

Dyeing of nylon with disperse dye					
Winch vat	40:1	–	656	–	78.7
Winch vat	20:1	192	777	1:4.1	46.4
Jigger	5:1	197	954	1:4.8	23.9

Dyeing of acrylic with basic dye					
Winch vat	40:1	–	281	–	33.7
Winch vat	20:1	218	362	1:1.7	28.9

Dyeing of polyester with disperse dye					
HTHP dyeing					
Jet	40:1	–	584	–	140.2
Jet	20:1	165	722	1:4.4	72.2
Carrier dyeing					
Winch vat	40:1	200	2043	1:10.2	408.6
Winch vat	20:1	189	1888	1:10.0	188.8

32.4   Quality of water for textile processing

Water used in processing sectors does not possess the exact standard. Successful results in the process houses are strongly dependent on clean and consistent supply of softened water, detailed in Table 32.2 (Sroka and Reineke, 1973; Gardiner and Borne, 1978). Elimination of undesirable impurities is one important objective. Water is not only a cheap and convenient medium for mass transfer and energy distribution within the machines used for processing, but is also a direct participant in the essential dye–fibre interaction mechanism. It is able to exert either favourable or adverse effects on these interactions, depending on the nature of the bonding forces concerned. Water is usually obtained by textile mills from three principal sources, viz. (i) surface water from reservoirs, rivers and canals, (ii) ground water from wells or bore-holes and (iii) mains water from the public supply in urban areas. Surface and ground water are cheaper sources and require little or no pre-treatment due to less dissolved solids; quality is highly consistent and reproducibility of dyeings is favoured. River water varies in composition with change in season. Mains water from public supply, though fit for human consumption, does not guarantee its suitability for textile processing; such water may contain traces of particulate matter which may well render it unacceptable for dyeing. Water with consistency in dissolved solids content favours reproducible results.

Treatment of fresh water supplies is usually done in textile mills itself due to: (i) removal of solid matter by sedimentation and filtration, (ii) correction of pH, (iii) elimination of residual chlorine, (iv) hardness removal and (v) removal of heavy metal ions. At the same time, used water must be treated through suitable ways before discharge to public sewage, criteria of such treated water is shown in Table 32.3 (Best, 1974; Desai, 1982).

Table 32.2 Specifications of water for use in textile processing

Parameter	Highest desirable level (mg/l)	Max. permissible level (mg/l)
Colour	colourless	colourless
Anionic surfactant	0.2	1.0
Mineral oil	0.1	0.3
Phenolic compounds	0.001	0.002
Total hardness (as $CaCO_3$)	100	500
Calcium	75	200
Chloride	200	600
Copper	0.05	1.5
Iron	0.1	1.0
Magnesium	30	150
Manganese	0.05	0.5
Sulphate (as SO_4)	200	400
Zinc	5	15
Fluoride	0.8	1.4
Arsenic	nil	0.5
Cadmium	nil	0.01
Lead	nil	0.05
Mercury	nil	0.01
Selenium	nil	0.01
pH	7.0–8.5	6.5–9.2

Table 32.3 Typical limits for discharge into public sewage*

Waste-water characteristics	Max. limit	Tolerance limit of		
		Pollution board	BIS-3306 1955	AMC-1975
pH	6–8	5.5–9.0	5.5–9.0	5.5–9.0
Temperature	45	40	45	45
Total suspended solids (mg/l)	300	100	600	600
Colour	–	–	100	–
Total dissolved solids (mg/l)	–	2600	2100	3500
Oil and grease (mg/l)	**	10	100	100
Compounds of Cr, Cu, Ni, Cd, Zn, Pb, Tin (mg/l)	20 (total)	–	–	–
Phenolic compounds (mg/l)	–	1	5	5
Fe (mg/l)	150	–	–	–
Sulphates (mg/l)	300	1000	1000	1000
Chlorides (mg/l)	–	600	600	600
BOD_5 (mg O_2/l)	600	31	500	500
COD (mg O_2/l)	–	30	500	500
Synthetic detergent (mg/l)	10	–	–	–

*colourless on visualization, **not visibly detectable

32.5   Alternate technologies to reduce load

Problems associated with degradation of unused dyes and chemicals in bath or processing section itself as well as limited availability of surface water has forced to innovate alternates of water or to replace or minimum use of it. These include padding technique, foam processing, spray techniques, kiss roll techniques, etc. Though use of water is kept at minimum, these processes have their inherent limitations. Stabilization of foam is impracticable for heavier processing. A part of solvent is always lost to atmosphere during processing, it spreads throughout shop-floor causing unhygienic surroundings and solvent recovery plant must be installed to reduce cost of processing. Padding technique is unsuitable for many types of materials, because of their construction or the economy of the process. However, these were not fully commercialized as these were continuous or semi-continuous in nature and not economical for short length fabric processing.

Waste-water load in textile processing can be reduced in various ways, viz. (i) short liquor processing, (ii) minimum use of hazardous chemicals, (iii) reuse of bath for multiple processing (follow-on technique), (iv) reduction in concentration of unused dye and chemicals through increase in affinity of fibre for dye (pre-treatment of fibre), (v) reduction in waste concentration by recovery and reuse, (vi) reduction in waste concentration through process modification, (vii) reduction in waste water load by chemical substitution (source-reduction technique) and (viii) devising suitable treatment based on point of ultimate disposal (effluent treatment plant) to meet the requirement. Reduction in waste water volume can also be improved by scientific production scheduling and housekeeping. However, different steps described above, have their own limitation too.

32.5.1 Short liquor processing

Reduction in volume of water can be affected by installing metering devices to ensure constant liquor ratio in all cases; washing-off is curtailed or shortened and overflow rinsing is eliminated. But reactive dyeing through exhaust technique necessitates thorough washing with overflow rinsing. Higher liquor ratio in vat dyebath ensures levelled dyeing through reduction in strike rate of dye. Heavy dyeing of polyester requires reduction cleaning succeeded by thorough washing. No doubt short liquor ratio reduces g COD per kg dyeings, but performance of dyeings in terms of levelling of shade, wash fastness, etc., are severely affected.

32.5.2 Minimum use of hazardous chemicals

To get desired depth of shade, concentration of chemicals used must be optimum for efficient processing and in doing that, reduction in dosing of chemicals may be exercised up to certain extent. Manufacturer's literature can be consulted. In spite of all these, use of hazardous chemicals becomes essential in selective cases, e.g. $Na_2S_2O_4$ in vat dyeing, which is applied quite beyond the stiochiometric requirement for quality dyeing. CH_3COOH used in most of dyeings has higher BOD_5, but required dose is a must to maintain specific pH. Concentration of dyes and other higher molecular weight chemicals, e.g. cationic surfactants mostly used in finishing can not be brought down below certain limit to achieve optimum process performance.

32.5.3 Follow on dyeing (reuse of bath)

Exhausted bath in dyeing is invariably drained out once the dyeing cycle is over and a fresh bath is prepared again even if dye–hue–fibre system remains unchanged for the successive operation. This is because the composition of exhausted bath is too complex in nature and is associated with class of dye used and permanent change in its structure during dyeing; loss in reactivity of chemicals is also another factor. Reuse of a bath, the so called follow-on technique, can only be ensured if the dye and chemicals retain their own structure during processing or if changed, do not interfere in next processing after compensation of consumed dye and chemicals in earlier operation (Cook et al, 1980; Pickford, 1981; Cook and Tincher, 1978; Carr and Cook, 1980). Reuse of baths is quite feasible in sulphur, vat and direct dyeing of cotton, dyeing of acrylic with basic dye, nylon, wool or silk with acid or metal complex dye, polyester with disperse dye, etc., as in all these cases dye structure does not get permanently changed. In contrast, reactive dyes are either hydrolysed or react with cotton and are not generally available in bath in original structure. This necessitates draining of bath after dyeing. A coupling bath in naphthol dyeing can not be reused on same ground. However, reuse of a bath is possible for specific cycles only. Each exhausted bath leaves some impurity in bath, which is slowly increased during subsequent cycles and retards rate and quality of dyeing. The impurity is mainly contributed by auxiliary chemicals as well as impurities in dye added during its manufacture. Number of reuse cycles in various dyeing systems is different which depends on extent of impurity generated in bath after each cycle. The follow-on concept is more suitable for dyeing of synthetics, as lesser number of chemicals is used which do not interact. A bath can be reused up to even 25–30 cycles and accuracy of reproduction of shades depends on how accurately spent baths are analyzed and fortified.

32.5.4 Reduction in unused dye and chemicals

This can be affected in dye bath through increase in affinity of fibre for dye through suitable pre-treatment of fibre prior to dyeing. Mercerisation of cotton is probably the simplest example to enhance affinity of dye for fibre. Cotton can be pre-treated with metal salts followed by conversion of water soluble salts to insoluble metal hydroxides, which remains embedded *in situ* fibre and enhances affinity of anionic dyes for cotton, resulting higher dye uptake and higher dye yield at lesser concentration of dye (Chavan and Chakraborty, 2000). The effect becomes highly effective if carried out through padding and with those dyes which gets oxidized to restore original structure and becomes non-ionic, e.g. vat dyes. The metal in hydroxide once used to attract more anionic dye now detaches its link with dye and is self-recharged for next attraction. Resins work in the same way. However, the technique has some complications, viz. incorporation of metal in situ fire develops stiffness in fibre and dyeings become acid sensitive. Mordanting of wool and silk prior to dyeing with acid dyes is another example. Rate of reaction of basic dyes on acrylic is proportional to extent of anionic co-monomers incorporated in fibre during polymerization; dye uptake can be enhanced for a given shade through increase in concentration of co-monomers as pure polyacrylonitrile not dyeable at all.

32.5.5 Recovery and reuse

Reduction in waste concentration through recovery and reuse is another concept to reduce waste-water load. NaOH can be recovered from mercerization as well as solvents can be recovered from spent baths. Reduced and solubilised sulphur dye can be precipitated in bath with H_2SO_4 and can be reused too (Spirkin, 1981). Unused vat dyes can be precipitated in exhaust liquor with excess of oxidising agent and can be filtered out. Unused disperse dye in polyester dyeing can be recovered by counter-acting its dispersion with cation active agents. Reactive and insoluble azoic dyebaths, on the other hand, neither can be recovered nor reused. In commercial sense, recovery is a costlier approach and in most cases, delays subsequent processing schedule.

32.5.6 Modification in technology

Modification in existing technology for a specific process may be done to reduce effluent load. Denim is customarily dyed in warp sheet or rope form using 6 different dye baths, in which consumption of indigo is compensated through on line monitoring – a cumbersome process. Draining of 6 different baths at the end of dyeing increases effluent load. In latest development, a new '1 in 6' loop dyeing method have been devised, according to which, the rope of warp is passed through the same indigo bath six times consecutively for development of shade (Godau, 1994). Compensation of indigo and chemicals becomes simpler due to one bath process, which enables to reduce floor space for dyeing.

32.5.7 Source reduction technique

Source reduction technique reduces waste water load by cutting the source. The disputed chemical in recipe is replaced with some suitable one and if required, the recipe and process conditions are also changed. If complete substitution is not feasible, certain percentage of the disputed chemical is substituted with another; process performance due to this change is tested and steps are taken accordingly (Kaushik et al, 1993a, 1993b). The method is very much flexible in nature with related advantages. It is one type of preventive measure and ensures decreased liability of waste, loss of materials, loss of energy, waste collection, handling and treatment costs and potential recycle and recovery value of waste. $K_2Cr_2O_7$ may be substituted with KIO_3 for oxidation of reduced sulphur dyes. The latter is relatively eco-friendly and produces a gentle black instead of a reddish one, generally produced with $K_2Cr_2O_7$. The harsh handle of sulphur dyed cotton can also be reduced by oxidizing it with KIO_3. Use of CH_3COOH increases BOD_5, COD of disperse dye bath and can be reduced by substituting CH_3COOH with $HCOOH$ but in exchange for lesser dye uptake. Due to increase in DOC in citric acid, it increases disperse dye uptake on polyester with simultaneous increase in waste water load. Carriers are to be omitted from recipe in disperse dyeing, if not essentially required. Reactive dyes are self-hydrolysable in nature and so if possible, application of non-degradable dyes may be restricted. Branched anionic detergents may be replaced with non-ionics.

32.5.8 Effluent treatment plant

A popular and scientific approach is to set up effluent treatment plant (ETP). Drained liquor from different processing zones is allowed to get mixed and settled in a settling tank, called equalization or neutralization. The suspended solids also get precipitated after which the waste liquor is processed through various treatments and recycled (Vaidya, 1988; Smethurst, 1989). The method is unique as (i) it is situated at a tolerable distance from the processing section and causes no obstruction in free working; also further expansion of the section is not disturbed, (ii) the land occupied for ETP is surrounded by low level area and hence can be used for direct sludge disposal, (iii) the plant is simple in design and operation, hence operating cost is low. However, it has some serious drawbacks too: (i) design of ETP is based on present composition of the effluent, any change in effluent composition , which comes from future expansion, complicates its efficient working, (ii) there is limited control over functioning of ETP which leads to inaccurate addition of chemicals, coagulants, inaccurate flow, etc., (iii) no facilities for testing and analysis of effluents are available, (iv) the plant pollutes the surrounding air, (v) initial investment is very high. The higher cost involved to set up ETP is a headache for all these industries, because it costs with no return.

32.6　Problem related to unused colour

It is evident from Table 32.4 that a dyebath is never exhausted completely (Desai, 1982). Unexhausted colour, whether water soluble or not, imposes negative effect on water body when drained. Most of colours have compact structure with high organic content and are applied in soluble condition. On draining, these are spreaded over the water body, do not allow sunlight to pass through and acts by depriving the stream of its dissolved oxygen. The asphyxial point for trout at 10°C is approximately 1.5 ppm of dissolved oxygen for coarse fish; this may be as low as 0.5 ppm. For every 10°C rise in temperature, a stream fish requires an approximately two-fold increase in dissolved oxygen and maximum oxygen that can be dissolved by water at 20°C water is about 9 mg/l.

Table 32.4　Standard percent exhaustion of various dyes

Dye class	Percent fixation According to OECD
Sulphur	60–70
Vat	80–95
Direct	70–95
Azoic	90–95
Reactive	50–80
Disperse dye	80–92
Basic	97–98
Acid	80–93

The primary effects produced by various colours on receiving streams are (i) toxicity to stream life, which is not only destructive to fish and aquatic life of that charter, but has effect of killing bacteria and various other forms of biological life upon which stream self-purification processes depend, with the result that the development of fish life is prevented, (ii) oxygen depletion in stream water: on complete consumption of dissolved oxygen, putrefaction sets in and a septic condition is created, accompanied by the characteristic effects of black unsightly colour, bubbling gases and noxious odours, (iii) physical impairment of stream condition: it is not of great hygienic importance, but utmost important to the layman and the man bringing suit against the industry – that of physical appearance. It is the yardstick by which the layman judges damages. The water becomes coloured or turbid, sunlight is shut out and the green aquatic plants, dependent on sunlight for growth, die off. As these plants give off oxygen to the water as part of their life activity, their absence decreases the available dissolved oxygen supply.

A quantitative measure of the nature and density of colour can be made by using 'Lovibonds Nessleriser'. The Hazen colour standards have been matched with Lovibond glasses and a disc containing nine colour standards in which values 5–70 are present (Purohit, 1986).

Removal of colour can be affected in various ways:

(i)   *Chlorination*: the cheapest and simplest method. Hypochlorite bleaching has also the same effect. But it has several limitations, viz. if the final effluent goes into a watercourse from which water for domestic consumption is drawn, chlorination should not be done in presence of tar acids, because chlorophenols are highly resistant to bacterial oxidation. Noxious odour of chlorine poisons surrounding air.

(ii) *Chemical coagulation*: most effective and economical method, even suitable for most of the dyes. Popular coagulants are lime, alum and iron salts – used alone or in combination; a dose of 300–600 mg/l removes around 75–90% of colour from dye effluent (Summers, 1967; Halliday and Beszedits, 1986). Coagulation method uses action of chemical adsorption or bonding to separate dissolved contaminants from effluents. This method has two drawbacks: (a) it is time consuming – special tests are required to establish which coagulating agent or combination of agents effective for each type of effluents and (b) produces sludge – different coagulant produces different amount of sludge (Halliday and Beszedits, 1986).

(iii) *Biological oxidation*: colour removal is done by activated sludge method, primarily by flocculation or adsorption of dyes by microorganisms. But in many instances, most colours simply pass through the activated sludge process unaffected regardless of how the treatment system is operated (Halliday and Beszedits, 1986).

(iv) *Activated carbon adsorption*: two forms of activated carbons are used in waste water treatment, viz. granular and powdered. The powder form is utilized mainly in conjunction with the activated sludge process. This method is quite effective for decolourising reactive, basic, azoic and 1:2 metal complex dyeing waste waters and for removing colour from mixed effluents. However, it is ineffective in removing some disperse, vat and pigment colours from their pure solution.

(v) *Chemical Oxidation*: this is done by using chlorine, ozone or H_2O_2. Effect of ozone is twice than that of chlorine, because it is unstable – having a half-life of only several minutes. Problem is that it cannot be transported and must be generated on-site from air or oxygen. Except insoluble azoics and disperse dyes, other colours and odours can be completely removed through this technique.

(vi) *Reverse osmosis and ultra filtration*: these are pressure driven membrane processes. The modern methods are based on flat membrane sheets in arrangements similar to that of a plate and frame filter process.

Other methods associated for removal of colour from effluent are irradiation with gamma rays, UV light from carbon arc, the Katox process and catalytic treatment.

32.7   Waste-water problem in various exhaust dyeing processes

32.7.1 Dyeing with vat dyes

A vat dyebath consists of dye, $Na_2S_2O_4$ and NaOH. During dyeing, $Na_2S_2O_4$ is consumed in different ways; necessitating use of at least three times dosing over that is required stiochiometrically (Nair and Trivedi, 1970). Decomposition of $Na_2S_2O_4$ produces a long list of hazardous chemicals which too contribute to the overall waste-water load (see 8.7.1). The acidic products from decomposition of $Na_2S_2O_4$ like Na_2SO_3, $NaHSO_3$ consume a part of NaOH, asking for higher dose of NaOH over that is required stiochiometrically and corrode concrete pipes during discharge.

In an analysis as shown in Table 32.5, a vat dyebath was set with dye (0.75 g/l), NaOH32.5% (5.25 ml/l), $Na_2S_2O_4$ (1.6 g/l). $Na_2S_2O_4.10H_2O$ (17 g/l) and ethylene oxide based levelling agent (0.25 g/l) followed by further additions during dyeing as NaOH–32.5% (5.25 ml/l) and $Na_2S_2O_4$ (13.3 g/l). Dyeings were thoroughly washed, oxidized with H_2O_2 30% (10 ml/l) and soaped with Na_2CO_3 (1 g/l) and soap (2 g/l) for 20 min at boil (Fiebig and Konig, 1977a). There was marked rise in both BOD_5 and COD in dye bath after dyeing, followed by a fall due to dilution during rinsing processes, a sharp increase during oxidation bath and particularly by the soaping bath, finally to fall once again during subsequent rinsing.

Table 32.5 Waste water load in different steps of vat dyeing

Steps	pH	BOD$_5$ (mg O$_2$/l)	COD (mg O$_2$/l)	Volume of water in1/kg of goods	g COD/ kg goods
Dye liquor before dyeing	10.2	280	1526	40	61.04
Exhausted dyebath	10.0	1350	2525	40	101.0
1st rinsing bath	9.4	–	290	40	11.6
2nd rinsing bath	8.5	–	118	40	4.72
Oxidation bath	7.4	–	700	40	28.0
Soaping bath	10.2	–	5890	40	235.60
1st rinsing bath	9.3	–	415	40	16.60
2nd rinsing bath	8.0	–	95	40	3.80
Total waste water load					
Measured	9.6	510	1030	280	288.40
Calculated	9.0	–	1433	–	410.24

32.7.2 Dyeing with sulphur dyes

Na$_2$S used in dyeing with sulphur dyes acts as reducing and solubilizing agent both and the dye bath is not so complex in nature (Klein, 1982). Excess of sulphide, electrolyte and unused dye create problem in discharged liquor. Due to poor exhaustion of dyebath, COD remain on very higher side. Hydrosulphite can be used for reduction of sulphurised vat dyes, e.g. hydron blue, but is only of minor importance in the application of granular sulphur dyes. There are a number of red-brown, green and olive dyes which decomposes in presence of hydrosulphite. In contrast, there are a few dyes which give better yield in presence of hydrosulphite but have not been used for dyeing sulphur blacks because of reduced colour yield and poor reproducibility. Solubilized sulphur blacks can successfully be dyed using hydrosulphite but only at higher cost.

Glucose generally produces unsatisfactory result in jiggers and winches. This is because dye yield is highly dependent on temperature. Only at constant temperature above 90°C, better results can be expected. Glucose works along with NaOH as solubilizing agent. It is a strong hydrocarbon and so though degradable, shows higher BOD$_5$ and COD.

Solubilized sulphur colours can be applied using mercaptoethanol [C$_2$H$_5$SH] in exhaust technique, but not suitable for normal dyes. The advantage is that there may be no sulphide effluent present, but costly and colour does not dissolve smoothly. Other reducing agents like thioglycolic acid, thiosalicylic acid causes change in tone.

K$_2$Cr$_2$O$_7$ in combination with CH$_3$COOH releases Cr(VI) compounds, which oxidize all reduced sulphur dyes rapidly. Colour yield and fastness properties are good. Dyeings become harsher and less hydrophilic due to deposition of small amount of chromium and so not recommended for dyeing of sewing thread to avoid needle cutting. Use of chromium compounds cause increase in suspended solids. KIO_3 can be used in the same way as that is used for $K_2Cr_2O_7$. It gives reproducible results with pleasant shades; handle of dyeings remain softer, but costlier. $K_2Cr_2O_7$ imparts an orange tone to dyeings and is not recommended for blue shades, whereas KIO_3 are the most suitable alternatives as it does not show such a problem. H_2O_2 is extensively used in yarn dyeing for softer handle. In acidic medium, its action is very high and wash fastness is reduced. Other compounds like sodium perborate, ammonium or potassium persulphate, sodium bromate all have limitations and are not recommended.

Waste-water load in terms of COD due to drained exhausted dyebath is 534 mg O_2 /l in winch vat at a liquor ratio of 1:40, which gets reduced to 447 mg O_2 /l in winch vat at a liquor ratio of 1:20. The total COD (mg O_2/l) are 2002 and 2477 respectively which is higher than those for vat dyeing (Fiebig and Konig, 1977b). Indeed, Na_2S and $K_2Cr_2O_7$ both are the chief source of such higher COD. Interestingly, Na_2S is the best reducing agent for sulphur dyes till date and so effective methods were developed to remove unused sulphides from effluents. Na_2S can be easily oxidized to thiosulphate by atmospheric oxygen. The large quantity of sulphur dye present in the effluent catalyse the process, which has to be carried out in aerated tanks or reactors in which different effluents can be collected and equalized. A second method is precipitation in the form of iron sulphide. In practice, conversion is quantitative and the dyes present are also precipitated. Sulphur dyes can be precipitated by adding H_2SO_4 to exhaust baths, filtered-off and reused. Analysis of dyeings using precipitated sulphur dyes show the same composition, give the consistent hues and levels of fastness (Spirkin et al, 1981). For small industries, it will be cheaper to dye with sulphur black using glucose–NaOH system.

32.7.3 Dyeing with reactive dyes

These dyes are relatively less harmful in waste water due to self-hydrolysable in nature with time, temperature, alkali, water, etc. Drained liquor consists of reasonable percent of colour, large amount of electrolyte and alkalis. Dyes are water soluble, so the streams become coloured, preventing sunlight to pass through. The suspended solids are minimum, dissolved solids are maximum due to good solubility of salt, alkali and colour altogether; strong alkaline pH. Washing bath contains sufficient colour, detergents, salts and alkalis.

Though NaCl is the ideal electrolyte commercially, glauber's salt, calcium chloride can also be used. In dyeing, salts and alkalis used are mostly sodium compounds, when pumped in after draining, reduces fertility of soil. Commonly applied alkali is soda ash; somewhere addition of little caustic soda shows better results. In order to remove such a load of sodium compounds, $Ca(OH)_2$ may be a good substitute. However, $Ca(OH)_2$ produces $CaCO_3$ during exposure to air producing turbidity by reaction with CO_2, suspended solid and pH of fixation bath falls rapidly. For good colour yield and fixation, combination of NaCl and $Ca(OH)_2$ may be useful. The COD for reactive dyeings were found to be 696 at a liquor ratio of 1:40, 1063 at a liquor ratio of 1:20 and 4177 at a liquor ratio of 1:5 (Fiebig and Konig, 1977b).

32.7.4 Dyeing with direct dyes

Dye bath consists of only colour and salt along with little soda ash. Adsorption of colour is also better than reactive dyes causing lesser problem with coloured water. Dissolved solid content is very high due to higher dose of salt. Volume of water required is quite less as soaping is omitted and net result is reduction in waste-water load. A COD (mg O_2/l) of 432 for liquor ratio of 1:40 and 597 for liquor ratio of 1:20 were reported (Fiebig and Konig, 1977b).

Though a few methods are in place for after-treatment of direct dyeings, application of cationic dye fixings agent is popular due to no change in colour of after-treated cotton. After-treatment with $K_2Cr_2O_7$ or with other metal salts cause liberation of a part of metal in waste water increasing its solid content.

32.7.5 Dyeing with disperse dyes

Dyeing of polyester with disperse dyes have been found to give rise to even much higher concentration of polluting substances and to higher 'harmful factor' of the waste water (Table 32.6). The COD is the same or even higher than the same for vat and sulphur dyeing on cotton. It is again the dispersing agent content in dyestuff and various dyebath additives to the dye liquor, as well as washing and cleaning processes of dyeings which bring about the high 'harmful factor'(COD / BOD_5). The COD of exhaust baths from carrier dyeing of polyester is considerably higher than those from all other exhaust dyeing processes, since in this case besides the ancillaries required for HT-dyeing, carriers are also used, which are highly toxic (Fiebig and Konig, 1977b).

Table 32.6 Waste water load in dyeing polyester

Dyeing method	Liquor ratio	BOD_5 (mg O_2/l)	COD (mg O_2/l)	COD/ BOD_5	g COD/kg of goods
HT dyeing					
Jet	40:1	–	584	–	140.2
Jet	20:1	165	722	1:4.4	72.2
Carrier dyeing					
Winch vat	40:1	200	2043	1:10.2	408.6
Winch vat	20:1	189	1888	1:10	188.8

Waste water load generated by some of the carriers individually is shown in Table 32.7 (Medilek et al, 1976). Many carrier formulations contain mixtures of active substances, together with auxiliary solvents and emulsifiers not absorbed significantly by polyester. In some instances, this necessary adjuvant may contribute more COD or TOC to the effluent than the carrier itself, as 70–90% of the later may be taken up by fibre. Since most of this is often released again during subsequent heat treatment, air pollution can be a greater problem than contamination of effluent by carriers. Many carriers have offensive odours and some of the chlorinated types are significantly toxic to fish and bacterial sludge (Vaidya and Trivedi, 1975).

Levelling agents based on alkyl-aryl sulphonates are absorbed more effectively than simple aryl or aliphatic anionic agents, and dye uptake is enhanced at lower pH and higher temperature.

The polyester dyeings examined, particularly those involving carriers are most unfavourable with regard to COD / BOD_5 in waste water, as illustrated below.

Vat, sulphur and reactive dyeings show a COD / BOD_5 ratio (harmful factor) of around 1:3–1:6, whereas all other dyeing waste-waters have a ratio of 1:2–1:3. The unfavourable COD and BOD_5 ratio in polyester dyeing with carrier is attributable to poor degradation of waste-water contents, toxicity of carrier and its property of hindering decomposition. In case of carriers examined in, the harmful effect of dyeing probably lies in the region of 100–200 g COD / kg goods with a liquor ratio of 20:1 and in a range of 200–460 g COD / kg goods with a liquor ratio of 40:1.

Table 32.7 Waste-water load generated by few carriers

Carrier type	COD (mg O_2/l)	BOD_5 (mg O_2/l)
o-phenylphenol	1000–2000	200–800
N-Alkylphthalimide	1000–2100	100–200
Arylcarbonate ester	900–1900	700–800
Methyl cresotinate	800–1700	200–800
Dichlorobenzene	500–1100	0
Trichlorobenzene	300–1000	0

CH$_3$COOH used in disperse dyeing has higher BOD$_5$ and other organic acids with lesser TOC may be recommended. Two such acids are oxalic acid (HOOC–COOH) and formic acid (HCOOH). HCOOH is reducing in nature and so its application may show adverse effect on dyeings (Durig at al, 1978). Oxalic acid has same TOC like acetic acid. TOC of 1 kg of 80% CH$_3$COOH is 320 g and that of HCOOH is only half, i.e.

75 g CH$_3$COOH (80%) (= 1 mol; mol wt 60) = 24 g TOC, corresponds to: 54 g HCOOH (85%) (= 1 mol; mol wt 46) = 12 g TOC.

To achieve a particular pH, dosing of HCOOH will be lesser than that with CH$_3$COOH, as the former has a greater degree of dissociation, but the dosing is somewhat critical and needs greater care. Dyeing in alkaline pH is possible with a few disperse dyes, but not used due to sensitivity of polyester in alkali, preferably at higher temperature.

In producing levelled shades through efficient dispersion of dye, dispersing agents play decisive role. These are not applied separately rather mixed with dyes by manufacturers before marketing. However, in dyeing textured polyester or to cover barre', introduction of an extra amount of dispersing agent is inevitable for levelled shades, while dyes retain one-third of dispersing agent out of their total weight (Slack, 1979). Final exhaustion depends strictly on type of dispersing agent used and so selective dispersing agents are only used in dyeing.

In HT-dyeing, specifically in jet-dyeing, both material and liquor circulate at higher speed. This causes foaming inside the machine and application of defoaming agent is essential to suppress or destroy foam for level dyeing. Although dosing remains on too lower side (0.05 g/l), still it contributes to waste-water load.

Since, the waste-water problem in disperse dyeing in higher, some precautions must be taken to ensure a reduced load. HTHP dyeing can be well done without carrier; CH$_3$COOH may be replaced with HCOOH at pH 5.5–6.5. Polyester–cotton blends may be dyed with cellestran dyes to dye both the components simultaneously. If carrier application becomes must, biodegradable carriers may be given preference.

32.8   Cleaning of dyeings with surfactants

Textile processing is leading consumer of surfactants, which are used for different purposes depending on the type. Out of three types, only anionics and non-ionics hold importance in removal of superficial dyes from dyeings but both possesses advantages and restrictions in use. Efficiency of non-ionics largely depends on their cloud points beyond which these get separated out from bath and are precipitated. Importance of anionics is due to mainly its multi-purpose use as emulsifier, wetting agent, washing, solubilizing agent, detergent, softening agents, antistatic agent and lubricants. Anionic detergents are efficient even at higher temperature. However, anionics produce foam and somewhere hard to biodegrade. In contrast, non-ionic are lesser efficient, has no tendency of foaming; no affinity for fibres and washing is easier with minimum water and are degradable, as listed in Table 32.8 (Fischer and Weber, 1979).

Table 32.8 Extent of degradability of various surfactants

Anionic surfactant	Biodegradability (%)
Tetrapropylenebenzenesulphonate	15–20
N-dodecylbenzenesulphonate	90–94
Alkane sulphonate	95–98
Alpha-olefinsulphonate	85–95
Branched sulphosuccinate	45–50
Unbrabched sulphosuccinate	80–95
Tallow alcohol sulphate	95–98
Isohexylnapthalenesulphonate	45–50
Alpha sulphofatty acid ester	85–95

Biodegradability of anionic surfactants is influenced by their structure. Branched dodecylbenzene sulphonate (DBS) is not biodegradable, whereas those possessing linear structure are readily degradable. The list in Table 32.8 describes the degradability of few anionic surfactants. Structure of hydrophobic chain determines the degradability – the chain must be linear.

In regard to biodegradability of different anionic surfactants, primary alcohol sulphates are degraded within a very short period, followed by secondary alcohol sulphate and linear alkyl benzene sulphonates. The branched chain alkyl benzene sulphonates are not completely biodegraded even after seven days.

Biodegradability of non-ionics exclusively depends on hydrophobic and hydrophilic parts of the same. Most of the commercially used non-ionic surfactants consist of ethylene oxide condensates of various hydrophobic compounds. The ethoxy chain itself is degraded through biological oxidation and the hydrophilic part of non-ionics is the site for biological attack. If the ethylene oxide chain is remarkably longer, degradability gets reduced in the same way like presence of benzene ring in anionics. Primary and secondary linear alcohol ethoxylates are the quickest biodegradable among non-anionics. Branched chain alkyl phenol ethoxylates, like nonyl and octyl phenol ethoxylates are hard to degrade, i.e. non-biodegradable. Substitution of straight alkyl chain for branched alkyl chain in the molecule of these non-ionics has only a minor effect on their biodegradability.

Before application, suitability of surfactants must be judged as a list of readily biodegradable non-ionic surfactants are available and in many instances are more efficient over the currently available non-biodegradable one.

32.9   Basics of assessing waste-water load

Various government and municipal agencies specify norms and methods of measuring waste-water load (Durig, 1976; Parish, 1976). In textile process houses, the effluent is primarily tested for 'biological oxygen demand' (BOD), 'chemical oxygen demand' (COD), dissolved and suspended solids, pH and harmful factor (COD:BOD$_5$). A calculation of weight of dyeings in respect to total COD of the exhausted and drained bath enables dyer to restrict extent of load for each kg of dyeings (g COD / kg dyeings). In some cases, dissolved organic carbon (DOC) and total organic carbon (TOC) are also assessed. Needless to say, it is the BOD, which is of paramount importance to ascertain susceptibility of drained out liquor towards its degradation.

32.9.1 Biological Oxygen Demand (BOD)

It is a measure of volume of oxygen required under natural circumstances to degrade pollutants for each l of liquor. In general, BOD is measured after treating the liquor at 20°C for 5 days and is denoted by BOD$_5$. It is a measure of rate of degradation of pollutant throughout a period of 5 days. However, if degradability of a chemical is very high, instead of measuring BOD$_5$, degradation for 1 or 2 days, e.g. BOD$_1$, BOD$_2$, etc., can be assessed. Similarly, if rate of degradation is too slow, a BOD value can be measured after a period of 10, 20 days, etc. However, assessment of consumed oxygen for 5 days has been accepted as a standard, in general. Though degradability of individual chemicals can be measured separately and follows a specific pattern, drained out liquor contains a list of chemicals, which does not show any specific trend during degradation. Easily degradable components are digested first at faster rate followed by slow digestion for relatively poor degradable one, even a part may not degrade at all. For example, if a specific chemical has a BOD$_5$ of 10mg O$_2$/l of its solution, the period and total oxygen required for complete degradation of the same can be calculated. This is true for any other chemical too. However, when these chemicals having differential degradability is mixed, which is the practice in textile process houses, the drained out liquor does not show any specific trend of degradation.

32.9.2 Chemical Oxygen Demand (COD)

It provides a measure of oxygen equivalent of that portion of the organic matter in a sample that is susceptible to oxidation by a strong chemical oxidant irrespective of its rate of degradation which under natural circumstances may take several weeks or even years. It is an important, rapidly measured parameter for stream and industrial waste studies and control of waste treatment plants. COD can be defined as milligrams of oxygen which one litre of effluent will absorb from a hot, acidic solution of potassium dichromate.

32.9.3 Prediction of load with BOD, COD and other factors

Assessment of chemical oxygen demand, in relation to BOD, gives first hand information on how much time the liquor may take for complete degradation. If the BOD_5 value of a liquor is 15 mg O_2/l and COD is, say, 75 mg O_2/l, the liquor will take $75 \div (15 \div 5)$ or 25 days for complete degradation. This is true for individual chemicals, but while in mixture, fails to trace out total time required for complete degradation. Forceful oxidation of a chemical in laboratory is possible which under natural circumstances, may be too hard to degrade.

In general, BOD helps to indicate rate of natural degradation, while COD predicts how much of oxygen is required and in combination with BOD how many days the liquor may take for complete degradation. BOD_5 of a given effluent is always lesser than its corresponding COD.

To assess BOD_5, the liquor is diluted and a known volume is added to water saturated with oxygen containing trace elements and inoculums of organism. Closed bottles are incubated precisely at 20°C for five days in incubators and the residual dissolved oxygen is compared with that on duplicates measured before incubation; the difference shows consumption of oxygen due solely to the biodegradable organic matter present. Trials are to be made at different dilutions to the test solution, both to observe the effect of dilution and to arrive at a level whereby about half the dissolved oxygen is consumed in the test. The method has several limitations, viz. (i) it is a time consuming method, (ii) it does not provide information of extent of organic pollutant present in water, that is not degraded by bacteria and (iii) selection of correct bacteria is a must to get consistent results.

To assess COD, an oxidant, like $K_2Cr_2O_7$ is mixed in diluted liquor under test in acidic conditions and boiled for two hours to oxidize a part of the liquor; unused dichromate is assessed to know how much has been consumed. The value indicates total organic matter, both degradable and non-biodegradable. Measurement of COD is more reproducible, but limited as some organic contaminants are resistant to dichromate oxidation.

Harmful factor is the ratio of COD and BOD_5. The lower the ratio, the higher is the rate of degradation of waste-water. A higher value indicates that degradation is too slow and may take longer time. This in turn, poses challenge on existence of bacteria in sewage which may not cope up with the toxicity of liquor and may die. For fairly biodegradable waste water, the ratio should be closer to 1, practically very difficult to achieve.

Solids present in drained out liquor is either in suspended or dissolved state or combination of these two. Dissolved solids must be separated out from liquor; otherwise diffusion of atmospheric oxygen in the sewage will be slowed down. Huge salt added in dyebaths, more specifically in direct, sulphur and reactive baths is a major concern in this regard. Other soluble auxiliaries are applied at very lower concentration and shares less load. Suspended solids, on the other hand, clog respiratory tracts or gills of aquatics to kill them. Separation of these is essential to preserve freshness of water body.

Dissolved organic carbon (DOC) is the amount of organic carbon present in dissolved phase after centrifuging or membrane filtration. Total organic carbon (TOC) is the organic contaminants, both suspended and dissolved and are expressed in terms of carbon content of the effluent. Higher the carbon content more will be oxygen required to oxidize that to CO_2, time for complete conversion will be lengthy and vice versa.

DOC and TOC both have achieved more importance in recent years, since these represent all organic material present in one or both phases, irrespective of relative ease of oxidation. It should be noted, however, that the above criteria are concerned essentially with organic pollution of effluents. These indicate nothing about the colour, temperature, pH, content of electrolytes or heavy metals present; any of these factors may prove critical in defining the acceptability of an effluent sample. Bio-elimination includes the material removed by adsorption on the biomass as well as that which undergoes biochemical decomposition.

32.9.4 Total Organic Carbon (TOC)

This is assessed using potassium persulphate ($K_2S_2O_8$), which releases – HO free radical on UV-radiation for oxidation of carbon impurities in effluent to CO_2. The TOC analyzer essentially consists of a septum (open end) to inject sample and reagents at the interior of the analyser, a container which retains waste water under test as well as other additives, two gas connections to sparge unoxidized and oxidized samples, a pump, an oxidizing chamber inside which an UV-lamp is fitted and finally an integrating CO_2 detector to measure extent of CO_2 formed which is directly proportional to TOC in effluent.

32.10 Methods for evaluation of waste-water load

32.10.1 Evaluation of suspended and dissolved solids

25 ml of the sample is filtered through 'Whatman 42' filter paper. The residue on paper is dried at $103\pm2°C$ in hot air oven till constant weight is obtained. The filter paper is previously weighed.

Suspended solid (mg/l) = weight of residue (mg) × 10^3 / 25

The filtered solution collected in porcelain desiccator is dried at $103\pm2°C$, weighed till constant weight is obtained and extent of dissolved solid is calculated using the same way as that is used for calculation of suspended solids.

32.10.2 Colour of the liquor and pH

The colour of the drained out and treated liquor is visually assessed against sunlight while pH is tested with the help of a pH paper or better with pH meter for better precision.

32.10.3 Chemical Oxygen Demand

A boiling mixture of chromic and sulphuric acid destroys most types of organic matters in effluent. A sample of waste water is refluxed with known concentration of $K_2Cr_2O_7$ and H_2SO_4; excess $K_2Cr_2O_7$ is titrated with ferrous ammonium sulphate; amount of oxidizable organic matter is measured as oxygen equivalent and is proportional to the $K_2Cr_2O_7$ consumed.

$$K_2Cr_2O_7 + 4H_2SO_4 = K_2SO_4 + Cr_2(SO_4)_3 + 4H_2O + 3[O]\uparrow$$

Potassium permanganate may be used in place of $K_2Cr_2O_7$ for the oxidation process.

$$2KMnO_4 + 3H_2SO_4 \text{ (dil)} = K_2SO_4 + 2MnSO_4 + 3H_2O + 5O\uparrow$$

But $KMnO_4$ is not free from MnO_2 as $KMnO_4$ is slowly auto-reduced to MnO_2 in presence of light or acid. Ordinary distilled water is likely to contain reducing substances, such as traces of organic matter, etc., which may reduce $KMnO_4$ to MnO_2. MnO_2 is standardized against oxalic acid of known strength. However, $K_2Cr_2O_7$ method is popular worldwide. A practical methodology to assess COD is described below in details to help readers to conduct such tests in laboratory.

Material: conical flask of 500ml capacity with suitable condenser, 0.25N $K_2Cr_2O_7$, ferroin indicator, 0.25N ferrous ammonium sulphate $[Fe(NH_4)_2(SO_4)_2.6H_2O]$, conc. H_2SO_4 and $AgSO_4$ crystals.

Ferroin indicator may be prepared in laboratory by dissolving 1.735g of 1,10 phenolthioline dihydrate along with 695 mg $FeSO_4.7H_2O$ in water followed by its dilution to 100 ml.

Standard 0.25N $K_2Cr_2O_7$ is prepared by dissolving 12.25 g of $K_2Cr_2O_7$ (previously dried at $103\pm2°C$ for 24 h) in distilled water and diluting to one litre. Standard 0.1N $Fe(NH_4)_2(SO_4)_2.6H_2O$ is prepared by dissolving 39 g of $Fe(NH_4)_2(SO_4)_2.6H_2O$ in 400 ml water. Conc. H_2SO_4 (20 ml) is added to it, cooled and diluted to one litre. This solution is standardized against standard $K_2Cr_2O_7$.

For standardization of $Fe(NH_4)_2(SO_4)_2.6H_2O$, 10 ml of standard $K_2Cr_2O_7$ solution is diluted to about 100 ml. 30 ml conc. H_2SO_4 is added to it and allowed to cool down. The solution is titrated with 0.1N $Fe(NH_4)_2(SO_4)_2.6H_2O$ using a few drops of ferroin indicator with constant stirring to blue–red end point.

Normality of $Fe(NH_4)_2(SO_4)_2.6H_2O$ = (ml $K_2Cr_2O_7$ × 0.25) / ml $Fe(NH_4)_2(SO_4)_2.6H_2O$

Procedure

25 ml of waste water is taken in 500 ml flask and diluted to 100 ml with cold water. 25 ml of 0.25N $K_2Cr_2O_7$ solution is added to it followed by addition of 75 ml conc. H_2SO_4 and 1g $AgSO_4$ respectively. The mixture is refluxed for at least 2 h followed by cooling. Alternately, the mixture may be digested at 15psi for 30 min in an autoclave.

The refluxed or digested mixture is diluted to about 350 ml with water and the excess $K_2Cr_2O_7$ is titrated against standard $Fe(NH_4)_2(SO_4)_2.6H_2O$ solution using ferroin indicator. The end point is blue–red.

The procedure is carried out for blank with 100 ml water (no effluent) along with 75 ml conc. H_2SO_4, 25 ml of 0.25N $K_2Cr_2O_7$ and the mixture is refluxed for 2 h or digested as stated earlier. Volume of $Fe(NH_4)_2(SO_4)_2.6H_2O$ solution (for blank) is determined.

COD (mg O_2 /l) = {(A-B) × C × 8 × 1000} / ml of sample used

[Sample =25 ml]

Where, A is ml of $Fe(NH_4)_2(SO_4)_2.6H_2O$ used for blank, B is ml of $Fe(NH_4)_2(SO_4)_2.6H_2O$ used for sample and C stands for normality of $Fe(NH_4)_2(SO_4)_2.6H_2O$.

The procedure of using 25 ml sample holds good for COD values up to 1000mg O_2 /l. For higher COD beyond that, all $K_2Cr_2O_7$ is consumed and the solution turns green. It is suggested to dilute the effluent under test with water in such a case and the obtained COD may be multiplied by the dilution factor to get actual COD of effluent.

32.10.4 Biological Oxygen Demand (BOD)

Decomposable material in effluent is consumed by various micro-organisms, naturally present in the stream with the help of sunlight and dissolved oxygen in water. The rate of degradation of decomposable material by micro-organisms cause decrease in amount of oxygen (free oxygen) present in the steam proportionately. Assessment of oxygen consumed during microbial degradation gives an indication of the amount of biodegradable matter degraded. The amount of decomposable material may be monitored at various stages of treatment of sewage through test of BOD and is expressed quantitatively as parts of oxygen per million parts of sewage or other wastes.

The test consists of incubation of dilutions of known volume of effluent in sealed bottles at 20°C for 5 days, followed by titration of residual oxygen. From the difference between amount of residual oxygen in original water and incubated effluent after titration, BOD_5 of effluent is calculated.

Apparatus and reagents

BOD incubator, BOD bottles of 300 ml capacity, pipettes, burettes, 250 ml Erlenmeyer flasks.

Manganese sulphates solution: It is prepared by dissolving 480 g of $MnSO_4.4H_2O$ in water followed by its dilution to one litre. This solution must not develop colour with starch when added to an acidified solution of KI.

Alkali–iodide–azide solution: 500 g of NaOH and 150 g of KI are dissolved in water and diluted. To this solution is added 10 g of NaN_3 (sodium azide) previously dissolved in 40 ml distilled water. The final solution is diluted to one litre. This solution must not develop colour with starch solution when diluted and acidified.

Starch indicator: 0.5 g starch powder is mixed in 100 ml distilled water, boiled for a few min and cooled.

Stock sodium thiosulphate ($Na_2S_2O_3$) solution (0.1N): 24.82 g $Na_2S_2O_3.5H_2O$ is dissolved in boiled and cooled distilled water and finally diluted to one litre. Preserve if necessary, by adding 5 ml $CHCl_3$ for one litre of thiosulphate solution.

Standard sodium thiosulphate solution (0.025N): 250 ml of stock $Na_2S_2O_3$ solution is diluted to one litre with freshly boiled and cooled distilled water and preserved in the same way by adding 5 ml $CHCl_3$. Standardize before each titration using standard $K_2Cr_2O_7$.

Another reagent required for this test is conc. H_2SO_4.

Procedure

(A) Preparation of sample
    (i)   Sample of waste water is filtered to remove other wastes and large suspended matter.
    (ii)   Four BOD bottles are taken and labeled as 1/100, 1/50, 1/33 and B (blank). Dilutions to the extent of 1/100, 1/50 and 1/33 of effluent were prepared by adding approximately 3, 6 and 9 ml of sample to labeled bottles. Each bottle is filled up with sufficient water that upon insertion of the glass stopper, water level will rise above the ground glass joint, inserting a liquid seal. To the bottle leveled B, only water is added. Filled up bottles must be free from air bubbles with the water seal maintained.
    (iii) All the four bottles are incubated for 5 days at 20°C in an incubator.
(B) Determination of residual oxygen
    (i)   After keeping 5 days in incubator, stoppers are removed from the bottles and 2 ml of $MnSO_4$ followed by 2 ml of alkali–iodide–azide reagent are added to each bottle keeping the tip of pipette below the surface of liquid in the bottle. After replacing stopper, the top of the bottle is rinsed with water.
    (ii)   Each bottle is vigorously shaked for about 30s to disperse any precipitate uniformly throughout each bottle; the action is to be repeated in case re-precipitation starts.
    (iii) The bottles are now allowed to stand until the precipitates have settled about half way down. Stoppers are removed and 2 ml of conc. H_2SO_4 is added to each bottle, allowing the acid to run down the inside of the neck. Stoppers are again placed on each bottle, the top of each bottle is rinsed followed by immediate vigorous shaking for 30 s.
    (iv) 200 ml of each sample is titrated with 0.025 N $Na_2S_2O_3$ with gentle agitation in the sample and the titration is continued till a faint yellow colour is observed. 1 ml of starch indicator solution is added, which causes change in colour to blue. The titration is further slowly continued with drop by drop addition of $Na_2S_2O_3$ till the blue colour disappears completely and the sample becomes colourless. Volume of $Na_2S_2O_3$ used is noted down which is directly proportional to the residual oxygen in each bottle.

Calculation

BOD_5 = [ppm residual oxygen of B (i.e ml $Na_2S_2O_3$) – ppm residual oxygen of diluted sample (i.e ml $Na_2S_2O_3$)] × dilution factor.

The BOD of only those sample dilutions are calculated where there has been a reduction of at least 2 ppm residual oxygen as judged by comparison with the blank (B) and in which there is at least 1 ppm oxygen remaining. Average of the BOD values those falls within these limits is calculated to determine BOD_5 of original sample.

32.11 References

BEST G A (1974) 'Water pollution and control, *J. Soc. Dyers Color.*, **90**, 11, 389–393.

CARR W W and COOK F L (1980) 'Savings in dyebath re-use depend on variations in impurity concentrations', *Textile Chem. Color.*, **12**, 5, 106–110.

CHAVAN R B and CHAKRABORTY J N (2000) 'Enhancement in indigo uptake on cotton through metal salt pretreatment', *Indian J. Fibre Textile Res.*, **25**, 2, June, 130–137.

COOK F L and TINCHER W C (1978) 'Dyebath reuse in batch dyeing', *Textile Chem. Color.*, **10**, 1, 1–5.

COOK F L, TINCHER W C, CARR W W, OLSON L H and AVERETTE M (1980) 'Plant trials on dyebath reuse show savings in energy, water, dyes and chemicals', *Textile Chem. Color.*, **12**, 1, 1–10.

COPPER S G (1978) '*The Textile Industry, Environmental control and Energy Conservation*', USA, Noyes Data Corporation.

DESAI S R (1982) 'Study of effluents', *Colourage*, **29**, 26, December 30, 10–14.

DURIG G (1976) 'Ecological aspects of water usage in the textile industry', *Rev. Prog. Color Related Topics*, **7**, 70–80.

DURIG G, HAUSMANN J P and O'HARE B J (1978) 'Review of the possibilities for recycling aqueous dyehouse effluent', *J. Soc. Dyers Color.*, **94**, 8, 331–339.

FIEBIG D and KONIG K (1977a) 'Examination of various dyeings by the exhaust method with regard to their pollution of waste waters', *Textile Praxis Int.*, **32**, 5, 577–578 and 585–587.

FIEBIG D and KONIG K (1977b) 'Examination of various dyeings by the exhaust method with regard to their pollution of waste waters', *Textile Praxis Int.*, **32**, 6, 694–696, 707.

FISCHER H and WEBER R (1979) 'Methods for reducing the pollution load in effluent by specific changes to textile auxiliaries', *Textile Praxis Int.*, **34**, 5, 578–582.

GARDINER D K and BORNE B J (1978) 'Textile waste waters: Treatment and environmental effects, *J. Soc. Dyers Color.*, **94**, 8, 339–348.

GODAU E T (1994) 'Dyeing denim warp threads by the Loopdye 1 for 6 process, *Int. Textile Bulletin*, **40**, 1, 23–24.

HALLIDAY P J and BESZEDITS S (1986) 'Colour removal from textile mill wastewaters', *Canadian Textile J.*, **103**, 4, 78–84.

KAUSHIK R C D, SHARMA J K and CHAKRABORTY J N (1993a) 'Source reduction technique to control effluent load in reactive dyeing', *Asian Textile J.*, **1**, 9, July, 33–40.

KAUSHIK R C D, SHARMA J K and CHAKRABORTY J N (1993b) 'Investigation of oxidiser/ reducer to control effluent load in dyeing with sulphur dyes', *Asian Textile J.*, **2**, 1, November, 47–50.

KLEIN R (1982) 'Sulphur dyes: today and tomorrow', *J. Soc. Dyers Color.*, **98**, 4, 106–113.

MARQUARDT K (1974) 'Fresh and used water treatment in the textile industry', *Melliand Textilber.*, **55**, 6, 569–576.

MEDILEK P, QUURKE M and JABLONOWSKI P (1976) 'Waste water pollution by carriers, *Melliand Textilber.*, **57**, 7, 583–587.

NAIR G P and TRIVEDI S S (1970) 'Sodium sulphide in vat dyeing', *Colourage*, **17**, 27, December 31, 19–24.

PARISH G J (1976) 'Effluent treatment in the textile industry', *Rev. Prog. Color Related Topics*, **7**, 55–69.

PARK J and SHORE J (1984) 'Water for the Dyehouse: Supply, Consumption, Recovery and Disposal', *J. Soc. Dyers Color.*, **100**, 383–398.

PICKFORD T (1981) 'Use of the follow-on in dyeing', *Textile Inst. Ind.*, **19**, 2, 55.

PUROHIT S K (1986) *Environmental Pollution Control Hand Book*, Bikaner, India, S K Publishers.

SCHEFER W (1978) 'Fundamentals of ecology for textile technologists', *Textilveredlung*, **13**, 5, 182–187.

SCHEFER W (1979) 'Water pollution control by critical product selection', *Melliand Textilber.*, **60**, 5, 434–435.

SCHEFER W (1982) 'Reducing environmental impairing in the textile industry, *Melliand Textilber.*, **63**, 8, 591–594.

SLACK I (1979) 'Effects of auxiliaries on disperse dyes', *Canadian Textile J.*, **96**, 7, 46–51.

SMETHURST G (1989) *Basic water Treatment–for application world-wide*, 2nd Edition, Delhi, IBT Publishers & Distributors.

SPIRKIN V G, GOVORUNOVA L V, RED'KO N V and CHERNYSHOV B D (1981) 'Removal of sulphur dyes in effluent treatment', *Tekstil'naya Promyshlennost'*, **41**, 8, 61–62.

SROKA P and REINEKE J (1973a) 'Effluent treatment over land, as applied in textile finishing for the ecology of inland waterways', *Textile Praxis Int.*, **28**, 3, 171–173.

SROKA P and REINEKE J (1973b) 'Effluent treatment over land, as applied in textile finishing for the ecology of inland waterways', *Textile Praxis Int.*, **28**, 4, 208–211.

SUMMERS T H (1967) 'Effluent Problems and their Treatment in the Textile Industry', *J. Soc. Dyers Color.*, **83**, 373–379.

VAIDYA A A and TRIVEDI S S (1975) *Textile Auxiliaries and Finishing Chemicals*, India, Ahmedabad Textile Industries Research Association.

VAIDYA A A (1988) *Production of Synthetic Fibres*, New Delhi, Prentice Hall of India Pvt Ltd.

USEPA (1974) '*Waste water treatment systems: Upgrading textile operations to reduce pollution*', Report 625/3-74-004, Washington D C, October, US Environmental Protection Agency.

Index